Lecture Notes in Electrical Engineering

Volume 942

The book series *Lecture Notes in Electrical Engineering* (LNEE) publishes the latest developments in Electrical Engineering - quickly, informally and in high quality. While original research reported in proceedings and monographs has traditionally formed the core of LNEE, we also encourage authors to submit books devoted to supporting student education and professional training in the various fields and applications areas of electrical engineering. The series cover classical and emerging topics concerning:

- Communication Engineering, Information Theory and Networks
- Electronics Engineering and Microelectronics
- Signal, Image and Speech Processing
- Wireless and Mobile Communication
- Circuits and Systems
- Energy Systems, Power Electronics and Electrical Machines
- Electro-optical Engineering
- Instrumentation Engineering
- Avionics Engineering
- Control Systems
- Internet-of-Things and Cybersecurity
- Biomedical Devices, MEMS and NEMS

For general information about this book series, comments or suggestions, please contact leontina.dicecco@springer.com.

To submit a proposal or request further information, please contact the Publishing Editor in your country:

China

Jasmine Dou, Editor (jasmine.dou@springer.com)

India, Japan, Rest of Asia

Swati Meherishi, Editorial Director (Swati.Meherishi@springer.com)

Southeast Asia, Australia, New Zealand

Ramesh Nath Premnath, Editor (ramesh.premnath@springernature.com)

USA, Canada:

Michael Luby, Senior Editor (michael.luby@springer.com)

All other Countries:

Leontina Di Cecco, Senior Editor (leontina.dicecco@springer.com)

** This series is indexed by EI Compendex and Scopus databases. **

More information about this series at https://link.springer.com/bookseries/7818

Zhihong Qian · M. A. Jabbar ·
Xiaolong Li
Editors

Proceeding of 2021 International Conference on Wireless Communications, Networking and Applications

Set 1

 Springer

Editors
Zhihong Qian
College of Communication Engineering
Jilin University
Jilin, Jilin, China

M. A. Jabbar
Department of AI & ML
Vardhaman College of Engineering
Hyderabad, Telangana, India

Xiaolong Li
College of Technology
Indiana State University
Terre Haute, IN, USA

ISSN 1876-1100   ISSN 1876-1119   (electronic)
Lecture Notes in Electrical Engineering
ISBN 978-981-19-2455-2   ISBN 978-981-19-2456-9   (eBook)
https://doi.org/10.1007/978-981-19-2456-9

This Springer imprint is published by the registered company Springer Nature Singapore Pte Ltd.
The registered company address is: 152 Beach Road, #21-01/04 Gateway East, Singapore 189721,
Singapore

Preface

WCNA2021 [2021 International Conference on Wireless Communications, Networking and Applications] will be held on December 17–19, 2021, Berlin, Germany (virtual conference). Due to the COVID-19 situation and travel restriction, WCNA2021 has been converted into a virtual conference, which will be held via Tencent Meeting.

WCNA2021 hopes to provide an excellent international platform for all the invited speakers, authors, and participants. The conference enjoys a wide spread participation, and we sincerely wish that it would not only serve as an academic forum but also a good opportunity to establish business cooperation. Any paper and topic around wireless communications, networking, and applications would be warmly welcomed.

WCNA2021 proceeding tends to collect the most up-to-date, comprehensive, and worldwide state-of-the-art knowledge on wireless communications, networking, and applications. All the accepted papers have been submitted to strict peer review by 2–4 expert referees and selected based on originality, significance, and clarity for the purpose of the conference. The conference program is extremely rich, profound, and featuring high-impact presentations of selected papers and additional late-breaking contributions. We sincerely hope that the conference would not only show the participants a broad overview of the latest research results on related fields but also provide them with a significant platform for academic connection and exchange.

The technical program committee members have been working very hard to meet the deadline of review. The final conference program consists of 121 papers divided into six sessions. The proceedings would be published on Springer Book Series Lecture Notes in Electrical Engineering as a volume quickly, informally, and in high quality.

We would like to express our sincere gratitude to all the TPC members and organizers for their hard work, precious time and endeavor preparing for the conference. Our deepest thanks also go to the volunteers and staffs for their long-hours work and generosity they have given to the conference. Last but not least, we would like to thank each and every of the authors, speakers, and participants for their great contributions to the success of WCNA2021.

WCNA2021 Organizing Committee

Organization

Committees

Honor Chair

Patrick Siarry — Laboratoire Images, Signaux et Systèmes Intelligents, University Paris-Est Cré, Paris, France

General Chair

Zhihong Qian — College of Communication Engineering, Jilin University, China

Co-chairs

Isidoros Perikos — Computer Engineering and Informatics, University of Patras, Greece

Hongzhi Wang — Department of Computer Science and Technology, Harbin Institute of Technology, China

Hyunsung Kim — School of Computer Science, Kyungil University, Korea

Editor in Chief

Zhihong Qian — College of Communication Engineering, Jilin University, China

Co-editors

M. A. Jabbar	Head of the Department, Department of AI &ML, Vardhaman College of Engineering, Hyderabad, Telangana, India
Xiaolong Li	College of Technology, Indiana State University, USA
Sivaradje Gopalakrishnan	Electronics and Communication Engineering Department, Puducherry Technological University, Puducherry, India

Technical Program Committee

Qiang Cheng	University of Kentucky, USA
Noor Zaman Jhanjhi	School of Computing and IT, Taylor's University, Malaysia
Yilun Shang	Department of Computer and Information Sciences, Northumbria University, UK
Pascal Lorenz	University of Haute Alsace, University of Haute Alsace, France
Guillermo Escrivá-Escrivá	Department of Electrical Engineering, Universitat Politècnica de València, Spain
Surinder Singh	Department of Electronics and Communication Engineering, Sant Longowal Institute of Engineering and Technology, India
Pejman Goudarzi	Iran Telecom Research Center (ITRC), Iran
Antonio Muñoz	University of Malaga, Spain
Manuel J. Domínguez-Morales	University of Seville, Spain
Shamneesh Sharma	School of Computer Science & Engineering, Poornima University, India
K. Somasundaram	Amrita Vishwa Vidyapeetham, India
Daniela Litan	Deployment & Delivery (Oracle Technology Center), Oracle Developer, Romania
Artis Mednis	Institute of Electronics and Computer Science, University of Latvia, Latvia
Hari Mohan Srivastava	Department of Mathematics and Statistics, University of Victoria, Canada
Chang, Chao-Tsun	Department of Information Management, Hsiuping University of Science and Technology, Taiwan
Sumit Kushwaha	Department of Electronics Engineering, Kamla Nehru Institute of Technology, India
Bipan Hazarika	Department of Mathematics, Gauhati University, India

Petko Hristov Petkov	Technical University of Sofia, Bulgaria
Pankaj Bhambri	Department of Information Technology, I.K.G. Punjab Technical University, India
Aouatif Saad	National School of Applied Sciences, Ibn Tofail University, Morocco
Marek Blok	Telecommunications and Informatics, Gdańsk University of Technology, Poland
Phongsak Phakamach	College of Innovation Management, Rajamangala University of Technology Rattanakosin, Thailand
Mohammed Rashad Baker	Imam Ja'afar Al-Sadiq University, Iraq
Ahmad Fakharian	Islamic Azad University, Iran
Ezmerina Kotobelli	Department of Electronics and Telecommunication, Faculty of Information Technology, Polytechnic University of Tirana, Albania
Nikhil Marriwala	Electronics and Communication Engineering Department, Kurukshetra University, India
M. M. Kamruzzaman	Department of Computer and Information Science, Jouf University, KSA
Marco Listanti	Department of Electronic, Information and Telecommunications Engineering (DIET), University of Roma "La Sapienza," Italy
Ashraf A. M. Khalaf	Electrical Engineering (Electronics and Communications), Minia University, Egypt
Kidsanapong Puntsti	Department of Electronics and Telecommunication Engineering, Rajamangala University of Technology Isan (RMUTI), Thailand
Valerio Frascolla	Director of Research and Innovation at Inte, Intel Labs Germany, Germany
Babar Shah	College of Technological Innovation, Zayed University, Dubai
DijanaIlišević	Department for Planning and Construction of Wireless Transport network
Xilong Liu	Department of Information Science and Engineering, Yunnan University, Yunnan University, China
Suresh Kumar	Computer Science and Engineering, Manav Rachna International University, India
Sivaradje Gopalakrishnan	Electronics and Communication Engineering Department, Puducherry Technological University, India
Kanagachidambaresan	Vel Tech University, India

Sivaradje   Department of Electronics and Communication
                                   Engineering, Pondicherry Engineering
                                   College, India
A. K. Verma   CSED, Thapar Institute of Engg.
                                   and Technology, India
Kamran Arshad   Electrical Engineering, Ajman University, UAE
Gyu Myoung Lee   School of Computer Science and Mathematics,
                                   iverpool John Moores University, UK
Zeeshan Kaleem   COMSATS University Islamabad, Pakistan
Fathollah Bistouni   Department of Computer Engineering, Islamic
                                   Azad University, Iran
Sutanu Ghosh   Electronics and Communication Engg., India
Sachin Kumar   School of Electronic and Electrical Engineering,
                                   Kyungpook National University, South Korea
Anahid Robert Safavi   Wireless Network Algorithm Laboratory Huawei
                                   Sweden, Sweden
Hoang Trong Minh   Telecommunications Engineering,
                                   Telecommunications Engineering, Vietnam
Devendra Prasad   CSE, Chitkara University, India
Hari Shankar Singh   Electronics and Communication Engineering,
                                   India
Ashraf A. M. Khalaf   Faculty of Engineering, Minia University, Egypt
Hooman Hematkhah   Electrical and Electronics Engineering, Chamran
                                   University (SCU), Iran
Mani Zarei   Department of Computer Engineering, Tehran,
                                   Iran
Jibendu Sekhar Roy   School of Electronics Engineering, KIIT
                                   University, India
Luiz Felipe de Queiroz   Computer Engineering and Automation
    Silveira   Department, Federal University of Rio Grande
                                   do Norte, Brazil
Alexandros-Apostolos A.   Digital Systems, University of Piraeus, Greece
    Boulogeorgos
Trong-Minh Hoang   Posts and Telecommunication Institute
                                   of Technology, Vietnam
Jagadeesha R. Bhat   Electronic Communication Engg., Indian Institute
                                   of Information Technology, India
Tapas Kumar Mishra   Computer Science and Engineering, SRM
                                   University, India
Zisis Tsiatsikas   Information and Communication Systems
                                   Engineering, University of the Aegean, Greece
Muge Erel-Ozcevık   Software Engineering Department, Manisa Celal
                                   Bayar University, Turkey
E. Prince Edward   Department of Instrumentation and Control
                                   Engineering, Sri Krishna Polytechnic College,
                                   India

Prem Chand Jain	School of Engineering, Shiv Nadar University, India
Vipin Balyan	Department of Electrical, Electronics and Computer Engineering, Cape Peninsula University of Technology, South Africa
Yiannis Koumpouros	Department of Public and Community Health, University of West Attica, Greece
Aizaz Chaudhry	Systems and Computer Engineering, Carleton University, Canada
Andry Sedelnikov	Department of Space Engineering, Samara National Research University, Russia
Alexei Shishkin	Faculty of Computational Mathematics and Cybernetics, Moscow State University, Russia
Sevenpri Candra	S.E., M.M., ASEAN Engg., BINUS University, Indonesia
Meisam Abdollahi	School of Electrical and Computer Engineering, University of Tehran, Iran
Sachin Kumar (Research Professor)	Kyungpook National University, South Korea
Thokozani Calvin Shongwe	Electrical Engineering Technology, University of Johannesburg, South Africa
Ganesh Khekare	Department of Computer Science and Engineering, Faculty of Engineering & Technology, Parul University, Vadodara, Gujrat, India
Nishu Gupta	ECE Department, Chandigarh University, Mohali, Punjab, India
Gürel Çam	Iskenderun Technical University, Turkey
Ceyhun Ozcelik	Muğla Sıtkı Koçman University, Turkey
Shuaishuai Feng	Wuhan University, China
W. Luo	School of Finance and Economics, Nanchang Institute of Technology, China
Y. Xie	Party School of CPC Yibin Municipal Committee, China
Thanh-Lam Nguyen	Lac Hong University, Vietnam
Nikola Djuric	University of Novi Sad, Serbia
Ricky J. Sethi	Fitchburg State University, USA
Domenico Suriano	Italian National Agency for new Technologies, Energy, and Environment, Italy
Igor Verner	Faculty of Education in Science and Technology Technion, Israel Institute of Technology, Israel
Nicolau Viorel	"Dunarea de Jos" University of Galati, Romania
Snježana Babić	Polytechnic of Rijeka, Rijeka, Croatia

Esmaeel Darezereshki   Department of Materials Engineering, Shahid
                                Bahonar University, Kerman, Iran
Ali Rostami   University of Tabriz, Iran
Hui-Ming Wee   Department of Industrial and Systems
                                Engineering, Chung Yuan
                                Christian University, Taiwan
Yongyun Cho   Dept. Information and Communication
                                Engineering, Sunchon National University,
                                Sunchon, Korea
Lakhoua Mohamed Najeh   University of Cathage, Tunisia
M. Sohel Rahman   Bangladesh University of Engineering
                                and Technology, Bangladesh
Khaled Habib   Materials Science and Photo-Electronics Lab.,
                                RE Program, EBR Center KISR, Kuwait
Seongah Chin   Sungkyul University, Korea
Ning Cai   School of Artificial Intelligence, Beijing
                                University of Posts and Telecommunications,
                                China
Zezhong Xu   Changzhou Institute of Technology, China
Saeed Hamood Ahmed   MSCA SMART 4.0 FELLOW, AIT, Ireland
   Mohammed Alsamhi
Lim Yong Kwan   Singapore University of Social Sciences,
                                Singapore
Imran Memon   Zhejiang University, China
Anthony Kwame Morgan   Kwame Nkrumah University of Science
                                and Technology, Ghanaian
Ali Asghar Anvary Rostamy   Tarbiat Modares University, Iran
Hasan Dincer   Istanbul Medipol University, Turkey
Prem Kumar Singh   Gandhi Institute of Technology and
                                Management-Visakhapatnam, India
Dimitrios A. Karras   National and Kapodistrian University of Athens,
                                Greece
Cun Li   Eindhoven University of Technology, Netherland
Natalia A. Serdyukova   Plekhanov Russian University of Economics,
                                Russia
Sylwia   Wroclaw University of Science and Technology,
   Werbinska-Wojciechowska   Poland
José Joaquim de Moura   University of A Coruña, Spain
   Ramos
Naveen Kumar Sharma   I.K.G. Punjab Technical University, India
Tu Ouyang   Case Western Reserve University, USA
Nabil El Fezazi   Sidi Mohammed Ben Abdellah University,
                                Morocco
Pedro Alexandre Mogadouro   University of Trás-os-Montes e Alto Douro,
   do Couto   Portugal

Sek Yong Wee	Universiti Teknikal Malaysia Melaka, Malaysia
Muhammad Junaid Majeed	AuditXPRT Technologies, SQA Engineer, Pakistan
Janusz Kacprzyk	Systems Research Institute, Polish Academy of Sciences, Poland
Cihan Aygün	Faculty of Sports Sciences, Eskişehir Technical University, Turkey
Ciortea Elisabeta Mihaela	"December 1, 1918" University of Alba Iulia, Romania
Mueen Uddin	University Brunei Darussalam, Negara Brunei Darussalam
Esingbemi Princewill Ebietomere	University of Benin, Benin City, Nigeria
Samaneh Mashhadi	Iran University of science and Technology, Iran
Maria Aparecida Medeiros Maciel	Federal University of Rio Grande do Norte, Brazil
Josefa Mula	Universitat Politècnica de València, Spain
Claudemir Duca Vasconcelos	Federal University of ABC (UFABC), Brazil
Katerina Kabassi	Head of the Department of Environment, Ionian University, Greece
Takfarinas Saber	School of Computer Science, University College Dublin, Ireland
Zain Anwar Ali	Beijing Normal University, China
Jan Kubicek	VSB-Technical University of Ostrava, Czech Republic
Amir Karbassi Yazdi	School of Management, Islamic Azad University, Iran
Sujata Dash	Dept. of Computer Science and Application, North Orissa University, India
Souidi Mohammed El Habib	Abbes Laghrour University, Algeria
Dalal Abdulmohsin Hammood	Middle Technical Education (MTU) Electrical Engineering Technical College, Iraq
Marco Velicogna	Institute of Legal Informatics and Judicial Systems, Italian National Research Council, Italy
Hamad Naeem	College of Computer Science Neijiang Normal University, China
Hamid Jazayeriy	Babol Noshirvani University of Technology, Iran
Rituraj Soni	Engineering College Bikaner, India

Qutaiba Abdullah Hasan Alasad	University of Tikrit, Iraq
Alexandra Cristina González Eras	Universidad Técnica Particular de Loja, Department of Computer Science and Electronics, Ecuador
Falguni Roy	Noakhali Science and Technology University, Bangladesh
Ioan-Lucian Popa	Department of Computing, Mathematics, and Electronics, "1Decembrie 1918" University of Alba Iulia, Romania

Keynote Speakers

Advanced Architectures of Next Generation Wireless Networks

Pascal Lorenz

University of Haute-Alsace, France

Abstract. Internet Quality of Service (QoS) mechanisms are expected to enable wide spread use of real-time services. New standards and new communication architectures allowing guaranteed QoS services are now developed. We will cover the issues of QoS provisioning in heterogeneous networks, Internet access over 5G networks, and discusses most emerging technologies in the area of networks and telecommunications such as IoT, SDN, edge computing, and MEC networking. We will also present routing, security, and baseline architectures of the Internet working protocols and end-to-end traffic management issues.

Biography: Pascal Lorenz received his M.Sc. (1990) and Ph.D. (1994) from the University of Nancy, France. Between 1990 and 1995, he was a research engineer at WorldFIP Europe and at Alcatel-Alsthom. He is a professor at the University of Haute-Alsace, France, since 1995. His research interests include QoS, wireless networks, and high-speed networks. He is the author/co-author of three books, three patents, and 200 international publications in refereed journals and conferences. He was Technical Editor of the IEEE Communications Magazine Editorial Board (2000–2006), IEEE Networks Magazine since 2015, IEEE Transactions on Vehicular Technology since 2017, Chair of IEEE ComSoc France (2014–2020), Financial chair of IEEE France (2017–2022), Chair of Vertical Issues in Communication Systems Technical Committee Cluster (2008–2009), Chair of the Communications Systems Integration and Modeling Technical Committee (2003–2009), Chair of the Communications Software Technical Committee (2008–2010), and Chair of the Technical Committee on Information Infrastructure and Networking (2016–2017). He has served as Co-Program Chair of IEEE WCNC'2012 and ICC'2004, Executive Vice-Chair of ICC'2017, TPC Vice Chair of Globecom'2018, Panel sessions co-chair for Globecom'16, tutorial chair of VTC'2013 Spring and WCNC'2010, track chair of PIMRC'2012 and WCNC'2014, symposium Co-Chair at Globecom 2007–2011, Globecom'2019, ICC 2008–2010, ICC'2014 and '2016. He has served as Co-Guest Editor for special issues of IEEE Communications Magazine, Networks Magazine, Wireless Communications Magazine, Telecommunications Systems, and LNCS. He is an associate editor for International Journal of Communication Systems (IJCS-Wiley), Journal on Security and Communication Networks (SCN-Wiley) and International Journal of Business Data Communications and Networking, Journal of Network and Computer Applications (JNCA-Elsevier). He is a senior member of the IEEE, IARIA fellow, and member of many international program committees. He has organized many conferences, chaired several technical sessions, and gave tutorials at major international conferences. He was IEEE ComSoc Distinguished Lecturer Tour during 2013–2014.

Role of Machine Learning Techniques in Intrusion Detection System

M. A. Jabbar

Department of AI and ML, Vardhman College of Engineering, Hyderabad, Telangana, India

Abstract. Machine learning (ML) techniques are omnipresent and are widely used in various applications. ML is playing a vital role in many fields like health care, agriculture, finance, and in security. Intrusion detection system (IDS) plays a vital role in security architecture of many organizations. An IDS is primarily used for protection of network and information system. IDS monitor the operation of host or a network. Machine learning approaches have been used to increase the detection rate of IDS. Applying ML can result in low false alarm rate and high detection rate. This talk will discuss about how machine learning techniques are applied for host and network intrusion detection system.

Biography: Dr. M. A. JABBAR is Professor and Head of the Department AI&ML, Vardhaman College of Engineering, Hyderabad, Telangana, India. He obtained Doctor of Philosophy (Ph.D.) from JNTUH, Hyderabad, and Telangana, India. He has been teaching for more than 20 years. His research interests include artificial intelligence, big data analytics, bio-informatics, cyber-security, machine learning, attack graphs, and intrusion detection systems.

Academic Research

He published more than 50 papers in various journals and conferences. He served as a technical committee member for more than 70 international conferences. He has been Editor for 1st ICMLSC 2018, SOCPAR 2019, and ICMLSC 2020. He also has been involved in organizing international conference as an organizing chair, program committee chair, publication chair, and reviewer for SoCPaR, HIS, ISDA, IAS, WICT, NABIC, etc. He is Guest Editor for the Fusion of Internet of Things, AI, and Cloud Computing in Health Care: Opportunities and Challenges (Springer) Series, and Deep Learning in Biomedical and Health Informatics: Current Applications and Possibilities–CRC Press, Guest Editor for Emerging Technologies and Applications for a Smart and Sustainable World-Bentham science, Guest editor for Machine Learning Methods for Signal, Image and Speech Processing –River Publisher.

He is a senior member of IEEE and lifetime member in professional bodies like the Computer Society of India (CSI) and the Indian Science Congress Association (ISCA). He is serving as a chair, IEEE CS chapter Hyderabad Section. He is also serving as a member of Machine Intelligence Laboratory, USA (MIRLABS) and USERN, IRAN , Asia Pacific Institute of Science and Engineering (APISE) Hong Kong , Member in Internet Society (USA), USA , Member in Data Science Society, USA, Artificial Intelligence and Machine Learning Society of India (AIML), Bangalore.

He received best faculty researcher award from CSI Mumbai chapter and Fossee Labs IIT Bombay and recognized as an outstanding reviewer from Elsevier and received outstanding leadership award from IEEE Hyderabad Section. He published five patents (Indian) in machine learning and allied areas and published a book on "Heart Disease Data Classification using Data Mining Techniques," with LAP LAMBERT Academic publishing, Mauritius, in 2019.

Editorial works

1. Guest Editor: The Fusion of Internet of Things, AI, and Cloud Computing In Health Care: Opportunities and Challenges (Springer)
2. Guest Editor: Deep Learning in Biomedical and Health Informatics: Current Applications and Possibilities (CRC)
3. Guest Editor: Emerging Technologies and Applications for a Smart and Sustainable World-Bentham science
4. Guest Editor: Machine Learning Methods for Signal, Image, and Speech Processing-River Publisher
5. Guest Editor: The Fusion of Artificial Intelligence and Soft Computing Techniques for Cyber-Security-AAP–CRC Press
6. Guest Editor Special Issue on Web Data Security: Emerging Cyber-Defense Concepts and Challenges Journal of Cyber-Security and Mobility-River Publisher

Data Quality Management in the Network Age

Hongzhi Wang

Computer Science and Technology, Harbin Institute of Technology, China

Abstract. In the network age, data quality problems become more serious, and data cleaning is in great demand. However, data quality in the network age brings new technical challenges including the mixed errors, absence of knowledge, and computational difficulty. Facing the challenge of mixed errors, we discover the relationships among various types of errors and develop data cleaning algorithms for multiple errors. We also design data cleaning strategies with crowdsourcing, knowledge base as well as web search for the supplement of knowledge. For efficient and scalable data cleaning, we develop parallel data cleaning systems and efficient data cleaning algorithms. This talk will discuss the challenges of data quality in network age and give an overview of our solutions.

Biography: Hongzhi Wang is Professor, PHD supervisor, the head of massive data computing center and the vice dean of the honors school of Harbin Institute of Technology, the secretary general of ACM SIGMOD China, outstanding CCF member, a standing committee member CCF databases, and a member of CCF big data committee. Research fields include big data management and analysis, database systems, knowledge engineering, and data quality. He was "starring track" visiting professor at MSRA and postdoctoral fellow at University of California, Irvine. Prof. Wang has been PI for more than ten national or international projects including NSFC key project, NSFC projects, and national technical support project, and co-PI for more than ten national projects include 973 project, 863 project, and NSFC key projects. He also serves as a member of ACM Data Science Task Force. He has won first natural science prize of Heilongjiang Province, MOE technological First award, Microsoft Fellowship, IBM PHD Fellowship, and Chinese excellent database engineer. His publications include over 200 papers in the journals and conferences such as VLDB Journal, IEEE TKDE, VLDB, SIGMOD, ICDE, and SIGIR, six books and six book chapters. His PHD thesis was elected to be outstanding PHD dissertation of CCF and Harbin Institute of Technology. He severs as the reviewer of more than 20 international journal including VLDB Journal,

IEEE TKDE, and PC members of over 50 international conferences including SIGMOD 2022, VLDB 2021, KDD 2021, ICML 2021, NeurpIS 2020, ICDE 2020, etc. His papers were cited more than 2000 times. His personal website is http://homepage.hit.edu.cn/wang.

Networking-Towards Data Science

Ganesh Khekare

Department of Computer Science and Engineering, Faculty of Engineering and Technology, Parul University, Vadodara, Gujrat, India

Abstract. For communication, network is a must. Nowadays, networking is generating big data. To handle and process this huge amount of data, data science is required. Due to the increase in connectivity, interactions, social networking sites, platforms like YouTube, then invention of big data, fog computing, edge computing, Internet of Everything, etc., network transactions have been increased. Providing the best network flow graph is a challenge. Researchers are working on various data science techniques to overcome this. Node embedding concept is used to embed various complex networking graphs. To analyze different nodes and graphs for embedding, KarateClub library is used with Neo4j. Neo4j Graph data science library analyzes multigraphs networks in a better way. When network information is required in a fixed size vector, node embedding is used. This information is used in a downstream machine learning flow. Pyvis library is used to Visualize Interactive Network Graphs in Python. It provides a customization facility by which the network can be arranged for user requirements or to streamline the data flow. Researchers are also looking for interactive network graphs through data science algorithms that are capable of handling real-time scenarios. To draw Hive plots, the open-source Python package Hiveplotlib is available. The intelligible and visual probe of data generated through networking can be done smoothly by using Hive Plots. A data science algorithm viz., DeepWalk, is used to understand relationships in complex graph networks using Gensim, Networkx, and Python. Undirected and unweighted network visualization is also possible by using Mercator graph layout/embedding for a real-world complex network. Visualization of high dimensional network traffic data with 3D 360-degree animated scatter plots is the need. A huge research scope is there in networking using data science for the upcoming generations.

Biography: Dr. Ganesh Khekare is currently working as an Associate Professor in the department of Computer Science and Engineering at Parul University, Vadodara, Gujrat, India. He has done Ph.D. from Bhagwant University India. He pursued Master of Engineering from G H Raisoni College of Engineering, Nagpur, in the year 2013, and Bachelor of Engineering from Priyadarshini College of Engineering, Nagpur, in 2010. He has published more than 25 research articles in reputed international journals and conferences including Thomson Reuters, IGI Global, Inderscience, Springer, IEEE, Taylor and Francis, etc. He has published one patent and three copyrights. He guided more than 50 research as well as industry projects. His main research work focuses on data science, Internet of everything, machine learning, computer networks, artificial intelligence, intelligent transportation system, etc. He has more than 12 years of teaching and research experience. He is an active member of various professional bodies like ACM, ISTE, IEEE, IAENG, IFERP, IERD, etc.

Contents

Wireless Sensor Networks

Wireless Communications

A Nonzero Set Problem with Aumann Stochastic Integral

Jungang Li[1](✉) and Le Meng[2]

[1] Department of Statistics, North China University of Technology, Beijing 100144, China
`jungangli@126.com`
[2] China Fire and Rescue Institute, Beijing 102202, China

Abstract. A nonzero set problem with Aumann set-valued random Lebesgue integral is discussed. This paper proves that the Aumann Lebesgue integral's representation theorem. Finally, an important inequality is proved and other properties of Lebesgue integral are discussed.

Keywords: Set-Valued · Random process · Aumann Representation Theorem · Lebesgue Integral

1 Introduction

In signal processing and process control, we often use set-valued stochastic integral (see [3, 4] e.g.). Fuzzy random Lebesgue integral is applied to equations and stochastic inclusions (see [8] e.g.). Some papers [1, 2] have studied the Aumann type integral. Jung and Kim [1] used decomposable closure to give definitions of the stochastic integral, we have the integral is measurable. Li et. al. [7] gave set-valued square integrable martingale integral. Kisielewicz discussed the boundedness of the integral in [2]. We discussed set-valued random Lebesgue integral in [2]. An almost everywhere problem is solved in [5]. Our paper is organized as following: a nonzero set problem is pointed out with the set-valued Lebesgue integral. Aumann integral theorem is proved. We shall also discuss its boundedness, convexity, an important integral inequality etc.

2 Set-Valued Random Processes

First, we provide some definitions and symbols of closed set spaces. A set of real numbers R, natural numbers set N, the d-dimensional Euclidean space R^d. $K(R^d)$ is the all non-empty, closed subsets family of R^d, and $K_k(R^d)$ $(resp. K_{kc}(R^d))$ the all nonempty compact (*resp.* compact convex) subsets family of R^d. For $x \in R^d$ *and* $A \in K(R^d)$,

$h(x, A) = \inf_{y \in A} ||x - y||$. Define the Hausdorff metric h_d *on* $K(R^d)$ as
$h_d(A, B) = \max\{\sup_{a \in A} h(a, B), \sup_{b \in B} h(b, A)\}$. For $A \in K(R^d)$, denote

$$||A||_K = h_d(\{0\}, A) = \sup_{a \in A} ||a||.$$

© The Author(s) 2022
Z. Qian et al. (Eds.): WCNA 2021, LNEE 942, pp. 3–10, 2022.
https://doi.org/10.1007/978-981-19-2456-9_1

Then some properties of set-valued random processes shall be discussed. From first to last, we assume $T > 0$, $W = [0, T]$ and $p \geq 1$. A complete atomless probability space (Ω, C, P), a σ-field filtration $\{C_t : t \in [0, T]\}$, and the topological Borel field of a topological space E is $\mathcal{B}(E)$. Assume that $f = \{f(t), C_t : t \in [0, T]\}$ is a R^d-valued adapted random process. If for any $t \in [0, T]$, the mapping $(s, \omega) \to f(s, \omega)$ from $[0, t] \times \Omega$ to R^d is $\mathcal{B}([0, t]) \times C_t$-measurable,

then f is sequential measurable.

If

$$D = \{B \subset [0, T] \times \Omega : \forall t \in [0, T], B \cap ([0, t] \times \Omega) \in \mathcal{B}([0, t]) \times C_t\},$$

we have that f is D-measurable if and only if f is sequential measurable.

Denote $SM(K(R^d))$ the set of all sequential measurable set-valued random process. Similarly, we know notations $SM(K_c(R^d))$, $SM(K_k(R^d))$ and $SM(K_{kc}(R^d))$. Sequential measurable F is adapted and measurable. For $f_1, f_2 \in SM(R^d)$, define metric $\Delta_M(f_1, f_2) = E \int_0^T \frac{\|f_1(s)-f_2(s)\|}{1+\|f_1(s)-f_2(s)\|} ds$, we have norm $\||f\||_M = E \int_0^T \frac{\|f(s)\|}{1+\|f(s)\|}$, then $(SM(R^d), \Delta_M)$ is a complete space (cf. [6]).

Definition 2.1 $g(t, \omega) \in G(t, \omega)$ for a.e. $(t, \omega) \in [0, T] \times \Omega$, we call the R^d-valued sequential measurable random process $\{f(t), C_t : t \in [0, T]\} \in SM(R^d)$ is a selection of

$$\{G(t), C_t : t \in [0, T]\}.$$

Let $S\{G(\cdot)\}$ or $S(G)$ denote the family of all sequential measurable selections, i.e. $S(G) = \{\{g(t) : t \in [0, T]\} \in SM(R^d) : g(t, \omega) \in G(t, \omega),$ for a.e.$(t, \omega) \in [0, T] \times \Omega\}$. There are many definitions and results on set-valued theory, we can read this paper [9]. In this paper, the Aumann type Lebesgue integral is given.

Definition 2.2 (cf. [4]): A set-valued random process $G = \{G(t), t \in W\} \in SM(K(R^d))$. Define $I_t(G)(\omega) = (A)\int_0^t G(s, \omega)ds = \{\int_0^t g(s, \omega)ds : f \in S(G)\}$, for $t \in W$, $\omega \in \Omega$,

where $\int_0^t g(s, \omega)ds$ is Lebesgue integral. We call $(A)\int_0^t G(s, \omega)ds$ Aumann type Lebesgue integral of set-valued random process G with respect to t.

Remark 2.3: The elements of $S(G)$ in Definition 2.2 are integrable. By the definition of $S(G)$, $g(t, \omega) \in G(t, \omega)$ is defined for a.e. $(t, \omega) \in [0, T] \times \Omega$, and the number of selections is uncountable. The union of uncountable a.e. zero measurable sets is NOT a zero measurable set in general, denoted by $A_{(F)[0, T] \times \Omega}$. This helps to solve the boundedness problems in stochastic integral (see [2]). In fact, it may be unmeasurable. Let $D_1 = \{B_{(F)[0, T] \times \Omega} \subset B \subset [0, T] \times \Omega : \forall t \in [0, T], B \cap ([0, t] \times \Omega) \in \mathcal{B}([0, t]) \times C_t\}$, denote $\min MB_{G[0, T] \times \Omega} = \bigcap_{i=1}^{\infty} B_i$, for any $B_i \in D_1$. Let $\Pr_{\Omega}(\min MB_{(G)[0, T] \times \Omega}) = B_{(G)}\Omega$, the project set on Ω of $\min MB_{(G)[0, T] \times \Omega}$, $\Pr_{[0, T]}(\min MB_{(G)[0, T] \times \Omega}) = B_{(G)[0, T]}$, the project set on $[0, T]$ of $\min MB_{(G)[0, T] \times \Omega}$. In the following, we denote $\min MB_{(G)[0, T] \times \Omega}$ as $B_{(G)[0, T] \times \Omega}$ for convenience. Thus, $B_{(G)[0, T] \times \Omega}, B_{(G)[0, T]}$ and $B_{(G)}\Omega$ are all measurable.

Definition 2.4 Let a set-valued random process $G = \{G(t), t \in W\} \in SM(K(R^d))$. $t \in [0, T] \backslash B_{(G)[0,T]}$, define the integral $L_t(G)(\omega)$ by.

$$L_t(G)(\omega) = \begin{cases} \{g(s)ds : g \in S_T(G)(\omega)\}, & (s, \omega) \notin B_{(G)[0,T] \times \Omega} \\ \{0\}, & (s, \omega) \in B_{(G)[0,T] \times \Omega} \end{cases}$$

We call it Aumann type Lebesgue integral.

Now let's discuss the following Auman theorem and representation theorem.

3 Theorem and Proof

Theorem 3.1: A set-valued random process $G \in SM(K(R^d))$, $t \in [0, T] \backslash B_{(G)[0,T]}$, (A) $\int_0^t G(s)ds$ is a nonempty subset of $SM(K(R^d))$.

Proof. $S(G)$ is not null, $g \in S(G)$, $\int_0^t g(s, \omega)ds$ is sequential measurable. So (A) $\int_0^t G(s)ds$ is nonempty.

In the following, a new definition will be given. First, we will define a decomposable closure.

Definition 3.2: Nonempty subset $\Xi \subset SM[[0, T] \times \Omega, C, \lambda \times \mu; R^d]$, $\overline{de}\Xi = \{\{g(s, \omega) : t \in [0, T]\}, \varepsilon > 0$, there exist a D-measurable finite partition $\{A_1, \cdots, A_n\}$ of $[0, T] \times \Omega$ and $f_1, \cdots, f_n \in \Xi$ such that $\left\| g - \sum_{i=1}^n I_{A_i} f_i \right\|_M < \varepsilon\}$ is called the decomposable closure of Ξ with respect to D,

Theorem 3.3: $\{G(t) : t \in [0, T]\} \in SM(K(R^d))$, $\Xi(t) = (A) \int_0^t G(s)ds$, there exists a D-measurable process $L(G) = \{L_t(G) : t \in [0, T]\} \in SM(K(R^d))$, we have $S(L(G)) = \overline{de}\{\Xi(t) : t \in [0, T]\}$. In addition, the decomposable of $\Xi(t) = (A) \int_0^t G(s)ds$ is bounded by a constant C using the norm in space $SM(R^d)$.

Proof From Theorem 3.1, we know, $t \in [0, T] \backslash B_{(G)[0,T]}$, $\Xi(t) = (A) \int_0^t G(s)ds$ is nonempty in space $SM(R^d)$. Let

$$M = \overline{de}\{\Xi(t) : t \in W \backslash B_{(G)[0,T]}\}$$

$$= \overline{de}\left\{ h = \{h(t) : t \in [0, t] \backslash B_{(F)[0,T]}\} : h(t)(\omega) = \int_0^t g(s, \omega)ds, g \in S(G), (s, \omega) \notin B_{(G)[0,T] \times \Omega} \right\}$$

M is a closed subset in $SM[W \times \Omega \backslash A_{(G)[0,T] \times \Omega}, D, \lambda \times \mu; R^d]$. According to Theorem 2.7 in [6], it shows that there is $L(G) = \{L_t(G) : t \in [0, T]\} \in SM(K(R^d))$, we have $S(L(G)) = M$.

Now we shall prove boundedness. That is,

$$\left\| \sum_{i=1}^n I_{A_i} \int_0^t g_i(s, \omega)ds \right\| \leq \sum_{i=1}^n \left\| I_{A_i} \int_0^t g_i(s, \omega)ds \right\|$$

$$\leq \sum_{i=1}^n I_{A_i} \int_0^t \|g_i(s, \omega)\|ds$$

$$\leq \sum_{i=1}^{n} I_{A_i} \int_0^t \|G(s, \omega)\|_K ds.$$

Since $\phi(r) = \frac{r}{1+r} > 0$ is increasing, we have

$$\left\| \sum_{i=1}^{n} I_{A_i} \int_0^t g_i(s, \omega) ds \right\|_M = E \int_0^T \frac{\left\| \sum_{i=1}^{n} I_{A_i} \int_0^t g_i(s, \omega) ds \right\|}{1 + \left\| \sum_{i=1}^{n} I_{A_i} \int_0^t g_i(s, \omega) ds \right\|} dt$$

$$\leq E \int_0^T \frac{\sum_{i=1}^{n} I_{A_i} \int_0^t \|G(s, \omega)\|_K ds}{1 + \sum_{i=1}^{n} I_{A_i} \int_0^t \|G(s, \omega)\|_K ds} dt$$

$$\leq E \int_0^T \frac{\int_0^T \|G(s, \omega)\|_K ds}{1 + \int_0^T \|G(s, \omega)\|_K ds} dt$$

$$\leq C$$

This constant C is not relative to n.

Theorem 3.4 (Aumann Representation Theorem):
$G = \{G(t) : t \in [0, T]\} \in SM(K(R^d))$, a sequence of R^d-valued random processes $\{g^i = \{g^i(t) : t \in [0, T]\} : i \geq 1\} \subset S(G)$ exists, we have

$$L_t(G)(\omega) = cl\left\{ \int_0^t g^i(s, \omega) ds : i \geq 1 \right\} a.e.(t, \omega), \ (s, \omega) \in [0, T] \times \Omega \backslash B_{(G)[0,T] \times \Omega}$$

In addition, we have

$$L_t(G)(\omega) = cl\left\{ \int_0^t g(s, \omega) ds : g \in S(G) \right\} a.e.(t, \omega), \ (s, \omega) \in [0, T] \times \Omega \backslash B_{(G)[0,T] \times \Omega}$$

Proof By Theorem 3.9 in [5], we know, a series of $\{\varphi_n = \{\varphi_n(t) : t \in I\} : n \geq 1\} \subset S(L(G))$ exist,
$L_t(G)(\omega) = cl\{\varphi_n(t, \omega) : n \geq 1\}$, $a.e.(t, \omega) \in W \times \Omega \backslash B_{(G)[0,T] \times \Omega}$ holds.
Since

$$S(L(G)) = \overline{de}\{\Xi(t) : t \in [0, T]\}$$

$$= \overline{de}\left\{ \{h(t) : t \in I\} : h(t) = \int_0^t g(s) ds, \{g(\cdot)\} \in S(G) \right\}$$

$$= cl\left\{ \{k(t) : t \in I\} : k(t) = \sum_{k=1}^{n} I_{A_k} \int_0^t g_k(s) ds, \{A_k : K = 1, 2, ..., l\} \subset D, \text{ is a finite partition of} \right.$$

$$W \times \Omega \backslash B_{(G)[0,T] \times \Omega} \text{and} \{\{g_k(\cdot)\} : k = 1, 2, .., l\} \subset S(G), 1 \leq l\},$$

then for any $1 \leq n$, there exists $\{k_n^i : 1 \leq i\}$ such that $\{\varphi_n = \{\varphi_n(t) : t \in I\} : 1 \leq n\} \subset S(L(G))$, $\left\| \varphi_n(t) - k_n^i(t) \right\|_M \to 0 (i \to \infty)$, and $k_n^i(t) = \sum_{k=1}^{l(i,n)} I_{A_k^{(i,n)}} \int_0^t g_k^{(i,n)}(s) ds$, where $\left\{ A_k^{(i,n)} : k = 1, ..., l(i, n) \right\} \subset D$ is a finite

partition of $[0, T] \times \Omega \backslash B_{(G)}[0, T] \times \Omega$, $\left\{\left\{g_k^{(i, n)}(t) : t \in I\right\} : K = 1, 2, \ldots, l(i, n)\right\} \subset S(G)$.
Therefore there is a subsequence $\{i_j : 1 \leq j\}$ of $\{1, 2, \ldots\}$ such that

$$\left\| \varphi_n(t, \omega) - k_n^{i_j}(t, \omega) \right\| \to 0 \, \text{a.e.} (t, \omega) \in [0, T] \times \Omega \backslash B_{(G)}[0, T] \times \Omega (j \to \infty)$$

Thus for a.e.$(t, \omega) \in [0, T] \times \Omega \backslash B_{(G)}[0, T] \times \Omega (j \to \infty)$, we have that

$$L_t(G)(\omega) = cl\left\{k_n^{i_j}(t, \omega) : n, j \geq 1\right\}$$

$$\subset cl\left\{\int_0^t g_k^{(i_j, n)}(s, \omega)ds : n, j \geq 1, k = 1, \ldots, l(i_j, n)\right\}$$

$$\subset L_t(G)(\omega)$$

This means that for a.e.(t, ω), $(s, \omega) \in [0, T] \times \Omega \backslash B_{(G)}[0, T] \times \Omega$, we have

$$L_t(G)(\omega) = cl\left\{\int_0^t g_k^{(i_j, n)}(s, \omega)ds : n, j \geq 1, k = 1, \ldots, l(i_j, n)\right\}$$

Without losing generality, we have

$$L_t(G)(\omega) = cl\left\{\int_0^t g^i(s, \omega)ds : g^i \in S(G), i \geq 1\right\}.$$

In addition,

$$cl\left\{\int_0^t g^i(s, \omega)ds : g^i \in S(G), i \geq 1\right\} \subseteq cl\left\{\int_0^t g(s, \omega)ds : g \in S(G)\right\}.$$

Since $\Gamma \subseteq \overline{de}\Gamma = S(L(F))$, then we have

$$cl\left\{\int_0^t g(s, \omega)ds : g \in S(G)\right\} \subseteq cl\left\{\int_0^t g^i(s, \omega)ds : g^i \in S(G), i \geq 1\right\}.$$

Therefore,

$$L_t(G)(\omega) = cl\left\{\int_0^t g(s, \omega)ds : g \in S(G)\right\}.$$

Corollary 3.5 (Representation Theorem):

$G = \{G(t) : t \in [0, T]\} \in PM(K(R^d))$. There is a sequence of R^d-valued random process $\{g^i = \{g^i(t) : t \in [0, T]\} : i \geq 1\} \subset S(G)$ such that

$$G(t, \omega) = cl\left\{g^i(t, \omega) : i \geq 1\right\} \quad \text{a.e.} (t, \omega) \in [0, T] \times \Omega \backslash B_{(G)}[0, T] \times \Omega,$$

$$L_t(G)(\omega) = cl\left\{\int_0^t g^i(s, \omega)ds : i \geq 1\right\} \text{a.e.} (t, \omega), (s, \omega) \in [0, T] \times \Omega \backslash B_{(G)}[0, T] \times \Omega.$$

Remark 3.6: Since $\{G(t) : t \in [0, T] \backslash B_{(G)}[0, T]\} \in SM(K(R^d))$ is measurable wih respect to $t \in \ominus B([0, T] \backslash B_{(G)}[0, T])$ for fixed $\omega \in \Omega \backslash B_{(G)}\Omega$. If.

$$s \in [0, t] \backslash B_{(G)[0, T]} \subset [0, T], \; G(s, \omega) \subseteq R_+^d,$$

by Remark 3.11 in [6], we have.

(A) $\int_0^t G(s, \omega)ds = $ (A) $\int_0^t \mathrm{conv}G(s, \omega)ds = \mathrm{conv}\left((A) \int_0^t G(s, \omega)\right)$.

Therefore, theAumann random Lebesgue integral (A) $\int_0^t G(s, \omega)$ is convex by using Aumann representation theorem.

Theorem 3.7: For $p \geq 1$, $F, G \in L^p([0, T] \times \Omega; K(R^d))$, a.e.
$(s, \omega) \in ([0, t] \times \Omega) \cap \overline{B}_{(F)[0, T] \times \Omega} \cap \overline{B}_{(G)[0, T] \times \Omega}$, we have.

$$h_d(L_t(F)(\omega), L_t(G)(\omega)) \leq \int_0^t h_d(F_s(\omega), G_s(\omega))ds. \tag{1}$$

Proof Since $F, \quad G \in L^p([0, T] \times \Omega; K(R^d))$, that is $\left(E \int_0^T \|F(s, \omega)\|^p ds\right)^{\frac{1}{p}} < +\infty$.

Thus, there exists Ω_F such that $P(\Omega_F) = 1$, for any $\omega \in \Omega_F$, $\left(\int_0^T \|F(s, \omega)\|^p ds\right)^{\frac{1}{p}} < +\infty$. In the same way, we have Ω_G. Assume $\omega \in \left(\Omega_F \backslash B_{(F)}\Omega\right) \cap \left(\Omega_G \backslash B_{(G)}\Omega\right)$ in the following proof. Take an $f \in S_T(F)(\omega)$. Then, for $t, s \in [0, T] \cap \overline{B}_{(F)[0, T]} \cap \overline{B}_{(G)[0, T]}$, we have.

$$h\left(\int_0^t f_s ds, L_t(G)(\omega)\right) = \inf_{g \in S_t^1(G)(\omega)} \left\| \int_0^t f_s ds - \int_0^t g(s)ds \right\|$$

$$\leq \inf_{g \in S_t^1(G)(\omega)} \int_0^t \|f_s - g_s\| ds$$

Further, by proving the same point of [8, Theorem 4],

$$\inf_{g \in S_t^1(G)(\omega)} \int_0^t \|f_s - g_s\| ds$$

$$= \int_0^t \inf_{y \in G(s, \omega)} \|f_s - y\| ds = \int_0^t h(f_s, G_s(\omega)) ds$$

$$\leq \int_0^t \sup_{x \in F_s(\omega)} h(x, G_s(\omega)) ds \leq \int_0^t h_d(F_s(\omega), G_s(\omega)) ds$$

Thus,

$$h\left(\int_0^t f(s)ds, L_t(G)(\omega)\right) \leq \int_0^t h_d(F_s(\omega), G_s(\omega))ds$$

We know $f \in S_T(F)(\omega)$, by Definition 2.4 we have that
$$\sup_{x \in L_t(F)(\omega)} h(x, L_t(G)(\omega)) \leq \int_0^t h_d(F_s(\omega), G_s(\omega))ds.$$

Similarly, we have

$$\sup_{x \in L_t(G)(\omega)} h(x, L_t(F)(\omega)) \leq \int_0^t h_d(F_s(\omega), G_s(\omega))ds.$$

The two inequalities above yield

$$h_d(L_t(F)(\omega), L_t(G)(\omega)) \leq \int_0^t h_d(F_s(\omega), G_s(\omega))ds.$$

We obtain (1).

Acknowledgment. This paper is supported by Beijing municipal education commission (No. KM202010009013), in part by Fundamental Research Funds for NCUT(No.110052972027/007).

References

1. Jung, E., Kim, J.: On set-valued stochastic integrals. Stoch. Anal. Appl. **21**, 401–418 (2003)
2. Kisielewic, M., Michta, M.: Integrably bounded set-valued stochastic integrals. J. Math. Anal. Appl. **449**, 1892–1910 (2017)
3. Levent, H., Yilmaz, Y.: Translation, modulation and dilation systems in set-valued signal processing. Carpathian Math. Publ. **10**, 143–164 (2018)
4. Li, J., Li, S.: Set-valued stochastic Lebesgue integral and representation theorems. Int. J. Comput. Intell. Syst. **1**, 177–187 (2008)
5. Li, J., Li, S., Ogura, Y.: Strong solution of Ito type set-valued stochastic differential equation. Acta Mathematica Sinica Eng. Ser. **26**, 1739–1748 (2010)
6. Li, J., Wang, J.: Fuzzy set-valued stochastic Lebesgue integral. Fuzzy Sets Syst. **200**, 48–64 (2012)
7. Li, S., Li, J., Li, X.: Stochastic integral with respect to set-valued square integrable martingales. J. Math. Anal. Appl. **370**, 659–671 (2010)
8. Michta, M.: On set-valued stochastic integrals and fuzzy stochastic equations. Fuzzy Sets Syst. **177**, 1–19 (2011)

Circular L(j,k)-Labeling Numbers of Cartesian Product of Three Paths

Qiong Wu[✉] and Weili Rao

Department of Computational Science, School of Science,
Tianjin University of Technology and Education, Tianjin, China
wuqiong@tute.edu.cn

Abstract. The circular $L(j, k)$-labeling problem with $k \geq j$ arose from the code assignment in the wireless network of computers. Given a graph G and positive numbers j, k, σ, and a circular σ-$L(j, k)$-labeling of a graph G is an assignment f from $[0, \sigma)$ to the vertices of G, for any two vertices u and v, such that $|f(u) - f(v)|_\sigma \geq j$ if $uv \in E(G)$, and $|f(u) - f(v)|_\sigma \geq k$ if u and v are distance two apart, where $|f(u) - f(v)|_\sigma = min\{|f(u) - f(v)|, \sigma - |f(u) - f(v)|\}$. The minimum σ such that graph G has a circular σ-$L(j, k)$-labeling of a graph G, which is called the circular $L(j, k)$-labeling number of graph G and is denoted by $\sigma_{j,k}(G)$. In this paper, we determine the circular $L(j, k)$-labeling numbers of Cartesian product of three paths, where $k \geq 2j$.

Keywords: Code assignment · Circular-*L(j,k)*-labeling · Cartesian product

1 Introduction

The rapid growth of wireless networks causes the scarcity of available codes for communication in Multihop *Packet Radio Network* (PRN) which was studied in 1969 at University of Hawaii [1] firstly. In a multihop PRN, it is an important design consideration to assign transmission codes to network nodes. Because of the finite number of transmission codes, the number of network nodes may be larger than the number of transmission codes. It may take place that the time overlap of two or more packet receptions at the destination station. That is called *interference* or *collision*. For example, there exist two types of interference in a PRN using code division multiple access (CDMA). *Direct* interference occurs when two adjacent stations transmitting to each other directly. *Hidden terminal* interference is due to two stations at distance two communicate with the same receiving station at the same time.

Two stations are *adjacent* if they can transmit to each other directly. If two stations are called *at distance two* if two stations are nonadjacent but they are adjacent to one common station.

The wireless network can be modeled as an undirected graph $G = (V, E)$, such that the set of stations are represented as a set of *vertices* $V = \{v_0, v_1, \cdots, v_{n-1}\}$, and two vertices are joined by an undirected *edge* in E if and only if their corresponding stations can communicate directly.

© The Author(s) 2022
Z. Qian et al. (Eds.): WCNA 2021, LNEE 942, pp. 11–21, 2022.
https://doi.org/10.1007/978-981-19-2456-9_2

Since the interference (or collision) lowers the system throughput and increases the packets delay at destination, it is necessary to investigate the problem of code assignment for interference avoidance in Multi-hop PRN. Bertossi and Bonuccelli [2] introduced a type of code assignment for the network whose direct interference is so weak that we can ignore it, that is, only two distance-two stations are required to transmit by different codes to avoid the hidden terminal interference. By abstracting codes as labels, the above problem is equivalent to an $L(0, 1)$-labeling problem. That is, the distance-two vertices should be labeled numbers with difference at least 1.

In the real world, the direct interference cannot be ignored. In order to avoid the direct interference and hidden terminal interference, the code assignment problem was generalized to $L(j, k)$-labeling problem by Jin and Yeh [3], where $j \leq k$. That is, to avoid direct interference, any two adjacent stations must be assigned with difference at least j, then any two distance-two apart stations are required to be assigned larger code differences to avoid hidden terminal interference, as well as to avoid direct interference.

For two positive real numbers j and k, an $L(j, k)$-labeling f of G is a mapping of numbers to vertices of G such that $|f(u) - f(v)| \geq j$ if $uv \in E(G)$, and $|f(u) - f(v)| \geq k$ if u, v are at distance two, where $|a - b|$ is called *linear difference*. The $L(j, k)$-*labeling number* of G is denoted by $\lambda_{j,k}(G)$, where $\lambda_{j,k}(G) = \min\limits_{f} \max\limits_{u,v \in V(G)} \{|f(u) - f(v)|\}$. For $j \leq k$, there exist some results on the $L(j, k)$-labeling of graphs. For example, Wu introduced the $L(j, k)$-labeling numbers of generalized Petersen graphs [4] and Cactus graphs [5], Shiu and Wu investigated the $L(j, k)$-labeling numbers of direct product of path and cycles [6, 7], Wu, Shiu and Sun [8] determined the $L(j, k)$-labeling numbers of Cartesian product of path and cycle..

For any $x \in \mathbb{R}, [x]_\sigma \in [0, \sigma)$ denotes the remainder of x upon division by σ. The *circular difference* of two points p and q is defined as $|p - q|_\sigma = min\{|p-q|, \sigma - |p-q|\}$.

Heuvel, Leese and Shepherd [9] used the circular difference to replace the linear difference in the definition of $L(j, k)$-labeling, and obtained the definition of circular $L(j, k)$-labeling as follows.

Given G and positive real numbers j and k, a circular σ- $L(j, k)$-labeling of G is a function $f : V(G) \rightarrow [0, \sigma)$ satisfying $|f(u) - f(v)|_\sigma \geq j$ if $d(u, v) = 1$ and $|f(u) - f(v)| \geq k$ if $d(u, v) = 2$. The minimum σ is called the *circular $L(j, k)$-labeling number* of G, denoted by $\sigma_{j,k}(G)$. For $j \leq k$, this problem was rarely investigated. For instance, Wu and Lin [10] introduced the circular $L(j, k)$- labeling numbers of trees and products of graphs. Wu, Shiu and Sun [11] determined the circular $L(j, k)$-labeling numbers of direct product of path and cycle. Furthermore, Wu and Shiu [12] investigated the circular $L(j, k)$-labeling numbers of square of paths.

Two labels are *t-separated* if the circular difference between them is at least t.

The *Cartesian product* of three graphs G, H and K, denoted by $G \Box H \Box K$, is the graph with vertices set $V(G \Box H \Box K) = V(G) \times V(H) \times V(K)$, and two vertices $v_{u,v,w}, v_{u',v',w'} \in V(G \Box H \Box K)$ are adjacent if $v_u = v_{u'}, v_v = v_{v'}$ and $(v_w, v_{w'}) \in E(K)$, or $v_u = v_{u'}, v_w = v_{w'}$ and $(v_v, v_{v'}) \in E(H)$, or $v_w = v_{w'}, v_v = v_{v'}$ and $(v_u, v_{u'}) \in E(G)$. For convenience, the Cartesian product of three paths P_l, P_m and P_n is denoted by $G_{l,m,n}$. For any vertex $v_{x,y,z} \in V(G_{l,m,n}), x, y, z$ are called *subindex* of vertex. If two vertices with one different subindex are called *at the same row*. For instance, $v_{a,y,z}$ and $v_{b,y,z}$ are at the same row, where $a \neq b, 0 \leq a, b \leq l-1, 0 \leq y \leq m-1$, and $0 \leq z \leq n-1$.

All notations not defined in this thesis can be found in the book [13].

2   Circular $L(j, k)$-Labeling Numbers of Cartesian Product of Three Paths

Lemma 2.1 [10]. Let j and k be two positive numbers with $j \leq k$. Suppose H is an induced subgraph of graph G. Then $\sigma_{j,k}(G) \geq \sigma_{j,k}(H)$.

Note that Lemma 2.1 is not true if H is not an induced subgraph of G. For example, $\sigma_{1,2}(K_{1,3}) = 6 > 4 = \sigma_{1,2}(K_4)$, where $K_{1,3}$ is a subgraph of K_4 instead of an induced subgraph.

Lemma 2.2 [5]. Let a, b and σ be three positive real numbers, then $|[a]_\sigma - [b]_\sigma|$ equals to $[a-b]_\sigma$ or $\sigma - [a-b]_\sigma$.

Lemma 2.3. Let a, b and σ be three positive real numbers with $0 \leq a < \sigma$, then $[a+b]_\sigma - [b]_\sigma = a$ or $a - \sigma$.

Proof: The conclusion can be obtained as following cases.

a) If $0 \leq a+b < \sigma$ and $0 \leq b < \sigma$, then $[a+b]_\sigma - [b]_\sigma = a+b-b = a$.
b) If $\sigma \leq a+b < 2\sigma$ and $0 \leq b < \sigma$, then $[a+b]_\sigma - [b]_\sigma = a+b-\sigma-b = a-\sigma$.
c) If $\sigma \leq b$, let $b = r+k\sigma$, where $0 \leq r < \sigma$ and $k \in \mathbb{Z}^+$, according to the above two cases, we have $[a+b]_\sigma - [b]_\sigma = [a+r]_\sigma - [r]_\sigma = a$ or $a - \sigma$.

Hence, the lemma is proved.

2.1   Circular $L(j, k)$-labeling Numbers of Graph $G_{2,m,n}$

This subsection introduces the circular $L(j, k)$-labeling numbers of $G_{2,m,n}$ for $m, n \geq 2$ and $k \geq 2j$.

Theorem 2.1.1 Let j and k be two positive numbers with $k \geq 2j$. For $n \geq 2$, Then $\sigma_{j,k}(G_{2,2,n}) = 4k$.

Proof: Given a circular labeling f for $G_{2,2,n}$ as follows:

$$f(v_{0,0,z}) = \left\lceil \frac{zk}{2} \right\rceil_{2k}, \quad f(v_{1,0,z}) = \left\lceil \frac{(z+3)k}{2} \right\rceil_{2k} + 2k,$$

$$f(v_{0,1,z}) = \left\lceil \frac{(z+1)k}{2} \right\rceil_{2k} + 2k, \quad f(v_{1,1,z}) = \left\lceil \frac{(z+2)k}{2} \right\rceil_{2k},$$

where $0 \leq z \leq n-1$.

Note that the labels of two adjacent vertices at the same row are $\frac{k}{2}$-separated ($k \geq 2j$), and the labels of distance-two vertices at the same row are k-separated. Let $\sigma = 4k$. For an arbitrary vertex $v_{x,y,z} \in V(G_{2,2,n})$, according to the symmetry of the graph $G_{2,2,n}$, we need to check the differences between the labels of vertices $v_{x,y,z}$ and $v_{1-x,1-y,z}$, $v_{x,1-y,z\pm1}$, $v_{1-x,y,z\pm1}$ (if they exist). That is, we need to make sure that f satisfies the following cases.

a) $|f(v_{1-x,y,z+1}) - f(v_{x,y,z})|_{4k} \geq k$, where $x, y \in \{0, 1\}, 0 \leq z \leq n - 1$.

By Lemma 2.3 and the definition of circular difference, we have the following four subcases.

$$|f(v_{1,0,z+1}) - f(v_{0,0,z})|_{4k} = \left| \left[\frac{(z+4)k}{2} \right]_{2k} + 2k - \left[\frac{zk}{2} \right]_{2k} \right|_{4k}$$

$$= |2k|_{4k} \geq k.$$

$$|f(v_{0,0,z+1}) - f(v_{1,0,z})|_{4k} = \left| \left[\frac{(z+1)k}{2} \right]_{2k} - \left(\left[\frac{(z+3)k}{2} \right]_{2k} + 2k \right) \right|_{4k}$$

$$= |-3k|_{4k} \text{ or } |-k|_{4k} \geq k.$$

$$|f(v_{1,1,z+1}) - f(v_{0,1,z})|_{4k} = \left| \left(\left[\frac{(z+3)k}{2} \right]_{2k} \right) - \left(\left[\frac{(z+1)k}{2} \right]_{2k} + 2k \right) \right|_{4k}$$

$$= |-k|_{4k} \text{ or } |-3k|_{4k} \geq k.$$

$$|f(v_{0,1,z+1}) - f(v_{1,1,z})|_{4k} = \left| \left(\left[\frac{(z+2)k}{2} \right]_{2k} + 2k \right) - \left(\left[\frac{(z+2)k}{2} \right]_{2k} \right) \right|_{4k}$$

$$= |2k|_{4k} \geq k.$$

Thus, $|f(v_{1-x,y,z+1}) - f(v_{x,y,z})|_{4k} \geq k$, for $x, y \in \{0, 1\}, 0 \leq z \leq n - 1$.

b) $|f(v_{1-x,y,z-1}) - f(v_{x,y,z})|_{4k} \geq k$, where $x, y \in \{0, 1\}, 0 \leq z \leq n - 1$.

By Lemma 2.3 and the definition of circular difference, we have the following four subcases.

$$|f(v_{1,0,z-1}) - f(v_{0,0,z})|_{4k} = \left| \left[\frac{(z+2)k}{2} \right]_{2k} + 2k - \left[\frac{zk}{2} \right]_{2k} \right|_{4k}$$

$$= |3k|_{4k} \text{ or } |k|_{4k} \geq k.$$

$$|f(v_{0,0,z-1}) - f(v_{1,0,z})|_{4k} = \left| \left[\frac{(z-1)k}{2} \right]_{2k} - \left(\left[\frac{(z+3)k}{2} \right]_{2k} + 2k \right) \right|_{4k}$$

$$= |-2k|_{4k} \geq k.$$

$$|f(v_{1,1,z-1}) - f(v_{0,1,z})|_{4k} = \left| \left(\left[\frac{(z+1)k}{2} \right]_{2k} \right) - \left(\left[\frac{(z+1)k}{2} \right]_{2k} + 2k \right) \right|_{4k}$$

$$= |-2k|_{4k} \geq k.$$

$$|f(v_{0,1,z-1}) - f(v_{1,1,z})|_{4k} = \left| \left(\left[\frac{zk}{2} \right]_{2k} + 2k \right) - \left(\left[\frac{(z+2)k}{2} \right]_{2k} \right) \right|_{4k}$$

$$= |3k|_{4k} \text{ or } |k|_{4k} \geq k.$$

Thus, $|f(v_{1-x,y,z-1}) - f(v_{x,y,z})|_{4k} \geq k$, for $x, y \in \{0, 1\}, 0 \leq z \leq n - 1$.

c) $\cdot\left|f\left(v_{x,1-y,z+1}\right)-f\left(v_{x,y,z}\right)\right|_{4k} \geq k$, where $x, y \in \{0, 1\}, 0 \leq z \leq n-1$.

By Lemma 2.3 and the definition of circular difference, we have the following four subcases.

$$\left|f\left(v_{1,1,z+1}\right)-f\left(v_{1,0,z}\right)\right|_{4k} = \left|\left[\frac{(z+3)k}{2}\right]_{2k}-\left[\frac{(z+3)k}{2}\right]_{2k}-2k\right|_{4k}$$

$$= |-2k|_{4k} \geq k.$$

$$\left|f\left(v_{1,0,z+1}\right)-f\left(v_{1,1,z}\right)\right|_{4k} = \left|\left[\frac{(z+4)k}{2}\right]_{2k}+2k-\left(\left[\frac{(z+2)k}{2}\right]_{2k}\right)\right|_{4k}$$

$$= |3k|_{4k} \text{ or } |k|_{4k} \geq k.$$

$$\left|f\left(v_{0,1,z+1}\right)-f\left(v_{0,0,z}\right)\right|_{4k} = \left|\left(\left[\frac{(z+2)k}{2}\right]_{2k}+2k\right)-\left(\left[\frac{zk}{2}\right]_{2k}\right)\right|_{4k}$$

$$= |3k|_{4k} \text{ or } |k|_{4k} \geq k.$$

$$\left|f\left(v_{0,0,z+1}\right)-f\left(v_{0,1,z}\right)\right|_{4k} = \left|\left(\left[\frac{(z+1)k}{2}\right]_{2k}\right)-\left(\left[\frac{(z+1)k}{2}\right]_{2k}+2k\right)\right|_{4k}$$

$$= |-2k|_{4k} \geq k.$$

Thus, $\left|f\left(v_{x,1-y,z+1}\right)-f\left(v_{x,y,z}\right)\right|_{4k} \geq k$, for $x, y \in \{0, 1\}, 0 \leq z \leq n-1$.

d) $\left|f\left(v_{x,1-y,z-1}\right)-f\left(v_{x,y,z}\right)\right|_{4k} \geq k$, where $x, y \in \{0, 1\}, 0 \leq z \leq n-1$.

By Lemma 2.3 and the definition of circular difference, we have

$$\left|f\left(v_{1,1,z-1}\right)-f\left(v_{1,0,z}\right)\right|_{4k} = \left|\left[\frac{(z+1)k}{2}\right]_{2k}-\left[\frac{(z+3)k}{2}\right]_{2k}-2k\right|_{4k}$$

$$= |-3k|_{4k} \text{ or } |-k|_{4k} \geq k.$$

$$\left|f\left(v_{1,0,z-1}\right)-f\left(v_{1,1,z}\right)\right|_{4k} = \left|\left[\frac{(z+2)k}{2}\right]_{2k}+2k-\left(\left[\frac{(z+2)k}{2}\right]_{2k}\right)\right|_{4k}$$

$$= |2k|_{4k} \geq k.$$

$$\left|f\left(v_{0,1,z-1}\right)-f\left(v_{0,0,z}\right)\right|_{4k} = \left|\left(\left[\frac{zk}{2}\right]_{2k}+2k\right)-\left(\left[\frac{zk}{2}\right]_{2k}\right)\right|_{4k}$$

$$= |2k|_{4k} \geq k.$$

$$\left|f\left(v_{0,0,z-1}\right)-f\left(v_{0,1,z}\right)\right|_{4k} = \left|\left(\left[\frac{(z-1)k}{2}\right]_{2k}\right)-\left(\left[\frac{(z+1)k}{2}\right]_{2k}+2k\right)\right|_{4k}$$

$$= |-3k|_{4k} \text{ or } |-k|_{4k} \geq k.$$

Thus, $\left|f\left(v_{x,1-y,z-1}\right)-f\left(v_{x,y,z}\right)\right|_{4k} \geq k$, for $x, y \in \{0, 1\}, 0 \leq z \leq n-1$.

e) $\left|f\left(v_{1-x,1-y,z}\right)-f\left(v_{x,y,z}\right)\right|_{4k} \geq k$, where $x, y \in \{0, 1\}, 0 \leq z \leq n-1$.

By Lemma 2.3 and the definition of circular difference, we have the following four subcases.

$$\left|f\left(v_{1,1,z}\right)-f\left(v_{0,0,z}\right)\right|_{4k} = \left|\left[\frac{(z+2)k}{2}\right]_{2k}-\left[\frac{zk}{2}\right]_{2k}\right|_{4k} = |-k|_{4k} \text{ or } |k|_{4k} \geq k.$$

$$\left|f\left(v_{1,0,z}\right)-f\left(v_{0,1,z}\right)\right|_{4k}$$

$$= \left|\left[\frac{(z+3)k}{2}\right]_{2k} + 2k - \left(\left[\frac{(z+1)k}{2}\right]_{2k} + 2k\right)\right|_{4k} = |k|_{4k} \text{ or } |-k|_{4k} \geq k.$$

Thus, $\left|f\left(v_{1-x,1-y,z}\right) - f\left(v_{x,y,z}\right)\right|_{4k} \geq k$, for $x, y \in \{0, 1\}$, $0 \leq z \leq n-1$.

Hence, f is a circular $4k$-$L(j, k)$-labeling of graph $G_{2,2,n}$, it means that $\sigma_{j,k}(G_{2,2,n}) \leq 4k$ for $n \geq 2$ and $k \geq 2j$.

Figure 1 shows a circular $4k$-$L(j, k)$-labeling of graph $G_{2,2,8}$.

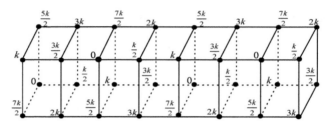

Fig. 1. A circular $4k$-$L(j, k)$-labeling of graph $G_{2,2,8}$

On the other hand, the vertices $v_{0,0,0}$, $v_{1,0,1}$, $v_{0,1,1}$, and $v_{1,1,0}$ are distance two apart mutually, the circular difference among their labels should be at least k, it implies that $\sigma_{j,k}(G_{2,2,n}) \geq 4k$ for $n \geq 2$.

Hence, $\sigma_{j,k}(G_{2,2,n}) = 4k$ for $n \geq 2$ and $k \geq 2j$.

Theorem 2.1.2. Let j and k be two positive real numbers with $k \geq 2j$. For $m, n \geq 3$, $\sigma_{j,k}(G_{2,m,n}) = 5k$.

Proof: Defined a circular labeling f for graph $G_{2,m,n}$ as follows:

$$f\left(v_{x,y,z}\right) = \left[\frac{(5x + y + 3z)k}{2}\right]_{5k},$$

where $x = 0, 1, 0 \leq y \leq m - 1$ and $0 \leq z \leq n - 1$.

Note that the labels of adjacent vertices at the same row are $\frac{k}{2}$-separated ($k \geq 2j$) and the labels of vertices with distance two apart at the same row are k-separated. Let $\sigma = 5k$. For an arbitrary vertex $v_{x,y,z} \in V(G_{2,m,n})$, according to the symmetry of the graph $G_{2,m,n}$, it is sufficient to verify the circular differences between $v_{x,y,z}$ and $v_{1-x,y+1,z}$, $v_{1-x,y,z+1}$, $v_{x,y+1,z\pm1}$(If they exist) are k-separated, respectively, where $x \in \{0, 1\}, 0 \leq y \leq m - 1$ and $0 \leq z \leq n - 1$. By Lemma 2.3 and the definition of circular difference, we have the following results.

a)
$$\left| f\left(v_{1-x,y+1,z}\right) - f\left(v_{x,y,z}\right)\right|_{5k}$$
$$= \left| \left[\frac{[5(1-x)+y+1+3z]k}{2} \right]_{5k} - \left[\frac{(5x+y+3z)k}{2} \right]_{5k} \right|_{5k}$$
$$= \left| \left[\frac{(6-5x+y+3z)k}{2} \right]_{5k} - \left[\frac{(5x+y+3z)k}{2} \right]_{5k} \right|_{5k}$$
$$= 2k \geq k.$$

b)
$$\left| f\left(v_{1-x,y,z+1}\right) - f\left(v_{x,y,z}\right)\right|_{5k}$$
$$= \left| \left[\frac{[5(1-x)+y+3(z+1)]k}{2} \right]_{5k} - \left[\frac{(5x+y+3z)k}{2} \right]_{5k} \right|_{5k}$$
$$= \left| \left[\frac{(8-5x+y+3z)k}{2} \right]_{5k} - \left[\frac{(5x+y+3z)k}{2} \right]_{5k} \right|_{5k}$$
$$= k \geq k.$$

c)
$$\left| f\left(v_{x,y+1,z+1}\right) - f\left(v_{x,y,z}\right)\right|_{5k}$$
$$= \left| \left[\frac{[5x+y+1+3(z+1)]k}{2} \right]_{5k} - \left[\frac{(5x+y+3z)k}{2} \right]_{5k} \right|_{5k}$$
$$= \left| \left[\frac{(4+5x+y+3z)k}{2} \right]_{5k} - \left[\frac{(5x+y+3z)k}{2} \right]_{5k} \right|_{5k}$$
$$= 2k \geq k.$$

d)
$$\left| f\left(v_{x,y+1,z-1}\right) - f\left(v_{x,y,z}\right)\right|_{5k}$$
$$= \left| \left[\frac{[5x+y+1+3(z-1)]k}{2} \right]_{5k} - \left[\frac{(5x+y+3z)k}{2} \right]_{5k} \right|_{5k}$$
$$= \left| \left[\frac{(5x+y+3z-2)k}{2} \right]_{5k} - \left[\frac{(5x+y+3z)k}{2} \right]_{5k} \right|_{5k}$$
$$= k \geq k.$$

Hence, f is a circular $5k$-$L(j, k)$-labeling of graph $G_{2,m,n}$, it means that $\sigma_{j,k}\left(G_{2,m,n}\right) \leq 5k$ for $m, n \geq 3$ and $k \geq 2j$.

For example, Fig. 2 is a circular $5k$-$L(j, k)$-labeling of graph $G_{2,3,3}$.

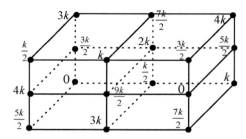

Fig. 2. A circular 5k-L(j, k)-labeling of graph $G_{2,3,3}$.

On the other hand, the vertices $v_{0,0,1}$, $v_{0,1,2}$, $v_{0,1,0}$, $v_{0,2,1}$ and $v_{1,1,1}$ are at distance two from each other, the circular difference among their labels should be at least k, it implies that $\sigma_{j,k}(G_{2,m,n}) \geq 5k$ for m,n $\geq$ 3.

Hence, $\sigma_{j,k}(G_{2,m,n}) = 5k$ for $m, n \geq 3$ and $k \geq 2j$.

2.2   Circular L(j, k)-Labeling Numbers of Graph $G_{2,m,n}$

This subsection introduces the general results on the circular $L(j, k)$-labeling numbers of $G_{l,m,n}$ for $l, m, n \geq 3$ and $k \geq 2j$.

Theorem 2.2.1. Let j and k be three positive real numbers with $k \geq 2j$. For $l, m, n \geq 3$, $\sigma_{j,k}(G_{l,m,n}) = 6k$.

Proof: Given a circular labeling f for $G_{l,m,n}$ as follows:

$$f\left(v_{x,y,z}\right) = \left\lceil \frac{(3x + y + 5z)k}{2} \right\rceil_{6k},$$

where $0 \leq x \leq l - 1, 0 \leq y \leq m - 1$ and $0 \leq z \leq n - 1$.

Note that the labels of adjacent vertices at the same row are $\frac{k}{2}$-separated ($k \geq 2j$) and the labels of distance-two vertices at the same row are k-separated. Let $\sigma = 6k$. For an arbitrary vertex $v_{x,y,z} \in V(G_{l,m,n})$, according to the symmetry of the graph $G_{l,m,n}$, it is sufficient to verify the circular differences between $v_{x,y,z}$ and $v_{x+1,y\pm1,z}$, $v_{x+1,y,z\pm1}$, $v_{x,y+1,z\pm1}$ (If they exist) are k-separated, respectively, where $0 \leq x \leq l-1, 0 \leq y \leq m-1$ and $0 \leq z \leq n - 1$. By Lemma 2.3 and the definition of circular difference, we have the following results.

a)
$$\begin{aligned}
&\left| f\left(v_{x+1,y+1,z}\right) - f\left(v_{x,y,z}\right) \right|_{6k} \\
&= \left| \left\lceil \frac{[3(x+1)+(y+1)+5z]k}{2} \right\rceil_{6k} - \left\lceil \frac{(3x+y+5z)k}{2} \right\rceil_{6k} \right|_{6k} \\
&= \left| \left\lceil \frac{(4+3x+y+3z)k}{2} \right\rceil_{6k} - \left\lceil \frac{(3x+y+5z)k}{2} \right\rceil_{6k} \right|_{6k} \\
&= 2k \geq k.
\end{aligned}$$

$$\left| f\left(v_{x+1,y-1,z}\right) - f\left(v_{x,y,z}\right)\right|_{6k}$$

b)
$$= \left| \left[\frac{[3(x+1) + (y-1) + 5z]k}{2} \right]_{6k} - \left[\frac{(3x+y+5z)k}{2} \right]_{6k} \right|_{6k}$$

$$= \left| \left[\frac{(2+3x+y+3z)k}{2} \right]_{6k} - \left[\frac{(3x+y+5z)k}{2} \right]_{6k} \right|_{6k}$$

$$= k \geq k.$$

$$\left| f\left(v_{x+1,y,z+1}\right) - f\left(v_{x,y,z}\right)\right|_{6k}$$

c)
$$= \left| \left[\frac{[3(x+1) + y + 5(z+1)]k}{2} \right]_{6k} - \left[\frac{(3x+y+5z)k}{2} \right]_{6k} \right|_{6k}$$

$$= \left| \left[\frac{(8+3x+y+3z)k}{2} \right]_{6k} - \left[\frac{(3x+y+5z)k}{2} \right]_{6k} \right|_{6k}$$

$$= 2k \geq k.$$

$$\left| f\left(v_{x+1,y,z-1}\right) - f\left(v_{x,y,z}\right)\right|_{6k}$$

d)
$$= \left| \left[\frac{[3(x+1) + y + 5(z-1)]k}{2} \right]_{6k} - \left[\frac{(3x+y+5z)k}{2} \right]_{6k} \right|_{6k}$$

$$= \left| \left[\frac{(3x+y+3z-2)k}{2} \right]_{6k} - \left[\frac{(3x+y+5z)k}{2} \right]_{6k} \right|_{6k}$$

$$= k \geq k.$$

$$\left| f\left(v_{x,y+1,z+1}\right) - f\left(v_{x,y,z}\right)\right|_{6k}$$

e)
$$= \left| \left[\frac{[3x + y + 1 + 5(z+1)]k}{2} \right]_{6k} - \left[\frac{(3x+y+5z)k}{2} \right]_{6k} \right|_{6k}$$

$$= \left| \left[\frac{(6+3x+y+3z)k}{2} \right]_{6k} - \left[\frac{(3x+y+5z)k}{2} \right]_{6k} \right|_{6k}$$

$$= 3k \geq k.$$

$$\left| f\left(v_{x,y+1,z-1}\right) - f\left(v_{x,y,z}\right)\right|_{6k}$$

f)
$$= \left| \left[\frac{[3x + y + 1 + 5(z-1)]k}{2} \right]_{6k} - \left[\frac{(3x+y+5z)k}{2} \right]_{6k} \right|_{6k}$$

$$= \left| \left[\frac{(3x+y+3z-4)k}{2} \right]_{6k} - \left[\frac{(3x+y+5z)k}{2} \right]_{6k} \right|_{6k}$$

$$= 2k \geq k.$$

Hence, f is a circular $6k$-$L(j, k)$-labeling of graph $G_{l,m,n}$, it means that $\sigma_{j,k}\left(G_{l,m,n}\right) \leq 6k$ for $l, m, n \geq 3$ and $k \geq 2j$.

For example, Fig. 3 is a circular $6k$-$L(j, k)$-labeling of graph $G_{3,3,3}$.

On the other hand, the vertices $v_{1,0,1}, v_{0,1,1}, v_{1,2,1}, v_{2,1,1}, v_{1,1,2}$ and $v_{1,1,0}$ are at distance two from each other, the circular difference among their labels should be at least k, this means $\sigma_{j,k}\left(G_{l,m,n}\right) \geq 6k$ for $l, m, n \geq 3$.

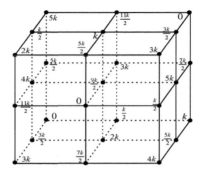

Fig. 3. A circular 6k-L(j, k)-labeling of graph $G_{3,3,3}$.

Hence, $\sigma_{j,k}\left(G_{l,m,n}\right) = 6k$ for $l, m, n \geq 3$ and $k \geq 2j$.

3   Conclusion

In this paper, we investigate the circular $L(j, k)$-labeling number of Cartesian product of three paths which arose from the code assignment of interference avoidance in the PRN. For $k \geq 2j$, we obtain that

$$\sigma_{j,k}\left(G_{l,m,n}\right) = \begin{cases} 4k, & \textit{if } l, m = 2 \textit{ and } n \geq 2, \\ 5k, & \textit{if } l = 2 \textit{ and } m, n \geq 3, \\ 6k, & \textit{if } l, m, n \geq 2. \end{cases}$$

Acknowledgements. This paper is partially supported by the NSF of Tianjin (Grant No. 18JCQNJC69700), and the Sci. and Tech. Develop. Fund of Tianjin (Grant No. 2020KJ115).

References

1. Abrahmson N.: The ALOHA system-Another alternative for computer communications. In: Proceedings of FJCC, pp. 281–285 (1970)
2. Bertossi, A.A., Bonuccelli, M.A.: Code assignment for hidden terminal interference avoidance in multihop packet radio networks. IEEE/ACM Trans. Networking **3**(4), 441–449 (1995)
3. Jin, X.T., Yeh, R.K.: Graph distance-dependent labeling related to code assignment in. computer networks. Naval Res. Logist. **52**(2), 159–164 (2005)
4. Wu, Q.: L(j, k)-labeling number of generalized Petersen graph. IOP Conf. Ser. Mater. Sci. Eng. **466**, 012084 (2018)
5. Wu, Q.: L(j, k)–labeling number of Cactus graph. IOP Conf. Ser. Mater. Sci. Eng. **466**, 012082 (2018)
6. Shiu, W.C., Wu, Q.: L(j, k)-number of direct product of path and cycle. Acta Mathematica Sinica, English Series **29**(8), 1437–1448 (2013)

7. Wu, Q.: Distance two labeling of some products of graphs. Doctoral Thesis, Hong Kong: Hong Kong Baptist University (2013)
8. Wu, Q., Shiu, W.C., Sun, P.K.: L(j, k)-labeling number of Cartesian product of path and cycle. J. Comb. Optim. **31**(2), 604–634 (2016)
9. Heuvel, J., Leese, R.A., Shepherd, M.A.: Graph labelling and radio channel assignment. J. Graph Theory **29**, 263–283 (1998)
10. Wu, Q., Lin, W.: Circular L(j, k)-labeling numbers of trees and products of graphs. J. Southeast Univ. **26**(1), 142–145 (2010)
11. Wu, Q., Shiu, W.C., Sun, P.K.: Circular L(j, k)-labeling number of direct product of path and cycle. J. Comb. Optim. **27**, 355–368 (2014)
12. Wu, Q., Shiu, W.: Circular L(j, k)-labeling numbers of square of paths. J. Combinatorics Number Theory **9**(1), 41–46 (2017)
13. Bondy, J.A., Murty, U.S.R.: Graph Theory with Applications, 2nd edn. MacMillan, New. York (1976)

Five Application Modes of Mobile Government

Gangqiang Yu, Dongze Li, and Jinyu Liu[✉]

School of Politics and Public Administration, South China Normal University, Guangzhou
510631, China
shayu425_2000@yeah.net, ldz15362180603@163.com, liujinyu@sina.com

Abstract. To solve the problem that traditional e-government tends to lose real-time control of content and process, mobile government was created, it has 5 main application modes, which are mG2G mode between government departments and other government departments, mG2E mode between government and internal staff, mG2B mode between government and business, mG2C mode between government and the public, and mG2V mode between government and organizations & people outside the country. Mobile government uses mG2C, mG2B and mG2V as the external service mode and mG2G and mG2E mode as the internal management mode to continuously improve the quality and level of external services through continuous optimization of internal management.

Keywords: Mobile government · mG2C · mG2B · mG2V · mG2G · mG2E

1 Introduction

Since the 1990s, government departments have been increasingly using e-government to improve the quality of public services. However, early e-government mainly used fixed, wired information networks to transmit data and provide services electronically. One of the inconveniences of such e-government is that both government staff and government service recipients rely on wired Internet and desktop computer to access government systems. Once step away from the office area, the government staff tends to lose control of the service content and process, which in turn affects the response speed and service effectiveness of certain matters. With the rapid development of wireless communication technology, more and more government departments are providing public services through mobile devices [1], which is also known as mobile government. In the 21st century, the large-scale use of mobile terminals such as smartphones and tablet PCs, as well as the popularization of wireless LAN, have not only made wireless offices possible within government departments, but also made it possible for the public to access convenient mobile government services.

2 The Nature of Mobile Government

Mobile Government (mGov), also known as mobile e-government, is simply an application model of mobile communication technology in government management and

Z. Qian et al. (Eds.): WCNA 2021, LNEE 942, pp. 22–27, 2022.
https://doi.org/10.1007/978-981-19-2456-9_3

service work. It is often regarded as an extension and upgrade of e-government [2]. It uses mobile devices instead of traditional electronic devices and its goal is to provide real-time access to government information and services when and where they are available from any location [3]. According to Ibrahim Kushchu, mGovernment is an e-government service provided through a mobile platform, a strategy and its implementation using wireless and mobile technologies, services, applications, and devices, which aims to enhance the various parties involved in e-government --- citizens, enterprises and government [4]. Mobile Government is considered by Chanana et al. as a public service provided through mobile devices (e.g., cell phones, PDAs, etc.) [5]. In simple terms, mobile government is an application mode of mobile communication technology in government management and service work. Mobile government indirectly solves the problem of time constraint and computer-based space constraint [6], and has shown its strong potential to provide public services "anytime, anywhere", expand government functions, and improve the quality and efficiency of government services [7], it has been applied in many government departments.

3   The Main Contents of the Five Application Modes of Mobile Government

According to the difference in the nature of the interacting subjects under the contextual theme, mobile government can be divided into the following five modes respectively, mG2G mode, mG2B mode, mG2V mode, mG2G mode and mG2E mode. Among them, the first three can be categorized as external service forms and the last two can be categorized as internal management forms (Fig. 1).

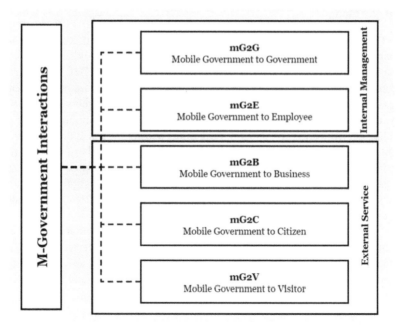

Fig. 1. Five application modes of mobile government

3.1 mG2G Mode

Mobile Government to Government, or mG2G for short, refers to the use of wireless network technology and mobile terminals between local governments, government departments at all levels, and their internal agencies to achieve internal management data push and business information processing. mG2G has its own application focus areas at the executive, management and decision-making levels. For the executive level, mG2G is mainly used to execute field operations. The executive can use the mobile government client to collect field data and send it back to the government data platform in real time. For management, the mobile government platform is mainly used to transmit data information to the executive level, for the support and coordination of front-line work. For the decision-making layer, the mobile government platform is mainly used to understand the overall situation at anytime, grasp the real-time statistics, to receive, issue documents, send work tasks even they are out of town. The development of mobile government makes communication between the decision-making layer and the executive layer more convenient and accelerates the speed of information transfer [8], thus improving the efficiency of internal management.

The focus of mobile government services among government departments is: first, to increase the efficiency of the common construction and sharing of government information resources, continuously optimize and improve electronic processes, promote seamless connections between services, and improve management efficiency; second, to promote the transformation of government functions, fully streamline and optimize the administrative approval process, provide quality and efficient management & services to the community, and enhance the overall image of government departments; third, to achieve mutual supervision and power balances among departments at the lowest cost.

3.2 mG2E Mode

Mobile Government to Employee, or mG2E for short, is a mobile government service that enables internal staff of government departments to work online using wireless communication technology. Compared with the mG2G mode, mG2E mainly provides management and service related to individuals and non-confidential management-type services to internal staff. Its content may include personal comprehensive information inquiry, email sending and receiving, internal control document inquiry and browsing, work schedule, work task reminder, online learning and continuing education, etc. Using the mobile government office platform, government staff can receive official information at any time and deal with pending matters in a timely manner without the restrictions of places and equipment. This reduces the government's administrative expenses and improves efficiency to a large extent [9], and in the meantime, effectively strengthens cooperation and communication among staff members.

The government's mobile government services for internal staff focus on, first, improving the energy efficiency of internal management and services to support the front-line work; second, implementing whole-process monitoring to continuously optimize and improve work quality and enhance job performance; and third, increasing professional proficiency, improving personal qualities, and promoting team building.

3.3 mG2B Mode

Mobile Government to Business, or mG2B for short, refers to the use of mobile communication technology between government and enterprises to achieve government-enterprise interaction in mobile government. The use of mG2B can further reduce the operating costs of enterprises in dealing with government departments, and can also save the cost of government expenditures incurred in providing public services. mG2B mode is mainly used in electronic license processing, electronic procurement and bidding, electronic taxation, public information consulting services, small and medium-sized enterprise e-services and other fields.

The government's mobile government services to enterprises focus on: first, creating a good social environment for enterprises and providing easy and low-cost management services, such as open public data; second, building a platform for mutual communication and legal access to public resources for the development of enterprises, such as establishing electronic trading platforms; third, effectively supervising enterprises under the framework of the rule of law to reduce negative effects, such as environmental pollution monitoring.

3.4 mG2C Mode

Mobile Government to Citizen, or mG2C for short, refers to government departments provide services to the public using mobile communication technology. Its main applications include education and training, employment, e-health, social security network, e-tax, social governance and public management information services. The main point of the current construction is to actively push online community service projects and online personal government service matters into the mobile government platform. This has greatly facilitated the two-way interaction between the government and the public, and the characteristics of public service provision centered on public demand are becoming more and more obvious [10], thus enabling "government departments or institutions to truly realize their mission of serving the public" [11] [12].

The government's mobile government services to the public focus on: first, "all-round" public affairs, government departments should publish all public services and work procedures to the public through the information network, so that people can understand the content of services in a timely manner. Second, "all-weather" government services, government departments should make full use of the mobile government services platform, so that the public can receive 7×24 h of service. Third, the "whole process" of supervision, while citizens enjoy mobile government services, they can also evaluate and supervise the service contents and effects in a timely manner to strengthen the supervision of the government by the society.

3.5 mG2V Mode

Mobile Government between government and foreign organizations and visitors, or mG2V for short, refers to mobile government services provided by foreign-related government departments to foreign organizations and personnel using mobile communication technology. The government has a diplomatic function. In the era of globalization, interactions between countries are becoming more frequent, and an open country always has a large number of international organizations and foreigners permanently stationed there. As a result, mG2V is increasingly of interest to modern governments.

The government's mobile government services for foreign organizations and personnel are, on the one hand, to provide foreign governments and the public with promotional information about various fields in the country, to introduce policies, regulations, finance, environmental and other issues to foreign enterprises and citizens interested in investing in the country, to introduce cultural resources of travel destinations and to explain laws and regulations such as visas and currency exchange to foreign tourists. On the other hand, it also has the function of handling immigration management and immigration services.

4 Conclusion

Technological and management innovations will continue to expand and deepen the content of the mobile government application modes, rather than being limited to those described in this paper. External services in the form of mG2C, mG2B and mG2V, and internal management in the form of mG2G and mG2E, constitute the basic application modes of mobile government. Handling the relationship between these two types is the guarantee of implementing, expanding and deepening mobile government services. First, we must always insist that the fundamental purpose of improving the internal management level of government affairs is to enhance the quality of external services. This is the prerequisite element for building a mobile government service model, otherwise it may lead to the blind introduction of various mobile information technologies, while ignoring the government service itself. Second, the relationship between mG2G and mG2E should be handled well. They are highly interrelated. mG2E directly serves specific "people", while mG2G directly serves seemingly abstract levels of government, but actually, both of them exist to meet the needs of the public. They ultimately converge in the specific service projects or events provided to the public. Third, mG2C, mG2B, and mG2V are deemed as the fundamental purpose of mobile government, and there is no priority among the three. Especially for mG2V, any free and open country should provide possible, equal and quality wireless government services to everyone in the world.

Acknowledgments. This work was supported by the following items: National Social Science Fund Project total "community-level data based on the authorization of a major community-level public health emergencies coordinated prevention and control mechanisms of innovative research" (20BGL217).

References

1. Ojo, A.F., Janowski, T.S., Awotwi, J.T.: Enabling development through governance and mobile technology. Government Inf. Q. **30**, S32–S45 (2013)

2. Lin Sitao, F.: Mobile e-government construction based on the public requirements. Chinese Public Adm. **4**, 52–56 (2015)
3. Zhou Pei, F., Ma Jing, S.: A study for the development of public service oriented mobile e-government. In: 2011 Third International Conference on Multimedia Information Networking and Security, pp. 545–549. IEEE, Shanghai (2011)
4. Kushchu, I., Kuscu, H.: From e-government to m-government: facing the inevitable. In: 3rd European Conference on eGovernment, pp. 253–260. Dublin (2004)
5. Chanana, L., Agrawal, R., Punia, D.K.: Service quality parameters for mobile government services in India. Global Bus. Rev. **17**(1), 136–146 (2016)
6. Su, C., Jing, M.: A general review of mobile e-government in China. In: 2010 International Conference on Multimedia Information Networking and Security, pp. 733-737 (2010)
7. Liu Shuhua, F., Zhan Hua, S., Yuan Qianli, T.: Mobile government and urban governance in China. E-Government **6**, 2–12 (2011)
8. Pan Wei, F., Su Lining, S.: Research on mobile government affairs and the construction of intelligent-service-government. J. Shanxi Youth Vocat. Coll. **34**(2), 42–45 (2021)
9. Song Zengwei, F.: Theory and Practice of Service-Oriented Government Construction, 1st edn. Economic Press China, Beijing (2012)
10. Bertot, J.C., Jaeger, P.T., Munson, S.: Social media technology and government transparency. Computer **43**(11), 53–59 (2010)
11. Jonathan, D., Breul, F.: Practitioner's perspective-improving sourcing decisions. Public Adm. Rev. **70**, 193–200 (2010)
12. Hilgers, D., Ihl, C.: Citizensourcing: applying the concept of open innovation to the public sector. Int. J. Public Participation **4**(1), 67–88 (2010)

Analysis of the Micro Implicit Feedback Behavior of User Network Exploring Based on Mobile Intelligent Terminal

Wei Wang[✉], Chuang Zhang, Xiaoli Zheng, and Yuxuan Du

School of Information and Electrical Engineering, Hebei University of Engineering, Handan 056038, China
wangwei83@hebeu.edu.cn

Abstract. In the face of the information recommendation requirements in mobile Internet applications, in order to better use the user micro implicit feedback behavior obtained by the mobile intelligent terminal to improve the recommendation efficiency, this paper intends to carry out the analysis of the implicit feedback behavior by analyzing the behavior distribution and behavior correlation. The analytical results reveal the particularity of the implicit feedback behavior in mobile intelligent terminal.

Keywords: Recommended system · Mobile intelligent terminal · Implicit feedback behavior · Behavior distribution

1 Introduction

The analysis of user network behavior characteristics is the design basis of many Internet products. Through in-depth analysis of user behavior, completing personalized recommendation can bring users a better application experience. In the field of market-driven software engineering, user behavior analysis also provides new ideas and improvement direction for application development to meet the requirements of the new situation.

User network behavior can be divided into two categories: explicit feedback behavior and implicit feedback behavior. The definition, characteristics, differences and types of the two types of behavior, relatively stable and unified views have been formed. Display feedback behavioral data can accurately express user intention, but because it interferes with the normal interaction process in the network, increases the cognitive burden and reduces the user experience, it is difficult to obtain data. On the contrary, for users' implicit feedback behavior data, it is much less difficult to obtain and has large information abundance. Therefore, although such information has low accuracy, large data noise and large context sensitivity, this research field is still getting more and more attention.

Z. Qian et al. (Eds.): WCNA 2021, LNEE 942, pp. 28–39, 2022.
https://doi.org/10.1007/978-981-19-2456-9_4

2   Related Studies

With the rapid development of social networks and e-commerce, the number of Internet users has increased and the demand for personalized recommendation services is growing. It is the focus and difficulty of current research to deal with the massive amount of multi-source heterogeneous data generated when users browse the mobile Internet.

The original personalized recommendation service is mainly for PC-based users, and the relevant research is mainly divided into the following four aspects: research on an application scenario, a kind or technology, recommendation system evaluation method, and a kind of common problems in the recommendation system.

The study of user network behavior was initially applied in the field of information retrieval, which significantly improves the performance of information filtering compared to other feedback, and quickly filters from massive information sets, providing the retrieval set with the highest correlation with their interest preferences[1]. Lots of researches show that user browsing time is important to find person's preference [2, 3]. Moreover, bookmarking, printing and saving could show users' interesting. Oard and Kim clustered them into three groups [4–6].

In addition, mobile network environment give a challenge. Researches such as [7, 8] focus on this condition. Implicit behaviors from user exploring website in this condition are hot [9–11]. Therefore, this paper conducts the analysis of the implicit feedback behavior of mobile intelligent terminals.

3   Problem Description and Behavioral Analysis

3.1   Problem Description

Users' network implicit behavior contains information about their preferences, but it is generally not clearly expressed, so it is more difficult to correctly judge their preferences, and the researchers have carried out more work in this regard. At present, there are many implicit studies on macro-network behavior, such as behavioral sequence analysis or item recommendation based on browsing, adding shopping carts, buying and other behaviors. For the implicit feedback behavior of user micro network, there are few studies and conclusions that are found due to small data scale, less data category and low data dimension. This paper plans to carry out implicit feedback behavior analysis, explore the characteristics of implicit feedback behavior data, and lay the foundation for the subsequent recommendation based on implicit feedback behavior.

3.2   Analysis of User Microscopic Implicit Feedback Behavior

Acquiring approach of users' micro implicit behavior includes two ways. The first one is direct acquiring way, which is conducted by running some software in background. The other is indirect way, generally speaking, which is acquired by questionnaire. In direct acquisition, there are problems of sparse data, less categories and low dimensions, which is not conducive to subsequent analysis and deterministic conclusions. In this paper, we use data in indirect acquisition mode to analyze the micro indirect feedback behavior, extracting part of the survey content (Q4-Q15) from the questionnaire, and mapping it to micro implicit behaviors, IFBn above, from user exploring in website, as below in Table 1.

Table 1. User micro implicit behavior.

Raw data (users' behavior)	Description	Corresponding behavior (micro implicit behavior)
Which app store do you use?(Q4)	Discrete, type: 10, Category mutual exclusion	Category selection of application market(IFB1)
How frequently do you visit the app store to look for apps?(Q5)	Discrete, type: 9, Category mutual exclusion	Access frequency of application market(IFB2)
On average, how many apps do you download a month?(Q6)	Discrete, type: 6, Category mutual exclusion	Number of monthly attention to items(IFB3)
When do you look for apps?(Q7)	Discrete, type: 6, Categories are not mutually exclusive	Query frequency of item(IFB4)
How do you find apps? (Q8)	Discrete, type: 9, Categories are not mutually exclusive	Query method for item(IFB5)
What do you consider when choosing apps to download?(Q9)	Discrete, type: 13, Categories are not mutually exclusive	Detail level of item browsing(IFB6)
Why do you download an app? (Q10)	Discrete, type: 15, Categories are not mutually exclusive	Focus on item (purchase possibility)(IFB7)
Why do you spend money on an app? (Q11)	Discrete, type: 12, Categories are not mutually exclusive	Purchase behavior of item(IFB8)
Why do you rate apps?(Q13)	Discrete, type: 7, Categories are not mutually exclusive	Evaluation behavior of item(IFB9)
What makes you stop using an app? (Q14)	Discrete, type: 15, Categories are not mutually exclusive	Cancel attention to item(IFB10)
Which type of apps do you download?(Q15)	Discrete, type: 23, Categories are not mutually exclusive	Category focus behavior on item(IFB11)

In order to facilitate the subsequent association analysis of various kinds of influence variables, the user micro implicit feedback behavior is divided into two categories according to the questionnaire data: 1) mutually exclusive micro implicit feedback behavior and 2) non-mutually exclusive micro implicit feedback behavior in the literature [7]. Among them, IFB1-IFB3 is category mutually exclusive micro implicit feedback behavior, each user corresponds to a micro implicit feedback behavior result, such as selecting only one application market class, a certain access frequency and attention frequency to item determined; IFB4-IFB11 is category non-mutually exclusive type micro implicit feedback behavior, each user can correspond to multiple micro implicit feedback behavior results, such as the query frequency to item when the user is depressed, when the user needs to complete the task, when the user is bored.

The variable $f_{\text{IFBn}}(C_m)$ is defined as the occurrence frequency of some implicit behavior IFBn. Then for mutually exclusive user behavior, $f_{\text{IFBn}}(C_m) = \sum_m C_m = 1$, and for non-mutually exclusive user behavior, $f_{\text{IFBn}}(C_m) = \sum_m C_m \geq 1$. Among these, C_m is the mth the category attribute values of the nth micro implicit feedback behavior IFBn.

Let the sample size of user micro implicit feedback behavior be N, then the behavior distribution is defined as $\left(\sum_N Cm\right)/N$ to clearly reflect the differences of various attributes of user micro implicit feedback behavior. At the same time, the correlation of the behavior by calculating the micro implicit feedback behavior. Due to the large numerical discretization, $f_{\text{IFBn}}(C_m)$, of the microscopic implicit feedback behavior IFBn and the inconsistent range of variation, it was normalized before the correlation analysis.

4 Experiments and Analysis

4.1 Microscopic Implicit Feedback Behavior Distribution

1) Users differ greatly in category selection (IFB1) for the application market. In Fig. 1, the top three are the differences in micro implicit feedback behavior of Android Market, Apple iOS App Store, Nokia Ovi Store, except from the context influence of user attributes discussed here, and more from the influence of software and hardware of mobile intelligent terminals, which will be discussed in subsequent studies.

2) The frequency of access (IFB2) in the application market is the reflection of user demand. This statistical data has not a strong relationship between the hardware and software of the mobile intelligent terminals used by the user, so the category is relatively evenly distributed, as shown in Fig. 2.

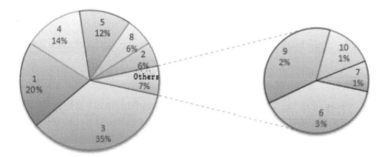

Fig. 1. Class distributions of microscopic implicit feedback behavior IFB1.

3) The number of attention to item per month (IFB3) reflects the strong willingness and choice tendency, but few users with high attention, as shown in Fig. 3, more users pay attention to item within 5 times a month, among which the number of attention to item is 0 or 1 is 40% and 2–5 for 36%, showing certain long tail characteristics.

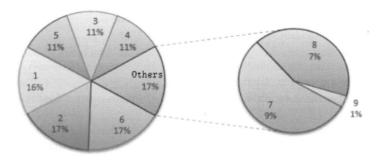

Fig. 2. Distributions of microscopic implicit feedback behavior IFB2.

4) The query frequency (IFB4) to item is also a microscopic implicit feedback behavior that reflects user willingness and choice propensity. According to the questionnaire data of literature [7], except for the last category (including data that cannot be classified to the top 5 categories), users with different needs, such as work demand, query demand, entertainment demand, etc., the query frequency fluctuates little, as shown in Fig. 4.

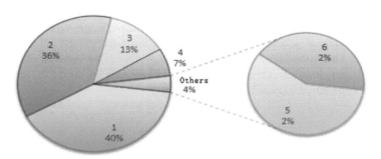

Fig. 3. Class distributions of microscopic implicit feedback behavior IFB3.

5) The query way of item (IFB5), from the questionnaire data in the literature [7], except the last category (including data that cannot be categorized to the top 8 categories), is shown in Fig. 5. the most way users use to query of item is keyword search, the most distrust way is list ranking.

6) Detail level of item browsing (IFB6). The most user attention to item information is price, features, detail description and comments, as shown in Fig. 6. From the implicit feedback behavior of mobile smart terminals, it is similar to PC-based user behavior.

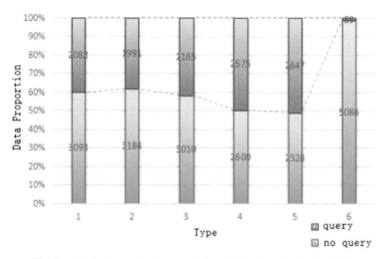

Fig. 4. Distributions of microscopic implicit feedback behavior IFB4.

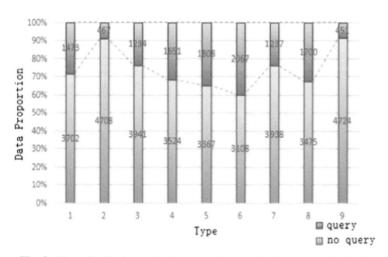

Fig. 5. Class distributions of microscopic implicit feedback behavior IFB5.

7) The intensity of attention on item (IFB7) also reflects user purchase possibilities for item. In addition to the last category (including data that cannot be classified to the top 14 categories), item with high intensity of user attention are entertainment, function and novelty, and lower ones are stranger communication, advertising effect and impulse purchase, reflecting users' rational attention, as shown in Fig. 7.

8) Purchases of item (IFB8). Except for the last category (including data that cannot be categorized to the top 11 categories), users preferred free item, unless there is no free version and similar features and requires increased functionality and performance, as shown in Fig. 8. Users don't tend to subscribe to a certain item and pay.

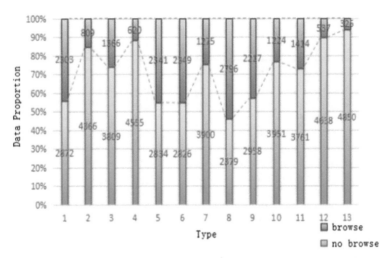

Fig. 6. Distributions of microscopic implicit feedback behavior IFB6.

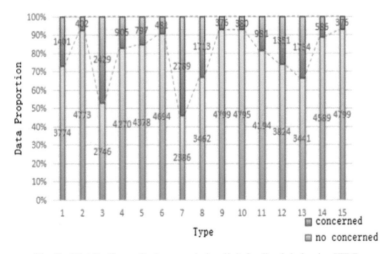

Fig. 7. Distributions of microscopic implicit feedback behavior IFB7.

9) Evaluation behavior (IFB9) for item. Except for the last category (including data that cannot be categorized to the top 6 categories), the data showed that the user did not like the evaluation, as shown in Fig. 9. Some existing reviews are given mainly to let others understand the merits of item. Mandatory evaluations are currently relatively few.

10) Cancel attention to item (IFB10). Except for the last category (including data that cannot be classified to the top 14 categories), causes users to dismiss item or find a better replacement, as shown in Fig. 10. The cancellation of attention is less affected by his family or friends.

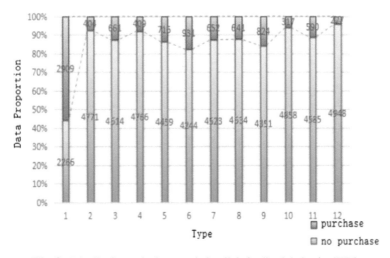

Fig. 8. Distributions of microscopic implicit feedback behavior IFB8.

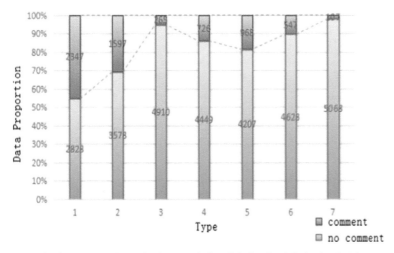

Fig. 9. Distributions of microscopic implicit feedback behavior IFB9.

11) Category focus behavior on item (IFB11). In addition to the last category (including data that cannot be classified to the top 22 categories), the item categories that users focus on are game category, social network category, music category, etc., and the item categories that users do not pay attention to are catalog category, medicine category and reference category, as shown in Fig. 11.

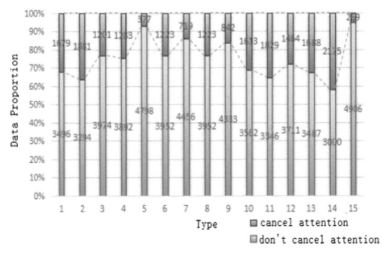

Fig. 10. Class of distributions of the microscopic implicit feedback behavior IFB10.

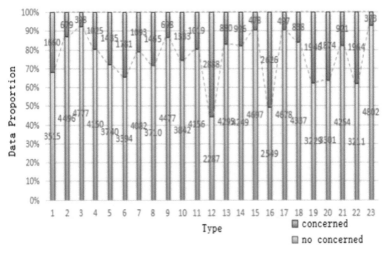

Fig. 11. Class of distributions of the microscopic implicit feedback behavior IFB11.

4.2 Microscopic Implicit Feedback Behavioral Correlations

The correlations between the implicit feedback behavior of non-mutually exclusive type microscopy are analyzed, as shown in Table 2.

Table 2. Microscopic implicit feedback behavioral correlations.

		IFB4	IFB5	IFB6	IFB7	IFB8	IFB9	IFB10	IFB11
IFB4	Pearson Correlation Coefficient	1	0.668^{**}	0.591^{**}	0.636^{**}	0.412^{**}	0.359^{**}	0.458^{**}	0.550^{**}
	Significance (two tailed)		0.000	0.000	0.000	0.000	0.000	0.000	0.000
IFB5	Pearson Correlation Coefficient	0.668^{**}	1	0.665^{**}	0.689^{**}	0.553^{**}	0.408^{**}	0.482^{**}	0.596^{**}
	Significance (two tailed)	0.000		0.000	0.000	0.000	0.000	0.000	0.000
IFB6	Pearson Correlation Coefficient	0.591^{**}	0.665^{**}	1	$0.714*^{*}$	0.486^{**}	0.419^{**}	0.618^{**}	0.594^{**}
	Significance (two tailed)	0.000	0.000		0.000	0.000	0.000	0.000	0.000
IFB7	Pearson Correlation Coefficient	0.636^{**}	0.689^{**}	0.714^{**}	1	0.578^{**}	0.461^{**}	0.579^{**}	0.651^{**}
	Significance (two tailed)	0.000	0.000	0.000		0.000	0.000	0.000	0.000
IFB8	Pearson Correlation Coefficient	0.412^{**}	0.553^{**}	0.486^{**}	0.578^{**}	1	0.430^{**}	0.337^{**}	0.484^{**}
	Significance (two tailed)	0.000	0.000	0.000	0.000		0.000	0.000	0.000
IFB9	Pearson Correlation Coefficient	0.359^{**}	0.408^{**}	0.419^{**}	0.461^{**}	0.430^{**}	1	0.385^{**}	0.413^{**}
	Significance (two tailed)	0.000	0.000	0.000	0.000	0.000		0.000	0.000
IFB10	Pearson Correlation Coefficient	0.458^{**}	0.482^{**}	0.618^{**}	0.579^{**}	0.337^{**}	0.385^{**}	1	0.526^{**}
	Significance (two tailed)	0.000	0.000	0.000	0.000	0.000	0.000		0.000
IFB11	Pearson Correlation Coefficient	0.550^{**}	0.596^{**}	0.594^{**}	0.651^{**}	0.484^{**}	0.413^{**}	0.526^{**}	1
	Significance (two tailed)	0.000	0.000	0.000	0.000	0.000	0.000	0.000	

[**] At the 0.01 level (two tailed), the correlation was significant

The significance value indicators in the table are all 0, less than 0.05, meeting the premise of correlation analysis.The Pearson correlation value of IFB4 with IFB 5, IFB 7 was greater than 0.6, indicating that the three microscopic implicit feedback behaviors are correlated and strongly correlated. Similarly, IFB 5 is associated strongly with IFB 6, IFB 7, IFB 6 with IFB 7, IFB 10, and IFB 7 with IFB 11. Purchase behavior (IFB8) for item and evaluation behavior for item (IFB9), showed a weak correlation with other behaviors.

5 Conclusions

This paper provides the analysis of the implicit feedback behavior of mobile intelligent terminal, establishes the micro implicit feedback behavior data set, and analyzes the behavior distribution and non-mutually exclusive micro implicit feedback behavior respectively, which lays the basis for further using the analysis results.

Acknowledgments. This work was supported by The National Natural Science Foundation of China (No. 61802107); Science and technology research project of Hebei University (No. ZD2020171); Jiangsu Planned Projects for Postdoctoral Research Funds (No. 1601085C).

References

1. Seo, Y.W., Zhang, B.T.: Learning user's preferences by analyzing Webbrowsing behaviors. In: Proceedings the 4th International Conference on Autonomous Agents, pp. 381–387 (2000)
2. Morita, M., Shinoda, Y.: Information filtering based on user behavior analysis and best match text retrieval. In: Proceedings the 17th Annual International ACM-SIGIR Conference on Research and Development in Information Retrieval, pp. 272–281 (1994)
3. Konstan, J.A., Miller, B.N., Maltz, D., et al.: GroupLens: applying collaborative filtering to Usenet news. In: Communications of the ACM, pp. 77–87 (1997)
4. Oard, D.W., Kim, J.: Implicit feedback for recommender systems. In: Proceedings of the AAAI Workshop on Recommender Systems, p. 83 (1998)
5. Kelly, D., Teevan, J.: Implicit feedback for inferring user preference: a bibliography. In: Acm Sigir Forum, pp. 18–28 (2003)
6. Yin, C.H., Deng, W.: Extracting user interests based on analysis of user behaviors. Comput. Technol. Dev. 18(5), 37–39 (2008)
7. Lim, S.L., Bentley, P.J.: Investigating country differences in mobile App user behavior and challenges for software engineering. IEEE Trans. Softw. Eng. 41(1), 40–64 (2015)
8. Zhou, G., Zhu, X., Song, C., et al.: Deep interest network for click-through rate prediction. In: Proceedings of the 24th ACM SIGKDD International Conference on Knowledge Discovery & Data Mining, pp. 1059–1068 (2018)
9. Xiao, Z., Yang, L., Jiang, W., et al.: Deep multi-interest network for click-through rate prediction. In: Proc. of the 29th ACM International Conference on Information & Knowledge Management, pp. 2265–2268 (2020)
10. Tang, H., Liu, J., Zhao, M., et al.: Progressive layered extraction (PLE): a novel multi-task learning (MTL) model for personalized recommendations. In: Fourteenth ACM Conference on Recommender Systems, pp. 269–278 (2020)
11. Qu, J.: Big data network user rowsing implicit feedback information retrieval simulation. Computer Simulation 430–433 (2019)

Design and Implementation of a Novel Interconnection Architecture from WiFi to ZigBee

Yu Gu[✉], Chun Wu, and Jiangan Li

School of Computer and Information, Hefei University of Technology, Hefei, China
yugu.bruce@ieee.org, {2019170971,2019170966}@mail.hfut.edu.cn

Abstract. The signal layer heterogeneous communication technology is a cross-technology communication (CTC) technology, which is a direct communication technology between different wireless devices. Since ZigBee and WiFi have overlapping spectrum distribution, the ZigBee transmission will affect the CSI sequence. We propose a CTC technology based on machine learning and neural network, from Zigbee to WiFi, leveraging only WiFi channel state information (CSI). By classifying WiFi CSI, we can distinguish whether there is ZigBee signal transmission in WiFi signal. This paper uses the machine learning method and neural network method to classify CSI sequence analyzes the importance of CSI sequence features to the classifier, improves the accuracy of machine learning classifier by extracting multiple CSI sequence features, and improves the classification accuracy by neural network classifier. In our experimental data set, the highest accuracy can reach 95%. The evaluation results show that our accuracy is higher than the existing methods.

Keywords: Heterogeneous communication · CSI · Machine learning · LSTM

1 Introduction

According to the prediction of the Global System for Mobile Communications assembly (GSMA), the number of global Internet of things (IoT) devices will reach about 24 billion in 2025. So many IoT devices bring challenges to the communication between different IoT devices. Traditionally, the method to realize the communication between heterogeneous IoT devices is to realize the indirect connection between heterogeneous IoT devices through IoT gateway. This will lead to an increase in cost, requiring Internet of things gateway equipment for transfer, slow data transmission and small traffic [1]. As a new research field, CTC has great application scenarios and good scientific research prospects [2]. According to different implementation schemes, CTC mainly includes packet-based CTC and signal-based CTC [3].

In packet-based CTC, the direct CTC of heterogeneous Internet of things devices is realized by embedding packet length, packet energy, and combined frame. Busybee [4] realized the CTC between WiFi devices and ZigBee devices and designed a scheme

© The Author(s) 2022
Z. Qian et al. (Eds.): WCNA 2021, LNEE 942, pp. 40–47, 2022.
https://doi.org/10.1007/978-981-19-2456-9_5

to encode channel access parameters. The system can correctly decode WiFi signals and ZigBee signals. Zifi [5] uses the unique interference signature generated by ZigBee radio through WiFi beacon to identify the existence of WiFi network. C-morse [6] It is the first to use traffic to implement CTC. When building recognizable wireless energy mode, c-morse slightly interferes with the WiFi packets. The packet-level CTC avoids hardware modifications, but it reduces the transmission rate and bandwidth.

Compared with packet-based CTC, the signal-based CTC will greatly improve throughput, which is conducive to improving throughput and expanding the application range of CTC [1]. TwinBee [7] realizes CTC by recovering chip errors introduced by imperfect signal simulation. LongBee [8] improves the reception sensitivity through new conversion coding, so as to realize CTC.

In this paper, the coding and decoding problem of the CTC signal is transformed into the classification problem of WiFi CSI. We extract several features of the WiFi CSI sequences, and then classify the CSI signal through machine learning classifiesr and neural network. We mark the CSI signal affected by ZigBee as "1" and the CSI signal not affected by ZigBee as "0". Specifically, our major contributions are as follows:

(1) We propose a CTC technology based on machine learning and neural network, from Zigbee to WiFi, using only WiFi CSI.
(2) We use a variety of machine learning methods to classify CSI sequences. We extracted eight CSI sequence features and analyzed the accuracy of machine learning classifier using six machine learning classifiers to improve the classification accuracy of CSI sequences.
(3) We use neural networks to classify CSI sequences, and neural network has a high accuracy. The experimental results show that the classification accuracy of CSI sequences by machine learning and neural network has reached a satisfactory level.

This paper consists of five sections, and the overall structure is as follows: The Sect. 2 introduces the preliminary work, the Sect. 3 introduces the system design, the Sect. 4 introduces the result analysis, and the Sect. 5 summarizes this paper.

2   Preliminary

2.1   The Spectrum Usage of ZigBee and WiFi

ZigBee is a new low-cost, low-power, and low-speed technology suitable for short-range wireless communication. It can be embedded in various electronic devices to support geographic positioning functions. This technology is mainly designed for low-speed communication networks. Different transmission speeds. WiFi and ZigBee use the 2.4 GHz wireless frequency band and adopt the direct sequence spread spectrum transmission technology (DSSS). ZigBee, transmission distance 50–300 m, rate 250 kbps, power consumption 5 mA. ZigBee is usually used in smart home. WiFi, fast speed (11Mbps), high power consumption, generally connected to the external power supply.

The spectrum usage of ZigBee and WiFi is shown in Fig. 1. Channel 1 of WiFi and channels 11, 12, 13, and 14 of ZigBee overlap, so we can try to achieve cross-technology communication from Zigbee to WiFi.

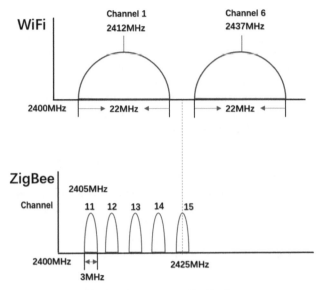

Fig. 1. The spectrum distribution

2.2 Channel State Information

In order to realize heterogeneous communication from Zigbee to WiFi, we need to analyze the changes of WiFi signals. Channel state information (CSI) is information used to estimate the channel characteristics of a communication link. Therefore, we use WiFi CSI information to analyze WiFi signals.

As shown in Fig. 2, the left figure shows the WiFi CSI signal when there is ZigBee, and the right figure shows the WiFi CSI signal when there is no ZigBee. It can be seen from the figure that ZigBee will affect the WiFi CSI signal. We can judge whether there is ZigBee by analyzing the WiFi CSI signal. Therefore, cross-technology communication from Zigbee to WiFi can be realized.

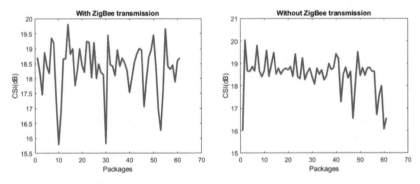

Fig. 2. The impact of ZigBee on WiFi CSI signal

2.3 The Support Vector Machines (SVM) Classifier

In this paper, we use machine learning classifiers to classify WiFi CSI signals. The experimental demonstrates that SVM classifier is the best classifier in our CSI sequence. Next, we introduce the SVM classifier.

Support vector machine (SVM) is a two class machine learning classifier. It is a supervised model, which is usually used for data classification of small samples. Support vector machine is the segmentation surface used to segment data points. Its position is determined by the support vector (if the support vector changes, the position of the segmentation surface will change). Therefore, this surface is a classifier determined by the support vector, that is, the support vector machine.

3 System Design

Figure 3 illustrates our system design, we first collect CSI data, then process the collected data, through the feature selection module and classification module, and finally analyze the classification results.

3.1 Hardware Setting

We conduct data acquisition on WiFi and ZigBee devices. We use the Intel 5300 network card as the WiFi device and the TelosB node as the ZigBee device. The transmission interval of WiFi packets is 0.5 ms and the length is 145 bytes. ZigBee packets are sent at an interval of 0.192 ms and 28 bytes in length. The experiment was conducted in a real environment. We extract some features of the WiFi CSI signal, and then classify the CSI signal through a machine learning classifier and neural network. We mark the CSI signal affected by ZigBee as "1" and the CSI signal not affected by ZigBee as "0".

3.2 Feature Extraction

The length of the classifier window is 16, which can obtain the optimal classification accuracy and transmission rate. In each window, we extract 8 features of CSI sequence: variance, peak to peak, kurtosis, bias, standard deviation, mean, mode and median. We classify the extracted features of CSI sequences with machine learning classifiers, and the classification results will be analyzed in Sect. 4.

3.3 Machine Learning Classification Selection and Neural Network Design

We use machine learning classifiers such as complex tree, quadratic discriminator, cubic SVM, fine KNN, medium tree, bagged trees and logistic regression. The classification results will be analyzed in Sect. 4.

Long short term memory network (LSTM) is a kind of time recurrent neural network (RNN), LSTM avoids long-term dependence through deliberate design. LSTM neural network is more suitable for dealing with timing problems. Our CSI sequences are timing problems, so we can use LSTM to classify them. Figure 4 illustrates the LSTM network structure we use.

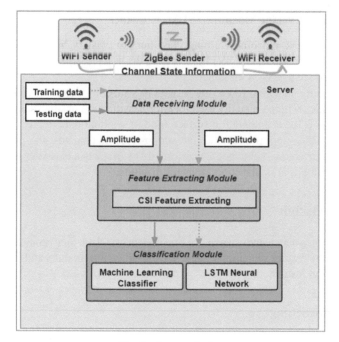

Fig. 3. System design

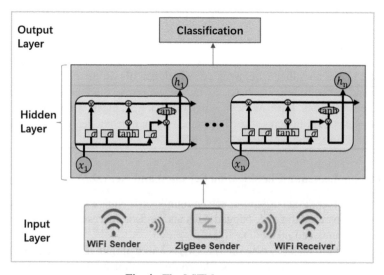

Fig. 4. The LSTM structure

4 Result Evaluation

4.1 Hardware

We experimented with off-the-shelf hardware. Figure 5 shows the placement of our transmitting antenna and receiving antenna. We used one WiFi transmitter and three WiFi receivers for the experiment. The distance between the transmitter and the receiver is about 100 cm, which can obtain better classification accuracy. ZigBee transmitter is between transmitting antenna and receiving antenna. The distance between the ZigBee transmitter and WiFi transmitting antenna and receiving antenna is about 50 cm.

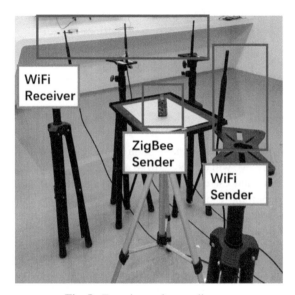

Fig. 5. Experimental setup diagram

4.2 Evaluation of Experiment Results

We extract 8 features of CSI sequence and train them with 10 machine learning classifiers. The classification results are shown in Table 1. Different machine learning classifiers have different classification accuracy, among which SVM classifier has the highest accuracy. The accuracy of Cubic SVM is 93.8%.This is the highest accuracy of machine learning classifier, reaching a high level.

Then we use the LSTM network introduced in Sect. 3 for training. The accuracy of LSTM is 94.2%, which is higher than that of SVM in machine learning classifier. LSTM is more suitable for training time series. Our CSI sequence is time series, which improves the accuracy of CSI sequence classification.

Table 1. Classification results our dataset.

Classifier	Accuracy
Complex Tree	88.9%
Quadratic Discriminant	72.8%
Cubic SVM	93.8%
Fine KNN	80.8%
Medium Tree	89.5%
RUSBoosted Trees	89.5%
Baggled Trees	90.6%
Boosted Trees	91.7%
Baggled Trees	90.6%
Logistic Regress	80.2%

5   Conclusion and Next Work

We realize the cross-technology communication from Zigbee to WiFi through CSI classification. In future work, we will explore how to realize cross-technology communication from WiFi to ZigBee, and use other neural networks to classify CSI sequences. CTC technology is an important technology in the Internet of things, which can realize the communication between different Internet of things devices. There is still a lot of work to be done in the future.

Acknoledgment. This work is supported by the National Key Research and Development Program Cyberspace Security Special Project "Research on Key Technologies for the Internet of Things and Smart City Security Assurance" under Grant No. 2018YFB0803403.

References

1. Xia, S., Chen, Y., Li, M., Chen, P.: A survey of cross-technology communication for iot heterogeneous devices. IET Commun. **13**(12), 1709–1720 (2019)
2. Zheng, X., He, Y., Guo, X.: StripComm: Interference-Resilient Cross-Technology Communication in Coexisting Environments, pp. 171–179 (2018)
3. Lu, B., Qin, Z., Yang, M., Xu, X., Lei, W.: Spoofing attack detection using physical layer information in cross-technology communication. In: 2018 15th Annual IEEE International Conference on Sensing, Communication, and Networking (SECON). IEEE (2018)
4. Croce, D., Galioto, N., Garlisi, D., Giaconia, C., Tinnirello, I.: Demo: unconventional WiFi-ZigBee communications without gateways. In: Proceedings of the 9th ACM international workshop on Wireless network testbeds, experimental evaluation and characterization. ACM (2014)

5. Zhou, R., Xiong, Y., Xing, G., Sun, L., Ma, J.: ZiFi: Wireless LAN Discovery via ZigBee Interference Signatures. In: International Conference on Mobile Computing & Networking. DBLP (2010)
6. Yin, Z., Jiang, W., Song, M. K., Tian, H.: C-Morse: Cross-technology communication with transparent Morse coding. IEEE INFOCOM 2017. In: IEEE Conference on Computer Communications. IEEE (2017)
7. Chen, Y.: TwinBee: Reliable Physical-Layer Cross-Technology Communication with Symbol-Level Coding Paper#1570385101 (2018)
8. Li, Z., He, T.: LongBee: Enabling Long-Range Cross-Technology Communication. pp. 162–170 (2018)

Cascaded GLRT Radar/Infrared Lidar Information Fusion Algorithm for Weak Target Detection

Peixuan Wu, Xiaoyong Du, and Weidong Hu$^{(\boxtimes)}$

National Key Laboratory of Science and Technology on Automatic Target Recognition, National University of Defense Technology, Changsha 410073, China
wdhu@nudt.edu.cn

Abstract. To deal with the problem of weak target detection, a cascaded generalized likelihood ratio test (GLRT) radar/infrared lidar heterogeneous information fusion algorithm is proposed in this paper. The algorithm makes full use of the target characteristics in microwave/infrared spectrum and the scanning efficiency of different sensors. According to the correlation of target position in the multi-sensor view field, the GLRT statistic derived from the radar measurements is compared with a lower threshold so as to generate initial candidate targets with high detection probability. Subsequently, the lidar is guided to scan the candidate regions and the final decision is made by GLRT detector to discriminate the false alarm. To get the best detection performance, the optimal detection parameters are obtained by nonlinear optimization for the cascaded GLRT Radar/Infrared lidar heterogeneous information fusion detection algorithm. Simulation results show that the cascaded GLRT heterogeneous information fusion detector comprehensively utilizes the advantages of radar and infrared lidar sensors in detection efficiency and performance, which effectively improves the detection distance upon radar weak targets within the allowable time.

Keywords: Radar · Infrared lidar · Heterogeneous fusion · Cascaded detector · False alarm discrimination

1 Introduction

Some important targets with collaborate design of shape and material are capable of backscattering the incidence electromagnetic wave weakly and the radar detection performance degrades a lot. Single-mode sensors are no longer satisfy the detection requirements, and multi-sensor fusion detection has become a development trend [1, 2], such as multi-radar sensors fusion [3], multi-infrared sensors fusion [4], radar/infrared fusion [5], radar/optical fusion [6], lidar point cloud/optical fusion [7], etc.

Since it is hard to control the targets characteristics in microwave and infrared frequency bands simultaneously, radar and infrared sensors have become an important combination mode for fusion detection. In [8], the infrared imaging/active radar fusion detection of weak target is realized through spatiotemporal registration and radar virtual

© The Author(s) 2022
Z. Qian et al. (Eds.): WCNA 2021, LNEE 942, pp. 48–57, 2022.
https://doi.org/10.1007/978-981-19-2456-9_6

detection image generation from infrared image. In [9], the relevance of radar/infrared characteristics is used for multi-target association. However, the maximum detection range of passive infrared sensors usually mismatches to that of radar. With the development of laser phased array technology, the combination of radar and infrared lidar will exhibits potential in aerial target detection.

Although it's easy to realize the spatiotemporal registration for co-platform radar/infrared lidar, the target characteristics in microwave/infrared spectrum has great difference which increased the difficulty of fusion detection. Besides, the mechanisms of radar and infrared lidar are different from each other. The wide beam of the radar can lead to quicker scanning but the detection angle resolution is low; the narrow beam of lidar can lead to higher detection angle resolution but the scanning and detection speed is low. Therefore, this paper proposed a cascaded GLRT radar/infrared lidar heterogeneous information fusion algorithm to solve the fusion detection problem of that radar/infrared lidar cross-spectrum sensors have difference on target characteristics and detection mechanisms.

Aiming at the problem of long distance and weak targets detection, the radar/infrared lidar heterogeneous fusion detection method is studied in this paper. Based on the target location prior constraint relationship of multi-sensor, a low detection threshold is set for radar detection firstly, and then the infrared lidar is guided by radar detection results for further detection and false alarm elimination. The organization of the paper is as follows: Sect. 2 describes the radar/infrared lidar measurement model, Sect. 3 provides the method of heterogeneous fusion detection, and the simulation experiments of typical scenarios are demonstrated in Sect. 4, and Sect. 5 concludes the paper.

2   Measurement Model of Radar and Infrared Lidar

2.1   Radar Echo Model

When the radar transmits a series of pulses with carrier frequency f_c, the echo of a target at distance R can be expressed as follows

$$S_r(t) = \sum_{k=0}^{CPI-1} \sqrt{P_{t_R} \cdot \sigma_R \cdot K} \cdot \text{rect}\left[\frac{t - \tau_{Ra} - kT_{r_R}}{T_{PR}}\right] \cdot \exp\{j2\pi f_c(t - \tau_{Ra})\} + e_R(t) \quad (1)$$

where P_{t_R} is emitted peak power, σ_R is the radar cross section of target, CPI is the number of pulses, T_{r_R} is pulse repetition period and T_{PR} is the pulse width. $\tau_{Ra} = 2R/c$ is the echo delay time of the target, $K = \frac{G^2\lambda_R^2}{(4\pi)^3 R^4}$ is the propagation decay factor, $e_R(t)$ is complex white Gaussian noise due to the receiver [11] with variance P_{n_R}, and

$$P_{n_R} = kT_0 B N_F \quad (2)$$

$k = 1.38 \times 10^{-23} J/K$ is the Boltzmann constant, $T_0=290\,K$, B is the bandwidth of receiver and N_F is the noise coefficient of receiver.

2.2 Lidar Echo Model

The lidar echo of a target at distance R can be expressed as follows [13].

$$S_r(t) = \sum_{k=0}^{CPI-1} \sqrt{P_{t_R} \cdot \sigma_R \cdot K} \cdot \text{rect}\left[\frac{t - \tau_{Ra} - kT_{rR}}{T_{PR}}\right] \cdot \exp\{j2\pi f_c(t - \tau_{Ra})\} + e_R(t).$$

(3)

$Tau = T_{half}/\sqrt{8\ln 2}$, P_{t_L} is emitted peak power, σ_L is the lidar cross section of target, T_{PL} is the pulse width, $\tau_{Li} = \frac{2R}{c}$ is target echo delay time, $K = \frac{G_T}{(4\pi R^2)^2} \cdot \frac{\pi D_r^2}{4}$ is the propagation decay, $e_L(t)$ is the background light noise including the sunlight reflected by the target and scattered by the atmosphere and the direct sunlight [14]. The noise variance

$$P_b = \frac{\pi}{16}\eta_{r_L}\Delta\lambda\theta_{t_L}^2 D_r^2[\rho T_a H_\lambda \cos\theta \cos\varphi + \frac{\beta}{4\alpha}(1 - T_a)H_\lambda + \pi L_\lambda]$$

(4)

In case of air-to-air lidar detection, by reviewing the paper [15], the angle between the sun ray and the target surface is taken $\theta = 0$, the angle between the normal line of the target surface and the receiving axis is taken $\varphi = 0$. In addition, the transmittance of receiving optical system is $\eta_t = 1$, receiving field angle $\theta_{t_L} = 1$ mrad, target reflection coefficient $\rho = 0.8$, the narrowband filter bandwidth $\Delta\lambda = 50$ nm, atmospheric transmittance $T_a = 0.87$. Atmospheric attenuation coefficient and scattering coefficient are $\alpha = 1$ and $\beta = 1$ combined with the detection requirements of more than 100 km [16]. The spectral radiance of atmospheric scattering and the spectral irradiance on the ground of sunlight are $L_\lambda = 3.04 \times 10^{-6}$ W/(cm$^2 \cdot$ sr $\cdot$ nm) and $H_\lambda = 6.5 \times 10^{-5}$ W/(cm$^2 \cdot$ nm) are simulated by MOTRAN4.0 software. When $\lambda = 1064$ nm, the $P_b \approx 1.1 \times 10^{-7}$ W.

3 Radar/Infrared Lidar Fusion Detection Algorithm

The Radar/Infrared lidar fusion detection algorithm proposed in this paper is asynchronously cascaded, the radar target detection is finished firstly, then based on the radar detection results and position correlation, the infrared lidar is used for further detection and false alarm discrimination. The algorithm flow is shown in Fig. 1

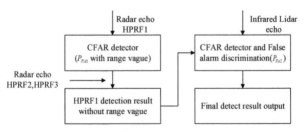

Fig. 1. Algorithm flow chart of cascade detection algorithm, the P_{FA1} and P_{FA2} are false alarm probability for radar detection and lidar detection

The radar/lidar heterogeneous fusion detection method includes two cascade target detection: the radar detection and the lidar false alarm discrimination. The received radar and lidar echo signals are converted into a discrete signal by the digital analogue digital converter (ADC), so the echo used for target detection is discrete sequence signal and the detection model [12] can be described as Eq. (5) uniformly.

$$\begin{cases} H_0: \mathbf{X} = \mathbf{w} \\ H_1: \mathbf{X} = A\mathbf{S} + \mathbf{w} \end{cases} \tag{5}$$

For radar detection, $\mathbf{X}$ and $\mathbf{S}$ are observation signal and signal wave with length N. $\mathbf{w} \sim CN(0, \sigma^2\mathbf{I}_N)$, and the probability density function (PDF) is $\mathbf{X} \sim CN(0, \sigma^2\mathbf{I}_N)$ for H_0 and $\mathbf{X} \sim CN(A\mathbf{S}, \sigma^2\mathbf{I})$ for H_1, A, σ^2 are both unknown parameters. The test statistics variable T can be constructed by GLRT [12].

$$T = \frac{(\mathbf{S}(\mathbf{S}^H\mathbf{S})^{-1}\mathbf{S}^H\mathbf{X})^H(\mathbf{S}(\mathbf{S}^H\mathbf{S})^{-1}\mathbf{S}^H\mathbf{X})/m}{((\mathbf{I} - \mathbf{S}(\mathbf{S}^H\mathbf{S})^{-1}\mathbf{S}^H)\mathbf{X})^H((\mathbf{I} - \mathbf{S}(\mathbf{S}^H\mathbf{S})^{-1}\mathbf{S}^H)\mathbf{X})/n} = \frac{(\mathbf{P_S X})^H(\mathbf{P_S X})/m}{((\mathbf{I} - \mathbf{P_S X})^H((\mathbf{I} - \mathbf{P_S X})/n} \tag{6}$$

The PDF of test statistics variable T is $T \sim F_{m,n}$ for H_0 and $T \sim F_{m,n}(\lambda)$ for H_1, $m = 2\text{rank}(\mathbf{P_S})$, $n = 2N - m$, $\lambda = \frac{2(A\mathbf{S})^H(A\mathbf{S})}{\sigma^2}$.

For lidar false alarm discrimination, $\mathbf{X}$ and $\mathbf{S}$ are two-dimensional observation signal and signal wave with size $M \times N$, M is the number of beams, N is the number of distance bins. $\mathbf{S} = [s_{mn}]_{M \times N}$ has an unknown parameter m_0 (m_0 is the index of a beam containing targets), $\mathbf{w} = [w_{mn}]_{M \times N}$, and $w_{mn} \sim \mathcal{N}(0, \sigma^2)$. Assuming that $\vec{\mathbf{X}}$ and $\vec{\mathbf{S}}$ are both one-dimensional vectors stretched from the two-dimensional matrix $\mathbf{X}$ and $\mathbf{S}$, the PDF is $\vec{\mathbf{X}} \sim \mathcal{N}(0, \sigma^2\mathbf{I}_{MN})$ for H_0 and $\vec{\mathbf{X}} \sim \mathcal{N}(A\vec{\mathbf{S}}, \sigma^2\mathbf{I}_{MN})$ for H_1, A, σ^2 are both unknown parameters. Then T constructed by GLRT [12] is as Eq. (7) shows.

$$T = \arg\max_{m_0 \in M} \frac{(\mathbf{P_S}\vec{\mathbf{X}})^T(\mathbf{P_S}\vec{\mathbf{X}})/p}{((\mathbf{I} - \mathbf{P_S})\vec{\mathbf{X}})^T((\mathbf{I} - \mathbf{P_S})\vec{\mathbf{X}})/(MN - p)} \tag{7}$$

$p = rank(\mathbf{P_S}) = rank(\vec{\mathbf{S}}(\vec{\mathbf{S}}^T\vec{\mathbf{S}})^{-1}\vec{\mathbf{S}}^T)$. Because the correlation of the M random variables is hard to analysis, it's difficult to calculate the PDF of T. Considering that the lidar echo of a point target for the different beam is independent, if we use $\vec{\mathbf{x}}_m$ (the beam echo that beam index is m) to substitute $\vec{\mathbf{X}}$, use $\vec{\mathbf{s}}_m$ (the beam wave with index m) to substitute $\vec{\mathbf{S}}$, the T is changed to

$$T = \arg\max_{m_0 \in M} \frac{(\mathbf{P}_{\vec{\mathbf{s}}_{m_0}}\vec{\mathbf{x}}_{m_0})^T(\mathbf{P}_{\vec{\mathbf{s}}_{m_0}}\vec{\mathbf{x}}_{m_0})/p_{m_0}}{((\mathbf{I}_N - \mathbf{P}_{\vec{\mathbf{s}}_{m_0}})\vec{\mathbf{x}}_{m_0})^T((\mathbf{I}_N - \mathbf{P}_{\vec{\mathbf{s}}_{m_0}})\vec{\mathbf{x}}_{m_0})/(N - p_{m_0})} \tag{8}$$

when m_0 is given, the $\vec{\mathbf{s}}_{m_0}$ will be definite. And for different values of m_0, $\vec{\mathbf{s}}_{m_0}$ are same, so the values of $p_{m_0} = rank(\mathbf{P}_{\vec{\mathbf{s}}_{m_0}}) = rank(\vec{\mathbf{S}}_{m_0}(\vec{\mathbf{S}}_{m_0}^T\vec{\mathbf{S}}_{m_0})^{-1}\vec{\mathbf{S}}_{m_0}^T)$ are same, and the observation echo in different beams are independent. Thus, the PDF of the test statistics variable is easy to analysis [17], the false alarm probability and detection probability are as Eq. (9) shows. F_t is the distribution function of t whose PDF is $F_{p,N-p}$, F_{t_2} is the distribution function of t_2 whose *pdf* is $F'_{p,N-p}(\lambda)$, $\lambda = \frac{(A\vec{s}_{m_0})^T (A\vec{s}_{m_0})}{\sigma^2}$.

$$
\begin{cases}
P_{FA} = \Pr\{T > \gamma|H_0\} = 1 - (F_t(\gamma))^M \\
P_D = \Pr\{T > \gamma|H_1\} = \Pr\{t_1 > \gamma, \, t_2 > \gamma|H_1\} = 1 - F_t(\gamma)^{M-1} F_{t_2}(\gamma)
\end{cases}
\tag{9}
$$

In summary, suppose the Radar test statistics variable is T_1, the infrared lidar test statistics variable is T_1, the total P_{FA} and P_D of the detection system can be calculated

$$
\begin{cases}
P_{FA} = \Pr\{T_1 > \gamma_1, \, T_2 > \gamma_2|H_0\} = \Pr\{T_1 > \gamma_1|H_0\} \cdot \Pr\{T_2 > \gamma_2|H_0\} = P_{FA1} \cdot P_{FA2} \\
P_D = \Pr\{T_1 > \gamma_1, \, T_2 > \gamma_2|H_1\} = \Pr\{T_1 > \gamma_1|H_1\} \cdot \Pr\{T_2 > \gamma_2|H_1\} = P_{D1} \cdot P_{D2}
\end{cases}
\tag{10}
$$

For a given P_{FA}, to get the best P_D and satisfy the engineering application requirements for algorithm complexity at the same time, the following nonlinear optimization strategy are given to get the optimized false alarm probability parameters for cascade detection

$$
\begin{cases}
P_D = \underset{P_{FA1}, P_{FA2}}{\arg\max} \, P_{D1} \cdot P_{D2} \\
0 < P_D \leq 1, 0 < P_{FA1} < a, 0 < P_{FA2} < 1 \\
P_{FA1} \cdot P_{FA2} = P_{FA}
\end{cases}
\tag{11}
$$

The value of a is related to the signal processing speed of the detection system. In actual engineering applications, it's necessary to minimize the time required for signal processing to achieve real-time updates of detection results. Assuming that the system need finish the lidar false alarm elimination within a given time $T_{\lim}$, the expected detection time can approximately satisfy the inequality that

$$
E(T_D) \approx N_{B_{Ra}} \cdot N_{R_{Ra}} \cdot P_{FA1}/n_{L'} \leq T_{\lim} \Rightarrow P_{FA1} \leq \frac{T_{\lim} \cdot n_{L'}}{N_{B_{Ra}} \cdot N_{R_{Ra}}} = a
\tag{12}
$$

$N_{B_{Ra}}$ is the number of Radar echo beams, $N_{R_{Ra}}$ is the number of Radar distance bins, P_{FA1} is the false alarm probability for radar detection, $n_{L'}$ is the number of false alarm eliminations completed by the signal processing system per unit time. $a = 10^{-1}$ in the paper, and the value of a can be changed for different detection situations.

In addition, the single radar detection model is the same as radar detection. Besides, for single lidar detection, the T can be constructed as $T = \frac{(P_S X)^T (P_S X)/p}{((I - P_S X)^T ((I - P_S X)/(N-p)}$, $p = \text{rank}(\vec{S}(\vec{S}^T \vec{S})^{-1} \vec{S}^T)$, $\lambda = \frac{(AS)^T (AS)}{\sigma^2}$.

4   Experiment and Analysis

4.1   Simulation Parameter Set

To verify the effectiveness of the cross-spectrum fusion detection algorithm, a stationary target detection simulation experiment is done in this part. (For moving targets, motion compensation can be used to convert target detection into an equivalent stationary target detection situation). The simulation parameters are as follows Table 1.

Table 1. The experiment simulation parameters

Radar parameters		Infrared lidar parameters	
Emitted peak power P_{t_R}	8000 W	Emitted peak power P_{t_L}	5000 W
Pulse width T_{PR}	200 ns	Pulse width T_{PL}	100 ns
Pulse repetition frequency $HPRF_1$	55 kHz	Pulse repetition frequency PRF	500 Hz
Pulse repetition frequency $HPRF_2, HPRF_3$	60 KH 66 kHz	Half maximum pulse width T_{half}	50 ns
Carrier frequency f_c	10 GHz	lidar wavelength λ_L	1064 nm
Pulse repetition number CPI	128	Normalized amplitude A	2×10^{-4}
Azimuth beam width θ_{t_R}	2°	Azimuth beam width θ_{t_L}	1 mrad
Beam scan interval θ_{D_R}	2°	Beam scan interval θ_{D_L}	0.1°
Radar antenna gain G	46 dB	lidar optical gain G_T	$G_T = 4\pi / \theta_T^2$
Receiver bandwidth	5 MHz	Receiver aperture	0.14 m
Receiver noise coefficient F_n	3.5 dB	Noise power P_b	110 nW
Sampling frequency f_{s_R}	50 MHz	Sampling frequency f_{s_L}	100 MHz
Radar cross section σ_R	0.2	lidar cross section σ_L	6.7241

Figure 2 shows the SNR varies with detection distance. It can be seen that the SNR of lidar echo is higher than that of the radar echo for the same detection distance.

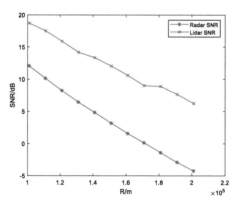

Fig. 2. The signal-to-noise ratio (SNR) of radar echo and lidar echo for different distances

4.2   Simulation Results and Analysis

Three comparative experiments are carried out to verify the effectiveness of fusion detection method, single radar detection, single lidar detection and radar/infrared lidar fusion detection. The number of Monte Carlo simulations are 1000, according to the evaluation method proposed in the paper [18], the multi-sensor information fusion performance is analyzed as follow.

Detection performance curve. Figure 3 shows the detection performance curve of the three detectors. The variations of P_D with detection distance when P_{FA} are 10^{-5} and 10^{-3} are given respectively. It shows that the detection probability of the fusion detection is obviously higher than single radar detection with the same detection distance; when the detection probability is 0.8, the combined detection result has the detection distance increment of 14 km and 12 km respectively compared with single-use radar detection in case of $P_{FA} = 10^{-5}$ and $P_{FA} = 10^{-3}$. Besides, when P_{FA} is low (corresponding to the case that $P_{FA} = 10^{-5}$), the detection result of fusion detection is close to that of single lidar detection.

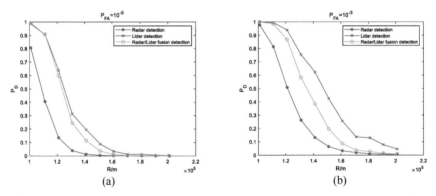

Fig. 3. The detection performance curve for single radar detection, single lidar detection and radar/infrared lidar fusion detection when P_{FA} is equal to 10^{-5} and 10^{-3}

Detection Time. The simulation scene is a two-dimensional plane with the azimuth angle range of $[-5°, 5°]$, and the number of detection units of the azimuth and distance dimension for radar and lidar detection is shown in Table 2.

Table 2. Detection unit parameters

	Radar detection	Infrared lidar detection	Radar/lidar fusion detection
Number of beams	$N_{BRa} = 5$	$N_{BLi} = 100$	$N_{BRa} = 5$
Number of range detection units	$N_{RRa} = 909$	$N_{RLi} = 66667$	$N_{RRa} = 909$ $N_{Li} = N_{BRa} \cdot N_{RRa} \cdot P_{FA1}$

(continued)

Table 2. (*continued*)

	Radar detection	Infrared lidar detection	Radar/lidar fusion detection
Signal length	$L_{Ra} = 183$	$L_{Li} = 183$	$L_{Ra} = 183, L_{Li'} = 183 \times 20$
Range window	[100 km,200 km]		
Azimuth range	[−5°, 5°]		

In addition, the detection time of different detection methods is analyzed in Table 3.

Table 3. The detection time of different detection methods

	Radar detection	lidar detection	Radar/lidar fusion detection
Detection time	$T_{D_{Ra}} = \dfrac{N_{B_{Ra}} \cdot N_{R_{Ra}}}{n_R}$	$T_{D_{Li}} = \dfrac{N_{B_{Li}} \cdot N_{R_{Li}}}{n_L}$	$T_{D_C} = \dfrac{N_{B_{Ra}} \cdot N_{R_{Ra}}}{n_R} + \dfrac{N_{B_{Ra}} \cdot N_{R_{Ra}} \cdot P_{FA1}}{n_{L'}}$

n_R, n_L and $n_{L'}$ are the number of detection times completed by the signal processing system per unit time for radar detection, lidar detection and radar/lidar fusion detection. $N_{B_{Li}}$ is the number of lidar beams and $N_{R_{Li}}$ represents the number of range units of the lidar echo. According to the simulation experiment, we can obtain that $n_R \approx n_L \approx 20 n_{L'}$, $P_{FA1} \leq 10^{-1}$, thus

$$T_{D_{Ra}} < T_{D_C} << T_{D_{Li}} \tag{13}$$

Figure 4 shows the variations of detection time with P_{FA} of the three detectors when R = 121 km. The time is calculated by MATLAB 2018b. The computer used in the experiment is a Lenovo Legion R7000 2020 notebook computer with 16G running memory, and the CPU is configured with an 8-core AMD Ryzen 7 4800 H.

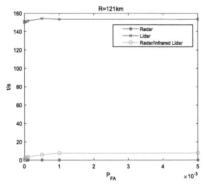

Fig. 4. The detection time of radar detection, lidar detection and radar/infrared lidar fusion detection for R = 121 km. The detection time is the average time for 1000 times simulation

It can be clearly seen that the detection time of the radar/infrared lidar fusion detection algorithm is much shorter than that of using a single lidar for detection.

5 Conclusion

Radar and infrared lidar are both active sensors, and they are complementary in working principle and detection performance. Based on the target characteristics and detection mechanism differences between radar and infrared lidar, this paper proposed a radar/infrared lidar cascade GLRT fusion algorithm for weak target detection and the optimal detection parameters are obtained by nonlinear optimization. The experimental simulation results show that the proposed fusion detection method has certain effectiveness: the heterogeneous information fusion detector comprehensively utilizes the advantages of radar and infrared lidar sensors in detection efficiency and performance, which effectively improves the detection distance upon radar weak targets within the allowable time. For further study, the joint statistics variable of radar/infrared lidar can be considered to constructed to make the best use of the target characteristics' correlation between microwave and infrared.

References

1. Luo, J.H., Yang, Y.: Overview of target detection methods based on data fusion. Control Decis. **35**(01), 1–15 (2020). (in Chinese)
2. He, Y.G.: Research on the key technologies of multi-sensor integration and information fusion. Sci. Eng. Res. Cent. **7**, (2015). (in Chinese)
3. Lei, B.: Research on Multi-Station Radar Cooperative Target Detection Method. Xidian University (2019). (in Chinese)
4. Zhang, H.B., Ju, Y.Q.: Helicopter multi-aircraft sensor cooperative detection method under radiation control conditions. Detect. Cont. **42**(05), 63–67 (2020). (in Chinese)
5. Zhang, W.L.: Research on image fusion and target detection algorithm based on multi-sensor infrared imaging system. Shan Dong University (2020). (in Chinese)
6. Wang, C.: Research on dangerous target detection method based on information fusion of millimeter wave radar and camera. Ji Lin University (2020). (in Chinese)
7. Hu, Z.Y., Liu, J.S., He, J.: Vehicle target detection method based on lidar point cloud and image fusion. Automobile Saf. Energy Conserv. (2019). (in Chinese)
8. Guo, M., Wang, X.W.: Infrared/active radar dim target fusion detection method based on infrared sensor parameters. Infrared Technol. **32**(8) (2010)
9. Liu, Z., Mao, H.X., Dai, C.M.: Research on association of dim and small targets based on multi-source data and multi-feature fusion. Infrared Laser Eng. **48**(05), 313–318 (2019). (in Chinese)
10. Fan, J.X., Liu, J.: Challenges and thoughts on intelligentized automatic target recognition of precision guidance. Aviation Weapon **26**(01), 30–38 (2019). (in Chinese)
11. Ding, L.F., Geng, F.L., Chen, J.C.: Radar Principle. Electronic Industry Publisher (2009). (in Chinese)
12. Kay, S.M.: The Basis of Statistical Signal Processing-Estimation and Detection Theory. Electronic Industry Publisher (2019). (in Chinese)

13. Ma, G.P., Yang, Y.: Multi-Pulse Lidar. National Defense Industry Publisher, p. 12 (2017). (in Chinese)
14. An, Y.Y., Zeng, X.D.: Photoelectricity Detection Principle. Xidian University Publishing House, Xi'an, pp. 42–45 (2004). (in Chinese)
15. Zheng, L.J.: Handbook of Optics (volume II). Shaanxi Science and Technology Press, Xi'an, p. 1802 (2010). (in Chinese)
16. Yan, D.K., Yan, P.Y., Huo, J., Guo, S., Jing, J.L.: Simulation research on maximum allowable noise of airborne long-range laser range finder. Laser Technol. (2018). (in Chinese)
17. Qu, T.Y.: The distribution of order statistics and its application in data analysis. Enterp. Technol. Dev. **11**, 127–129 (2018). (in Chinese)
18. Song, J., Ke, T., Zhang, H.: A performance evaluation method of multi-sensor information fusion system. Ship Electron. Countermeasures **43**(06), 60–64 (2020). (in Chinese)

Ship Encounter Scenario and Maneuvering Behavior Mining Based on AIS Data

Yinqiu Zhao, Yongfeng Suo$^{(\boxtimes)}$, and Bo Xian

Navigation College, Jimei University, Xiamen 361012, Fujian, China
yfsuo@jmu.edu.cn

Abstract. In order to gain a deep understanding of the operation of different ships in different time states and understand the geographical distribution of the encounters of ships near Gulei Port and the maneuvering behavior patterns of ships in the port area, this essay is different from the traditional single ship versus multi-target ship research. Through the comprehensive processing and data regulation of Gulei Port AIS (Automatic Identification System) data, the ships with consistent temporal and spatial characteristics are found, and the time and geographical position of the voyage data are revised, which solves the problem of asynchronous data processing of multi-target ships at different times. By ship navigation data mining, obtaining the trajectory distribution of the ship under a certain time condition, the distribution of the encounter area, the geographical distribution of the speed, and the law of ship speed and heading changes triggered by the formation of the encounter, summing up the same behavioral characteristics of different ship maneuvering modes in the port area at low speed.

Keywords: Ship encounter · AIS data · Maritime transportation

1 Introduction

Due to the complex characteristics of maritime transport itself, it is often necessary to comprehensively consider various aspects in the study of maritime transport, such as navigation waters, natural conditions, traffic conditions and other complex factors. In addition, the basic data collection and investigation of maritime transport also need to consider many data characteristics, such as ship density distribution, track distribution, traffic flow, traffic volume, speed distribution, ship arrival law, encounter rate and collision avoidance behavior. At the same time, due to the lack of AIS data, abnormal data, asynchronous broadcast time and large span, the availability and effectiveness of data are greatly reduced, and the subsequent data processing problem increasingly becoming the focus of research. The AIS data are used to realize ship behavior recognition based on multi-scale convolution [1]. The AIS data are mined, the complex and changeable ship routes are analyzed, and the behavior characteristics of ships are analyzed [2]. Research on ship behavior based on semantic level [3] and AIS data visualization [4] are exploring how to maximize the function of AIS data. Therefore, the use of AIS data mining

© The Author(s) 2022
Z. Qian et al. (Eds.): WCNA 2021, LNEE 942, pp. 58–65, 2022.
https://doi.org/10.1007/978-981-19-2456-9_7

for effective information on the regional distribution of multi-objective ships encountering and the characteristics of ship maneuvering behavior can help relevant personnel to understand the ship maneuvering law under realistic conditions and make corresponding adjustments according to the characteristics of ship maneuvering behavior. At the same time, it is of great significance for the deployment of maritime navigation aid facilities.

2   AIS Data Preprocessing

The data of ship automatic identification system includes the dynamic data and static data of the ship. Under realistic conditions, due to the influence of ship operation conditions and signal processing errors, AIS data are missing, repetitive and abnormal, which brings some difficulties to AIS data processing and analysis. The time asynchronous problem of AIS data between ships leads to further improvement of data processing difficulty. In order to improve the accuracy and reliability of the data, the missing value and abnormal value of the original AIS data are processed in advance. The data with the interval time span of data items greater than 30 min in the AIS data are deleted, and the data with abnormal speed are deleted. In order to facilitate the analysis of the actual navigation data with relatively large capacity, the ship navigation data with AIS data items greater than 300 are extracted, and the extracted 308 ship data are statistically analyzed. The data processing flow is show in Fig. 1.

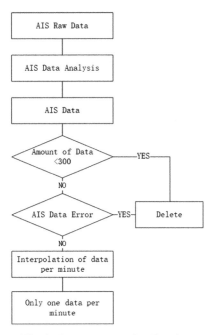

Fig. 1. Data preprocessing flowchart

3 Identification of Ship Encounter Area

3.1 Ship Distance Correction

Due to the asynchronous broadcast time of AIS data of different ships, there is a negative obstacle to the calculation of spherical distance between different ships. The spherical triangle sine theorem is used to correct the position of different ships. The distance between the target ship and the ship is corrected, and the navigation state of the two ships is compared at the same time. The spherical distance of the two ships is corrected to the same time. The distance between the two ships at the same time is used as the basis for detecting the occurrence of the encounter situation. At the same time, the distance between the two ships before and after the correction is recorded. When the distance between the two ships is small and less than a certain threshold, it is considered that the two ships have a potential encounter situation. The behavior mode of the ship before and after the time point is used to judge the steering and speed change measures after the encounter of the ship. The extracted AIS data of ships are shown in Table 1.

Table 1. AIS data processing items

MMSI	Postime	Course	Speed	Longitude	Latitude
813021827	1528143873	86.8	6.8	117.4514	23.60122
813021827	1528143875	86.8	6.8	117.4514	23.60122
568767867	1528782369	335.2	19.5	117.5596	23.67459
813021827	1528143933	88.7	6.6	117.4534	23.60125

By selecting two different ships, the distance of the point with the closest time difference is calculated. Due to the phenomenon of time asynchronous, the position of the ship A in time T_1 and the ship B in time T_2 is shown in Fig. 2. Due to the T_1 and T_2 is inequality, there is a certain time difference. Assuming T_1 is greater than T_2, to compare the distance of the two ships at the same time T_2, it is necessary to correct the position of the ship at the moment T_2, and move in the opposite direction along the existing course and speed of the ship A. The motion time is δ_t, the distance between the ship A and the ship B is corrected to the distance at the same time T_2, that is, the distance between the ship A and the ship B at the moment T_2.

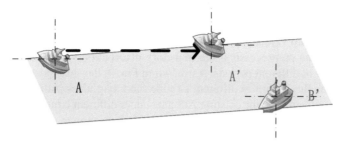

Fig. 2. Ship encounter distance correction

Spherical distance formula:

$$L = R \cdot \theta = Rar\cos[\cos(\alpha_1 - \alpha_2)\cos\beta_1\cos\beta_2 + \sin\beta_1\sin\beta_2] \tag{1}$$

$$\theta = \arccos[\cos(\alpha_1 - \alpha_2)\cos\beta_1\cos\beta_2 + \sin\beta_1\sin\beta_2] \tag{2}$$

where R is the radius of the earth, and the geographical coordinates of the two ships are $A(\alpha_1, \beta_1)$, $B(\alpha_2, \beta_2)$. Where α_1 and α_2 is the longitude of the ship and the target respectively, β_1 and β_2 is the latitude of the two ships, θ is the center angle of the large circle of the two points A and B, and L is the spherical distance of the two ships.

Since the navigation state of the ship on the water surface is constantly changing, the applicable condition of the correction method is that the change of ship heading and speed is relatively small under the condition of small-time difference, and the distance between the ship and the target ship is approximately linear. Therefore, in this paper, the time difference of the correction method is controlled to be less than or equal to 30 min, and the corrected distance is less than 3.8 nm [5] as the condition for the occurrence of the ship encounter situation. Thus, in the range of the existing AIS data, different ships with the corrected distance lower than the threshold and the close position and time are obtained. These two ships are considered as potential encounter ships, and the navigation data of these two ships are analyzed. The statistics of some encounter ships are shown in Table 2:

Table 2. Encounter ship list

MMSI of Ship A	MMSI of Ship B	Course of A	Speed of A	Course of B	Speed of B
413439530	416000147	226.5	2.5	30.1	5.2
413439530	416004349	317.1	0.1	297.5	4.9
900705594	416000147	298.3	7.8	51.5	0.5
413439530	814021779	280.1	7.3	294.6	6

3.2 Statistical Characteristics of AIS Data

Ship trajectory and velocity distribution in the region. Through the AIS data obtained after data preprocessing, the trajectory distribution of the ship in the region can be obtained in different periods. As shown in Fig. 3, the ship trajectory is dense in the triangle area that identifies different latitudes and longitudes. At the same time, the speed feature extraction of the existing AIS data under different latitudes and longitudes is carried out. Through three fittings, the speed distribution map of ship navigation in the region is obtained. Compared with the left and right parts, it can be seen that in the triangle area, the ship navigation speed is slow and the ship trajectory is dense. In this range, the maritime traffic volume is large and the frequency of ship encounters is relatively high.

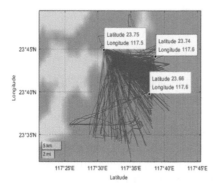 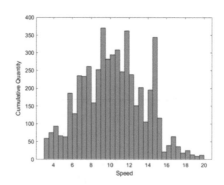

Fig. 3. Distribution of ship encounter areas

Ship encounter area mining. Through the historical AIS data information, the ships with relatively close distance at the same time and less than the threshold are selected. The speed change and heading change of each ship before and after the formation of the encounter situation are analyzed, and the latitude and longitude coordinates of the nearest encounter distance between the two ships are analyzed. From the existing statistical data, the geographical coordinates of all ships in the data range can be obtained, and the ship encounter area distribution near Gulei Port is obtained. As shown in Fig. 4, the ship encounter area is basically concentrated in the triangle area shown in Fig. 3, and at longitude 117.5, latitude 23.74. The zonal area formed by longitude 117.6, latitude 23.65 and longitude 117.6, latitude 23.69 shows that the natural conditions, traffic conditions and hydrological conditions of the three nearby areas have great influence on the ship. At the same time, the area should also be the place where the relevant departments set up navigation aids and focus on monitoring navigation safety.

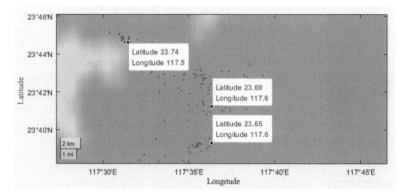

Fig. 4. Distribution of ship encounter areas

4   Feature Mining of Ship Maneuvering Behavior

The ship's navigation behavior is affected by the water period and the current maritime traffic facilities and equipment conditions, showing the adaptive navigation law of the ship itself to the environmental conditions [6, 7]. After the encounter ship identification and navigation data extraction of the existing ship navigation data, the course change rate and speed change rate of the ship near the nearest encounter point are calculated through the information of the ship's longitude and latitude, course and speed. The encounter ships whose course change rate and speed change rate fluctuate near zero are screened. It is considered that they are non-avoidance ships in the whole process of encounter, and their course and speed are basically unchanged. In addition, after in-depth analysis of the course and speed change rate of all encounter ship navigation data, it is concluded that the speed of most ships with a speed of less than 10 sections is basically unchanged

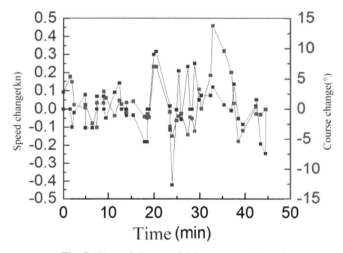

Fig. 5. Rate of change of ship course and speed

in the avoidance process, but the course change is relatively large. Taking one of the avoidance ships as an example, as shown in Fig. 5, the fluctuation range of the course change rate of the ship is basically maintained in the range of $[-15, +15]$, and the speed change rate fluctuates near zero and close to zero, which indicates that the ship with low speed does not adjust the ship speed frequently in the actual navigation process, but the ship course control is more frequent in the operation process. The steering operation of some ships may be guided by the tug near the port, but in most cases, the ship in the low-speed state is more dependent on steering for ship control, which is not the same as the frequent change of the direction and the acceleration in the road driving.

5   Conclusions

By processing the AIS data of ships in Gulei Port, the distribution of the encounter area, the trajectory distribution and the velocity distribution of the ships are excavated, and the conflict area of ship navigation in reality is obtained. Due to the temporal and spatial uncertainty of ship maneuvering and the adaptability to hydrology and geographical environment in the real navigation state, the ship maneuvering mode does not only consider the maritime navigation rules and the interference of other ships. Therefore, this paper mainly analyzes and calculates the low-speed ships near the port. Through the comparison of data, it is found that the common ship maneuvering behavior mainly depends on a large number of steering movements to complete the avoidance between ships, and the speed change is small. This conclusion is the same as the daily observation results of ship maneuvering near the port in life. At the same time, under certain conditions, the relevant personnel can effectively predict the behavior characteristics of ships near the port and complete their daily port work.

References

1. Li-lin, W., Jun, L.: Ship behavior recognition method based on multi-scale convolution. J. Comput. Appl. **39**(12), 3691–3696 (2019)
2. Wan Hui, X., Shan-shan, M.-Q.: A behavior analysis of ship characteristics in arctic Northeast Passage. J. Transp. Inf. Saf. **38**(02), 89–95 (2020)
3. Wen, Y.-Q., Zhang, Y.-M., Huang, L., Zhou, C.-H., Xiao, C.-S., Zhang, F.: Mechanism of ship behavior dynamic reasoning based on semantics. Navig. China **42**(03), 34–39 (2019)
4. Jia, L.: Research on Visualization of Inland Waterway Transportation Information Based on Massive AIS Data. Wuhan University Of Technology (2018)
5. Xiao, X., Qiang, Z., Zhe-peng, S., Xian-biao, J., Jia-cai, P.: Specific ship's encounter live distribution based on AIS. Navig. China **37**(03), 50–53 (2014)
6. Yang, T., Zhe, M., Ping, S., Bing, W.: A study of regularity of navigation patterns of cargo ships at the waterways near Wuhan Yangtze River Bridge based on ship manoeuvring behavior. J. Transport Inf. Safety **36**(01), 49–56 (2018)
7. Huan-huan, G., Hai-guang, H., Si-ning, J.: Study on the division of fishing vessel behavior based on VMS trajectory data analysis. Chinese Fisheries Econ. **38**(02), 119–126 (2020)

Cross-Knowledge Graph Entity Alignment via Neural Tensor Network

Jingchu Wang[1], Jianyi Liu[2(✉)], Feiyu Chen[2], Teng Lu[1], Hua Huang[3], and Jinmeng Zhao[1]

[1] State Grid Information and Telecommunication Branch, Beijing, China
[2] Beijing University of Posts and Telecommunications, Beijing, China
liujy@bupt.edu.cn
[3] Information and Telecommunication Company, State Grid ShanDong Electric Power Corporation, Jinan, China

Abstract. With the expansion of the current knowledge graph scale and the increase of the number of entities, a large number of knowledge graphs express the same entity in different ways, so the importance of knowledge graph fusion is increasingly manifested. Traditional entity alignment algorithms have limited application scope and low efficiency. This paper proposes an entity alignment method based on neural tensor network (NtnEA), which can obtain the inherent semantic information of text without being restricted by linguistic features and structural information, and without relying on string information. In the three cross-lingual language data sets DBP_{FR-EN}, DBP_{ZH-EN} and DBP_{JP-EN} of the DBP15K data set, Mean Reciprocal Rank and Hits@k are used as the alignment effect evaluation indicators for entity alignment tasks. Compared with the existing entity alignment methods of MTransE, IPTransE, AlignE and AVR-GCN, the Hit@10 values of the NtnEA method are 85.67, 79.20, and 78.93, and the MRR is 0.558, 0.511, and 0.499, which are better than traditional methods and improved 10.7% on average.

Keywords: Knowledge representation · Entity alignment · Neural tensor network

1 Introduction

The development of knowledge graph research has developed a variety of methods for the alignment of knowledge graph entities. Traditional entity alignment methods can only use the symbolic information on the surface of the knowledge graph data. The entity alignment between knowledge graphs can be realized efficiently and accurately.

This paper proposes a method for entity alignment based on joint knowledge representation and using improved NTN. We regard entity alignment as a binary classification problem, improve the evaluation function of NTN, and use the aligned entity pair vector as the input of alignment relationship model. If the "the Same As" relationship exists between the input entity pairs, the evaluation function of the model will return a high score, otherwise it will return a low score, based on the scores of the candidate entities to complete the entity alignment task.

© The Author(s) 2022
Z. Qian et al. (Eds.): WCNA 2021, LNEE 942, pp. 66–74, 2022.
https://doi.org/10.1007/978-981-19-2456-9_8

2   Related Work

2.1   Joint Knowledge Represents Learning

The purpose of knowledge representation learning is to embed entities and relationships into a low-dimensional vector space, and to maximize the preservation of the original semantic structure information. The TransE method opens a series of translation-based methods that learn vectorized representations of entities and relationships to support further applications, such as entity alignment, relationship reasoning, and triple classification. However, TransE is not very effective in solving many-to-one and one-to-many problems. In order to improve the effect of TransE learning multiple mapping relations, TransH, TransR and TransDare proposed. All variants of TransE specifically embed entities for different relationships, and improve the knowledge representation learning method of multi-mapping relationships at the cost of increasing the complexity of the model. In addition, there are some non-translation-based methods, including UM [1], SE, DistMult, and HolE [2], which do not express relational embedding.

2.2   Evaluation of the Similarity of the Neural Tensor Network

The goal of similarity evaluation is to measure the degree of similarity between entities. The BootEA model [3] designed a method to solve the problem that the training data set is very limited in the process of knowledge representation learning, iteratively marked out the possible entity alignment pairs, added them into the training of knowledge embedded model, and constrained the alignment data generated in each iteration. The similarity evaluation methods of these models belong to the traditional string text similarity calculation method. For example, KL divergence [4] is used to measure the amount of information lost when one vector approximates to another; There are also Euclidean distance, Manhattan distance [5] and other distance evaluation functions for mapping entities to vector space; There are many models using cosine similarity [6] as entity similarity calculation. Entity alignment algorithm.

3   Entity Alignment Algorithm

3.1   Algorithm Framework

This paper proposes an entity alignment method based on neural tensor network, which consists of two parts: Joint knowledge representation and neural tensor network similarity evaluation. The whole framework of this method is illustrated in Fig. 1. We use G to represent a set of knowledge maps, and G^2 to represent the combination of kgs (that is, the set of unordered knowledge pairs). For G_1 and G_2 is defined as the entity set in knowledge graph G, and R is defined as the relationship set in knowledge map G. $T = (h, R, t)$ denotes the entity relation triple of a positive example in the knowledge graph G, let h, t $\in$ E; r $\in$ R, vector_ h, vector_ r, vector_ T represents the embedding vectors of head entity h, relation R and tail entity t respectively.

We regard the alignment relationship "the Same As" as a special relationship between entities, as shown in Fig. 2, and perform alignment specific translation operations between aligned entities to constrain the training process of two knowledge maps to learn joint knowledge representation.

Formulaic given two aligned entities $e_1 \in E_1$ and $e_2 \in E_2$. We assume that there is an alignment relation r^{same} between two aligned entities, so $e_1 + r^{Same} \cong e_2$. The energy function of joint knowledge representation is defined as:

$$E\left(e_1, r^{Same}, e_2\right) = \|e_1 + r^{Same} - e_2\| \tag{1}$$

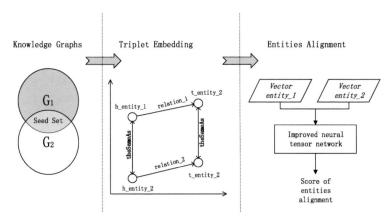

Fig. 1. NtnEA method framework

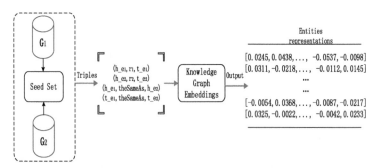

Fig. 2. Learning process of joint knowledge representation

The similarity evaluation model in 2.2 does not use the underlying semantic and structural information of the entity vector, and then considers that the neural tensor network is used in knowledge reasoning. This is in modeling the relationship between two vectors and inferring the relationship that exists between entities. A task has a very good effect, as shown in Fig. 3. Inspired by this, this article uses the NTN method as an alignment model to infer and judge whether there is a "the Same As" alignment relationship between two entities to be aligned. This method uses The tensor function regards entity alignment as a binary classification problem, and the evaluation function of the neural tensor network is:

$$S(e_1, e_2) = u^T f(e_1{}^T W^{[1:k]} e_2 + V \begin{pmatrix} e_1 \\ e_2 \end{pmatrix} + b) \tag{2}$$

Where $f = \tanh$ is a nonlinear function; $W^{[1:k]} \in R^{d \times d \times k}$ is a three-dimensional tensor; D is the dimension of entity embedding vector, k is the number of tensor slices; $V \in R^{2d \times k}$ And $b \in R^k$ is the parameter of the linear part of the evaluation function; $u \in R^k$.

In the legal triples, the relationship between the head entity and the tail entity is irreversible and directional for the current triple; However, for the alignment of entities to triples, the alignment relationship between entities is undirected, that is, there is such a triple relationship between aligned entity pairs (A, B):(A, theSameAs, B), (B, theSameAs, A),

The triplet embedding section in Fig. 1 shows this very well. We optimize the evaluation function:

$$S(e_1, e_2) = u^T f \begin{pmatrix} mean(e_1{}^T W^{[1:k]} e_2 + V \begin{pmatrix} e_1 \\ e_2 \end{pmatrix}, \\ e_2{}^T W^{[1:k]} e_1 + V \begin{pmatrix} e_2 \\ e_1 \end{pmatrix}) + b \end{pmatrix} \tag{3}$$

The final loss function is as follows:

$$L(\Omega) = \sum_{i=1}^{N} \sum_{c=1}^{C} max\left(0, 1 - S\left(T^i\right) + S\left(T_c^i\right)\right) + \lambda \|\Omega\|_2^2 \tag{4}$$

where Ω is the set of all parameters. T_c^i is the cth negative example of the ith positive example.

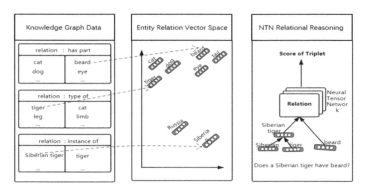

Fig. 3. Neural tensor network relational reasoning process

3.2 Algorithm Flow

The algorithm description of the specific NtnEA model is shown in Algorithm 1.

Algorithm 1 Entity alignment algorithm based on neural tensor network model

---

Input: the Seed Sets from two KGs as SS ,KG1,KG2,;

Output: the scores of entity pairs;

1: **while** *not converge* **do**

2:         **if** $i= 1$ **then**

3:                   Initialize the embeddings randomly. Model(h, r, t) $\in KGs$ to get the embeddings of entities, relations and "the Same As";

4:         **else**

5:                   Initialize the embeddings with the results from the(i-1) iteration. Model $(h, r, t) \in$ KGs (e1,theSameAs,e2) $\in SS$ to update all the embeddings;

6:         **end if**

7: **end while**

8:Use embeddings of seed sets to train a NTN evaluation model for "the Same As";

9: **for** *entity* $\in KGs$ **do**

10:         For each entity in the group, calculate the score of pairs with other entity in the group according to NTN(neural tensor network);

11: **end for**

---

4 Experiment

4.1 Datasets

This experiment is aimed at the comparison of entity alignment methods based on knowledge representation learning, in order to facilitate the horizontal comparison of multiple entity alignment methods, and evaluate the NtnEA method in the context of cross-language entity alignment tasks. This experimental data set uses a more general paper data, the DBP15K [7] data set, which contains three cross-language data sets. These data sets are constructed based on the multilingual version of the DBpedia knowledge base: DBP_{ZH-EN} (Chinese and English), DBP_{JP-EN} (Japanese and English) and DBP_{FR-EN} (French and English). Each data set contains 15,000 aligned entities.

4.2 Training and Evaluation

In order to verify the effectiveness of this research method on the task of knowledge map alignment, the following relatively common method pairs were selected as experimental reference comparisons:

- MTransE, the linear transformation between two vector spaces established by TransE;
- IPTransE, which embeds entities from different knowledge graphs into a unified vector space, and iteratively uses predicted anchor points to improve performance;
- AlignE [6] uses ε-truncated uniform negative sampling and parameter exchange to realize the embedded representation of the knowledge graph. It is a variant of BootEA method without bootstrapping;
- AVR-GCN uses VR-GCN as a network embedding model to learn the representation of entities and the representation of relations at the same time and use this network in the task of multi-relational network alignment based on this network;

To experimentally verify the algorithm in this paper, first learn the vectorized representation of entity relationships in the low-dimensional embedding space in the DBP15K data set. In the entire training process, the dimension d of the vector space is selected from the set $\{50, 80, 100, 150\}$, and the learning rate λ is selected from the set $\{10^{-2}, 10^{-3}, 10^{-4}\}$, the number of negative samples n is selected from the set $\{1, 3, 5, 15, 30\}$. Three sets of data sets are trained separately, and the final optimal parameter configuration is selected as follows: 1. ZH-EN data set, d $= 100$, $\lambda = 0.001$, n $= 5$; 2. JP-EN data set, d $= 100$, $\lambda = 0.001$, n $= 3$; 3. FR-EN data set, d $= 100$, $\lambda = 0.003$, n $= 5$.

The alignment entity data of each cross-language data set is divided according to the ratio of 3:7. As shown in Fig. 4, as the number of tensor slices k increases, the complexity of the model becomes larger, and its performance also improves, but considering that the parameter complexity will increase with the increase of tensor slice parameters. Therefore, the optimal parameter configuration of the neural tensor network model in this process is: $\lambda = 0.0005$, k $= 200$(tensor).

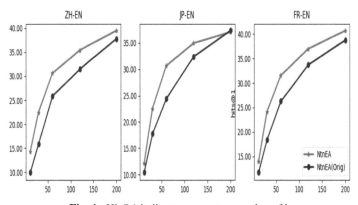

Fig. 4. Hit@1 indicator curve at any value of k

4.3 Experimental Results and Analysis

According to the experimental settings in the experimental method in the previous section, entity alignment experiments were performed on the three sets of cross-language data sets of DBP15K. The results of entity alignment are shown in Table 1. Through the experimental results, it can be seen that in the data sets DBP_{FR-EN}, DBP_{ZH-EN} and DBP_{JP-EN}, compared with the traditional entity alignment method on Hit@k and MRR indicators, The experimental results are shown in the table. The experimental results of MTransE, IPTransE, AlignE and AVR-GCN are obtained from the literature [8]. It can be seen from the table that the experimental results of the two NtnEA methods are significantly improved compared to the benchmark methods MTransE and IPTransE. For example, the Hit@10 values of NtnEA on the three cross-language data sets of DBP15k are 82.00, 78.07 and 77.10, respectively. Compared with the experimental indicators of the AlignE model, an average increase of 10.7%.

This paper uses the semantic structure information of triple data, and through joint knowledge indicates that more alignment information is integrated, so the results show that its alignment effect is significantly improved compared to the alignment methods based on knowledge representation learning such as MTransE and IPTransE. Among the two NtnEA entity alignment methods, the NtnEA model performs better than the NtnEA(Orig) model. This verifies the fact that the head entity and the tail entity in the triples of the alignment relationship are undirected graph structures under the relationship "the same As". On the three cross-language data sets, the Hit@10 and MRR indicators of the NtnEA(Orig) and NtnEA models proposed in this paper exceed the MTransE and IPTransE methods. However, there is no obvious advantage over the current more advanced AVR-GCN model in the Hit@1 indicator, which represents the alignment accuracy.

Table 2 shows that when using the similarity evaluation model for training, the more priori seed set training set alignment relationship data, the better the effect of the model on the entity alignment task.

Table 1. Comparison of entity alignment results

Method	DBP_{FR-EN}			DBP_{ZH-EN}			DBP_{JP-EN}		
	Hit@1	Hit@10	MRR	Hit@1	Hit@10	MRR	Hit@1	Hit@10	MRR
MTransE	7.0	31.81	0.146	13.46	41.45	0.232	13.02	38.80	0.218
IPTransE	12.46	43.51	0.225	21.94	45.90	0.328	17.02	48.74	0.275
AlignE	32.60	74.92	0.466	31.78	69.43	0.452	31.78	69.88	0.433
AVR-GCN	36.06	75.14	0.494	37.96	73.27	0.501	35.15	72.15	0.470
NtnEA(Orig)	38.00	82.00	0.533	37.60	78.07	0.504	35.36	77.10	0.487
NtnEA	40.81	85.67	0.558	39.27	79.20	0.511	35.47	78.93	0.499

Table 2. Comparison results under different seed set partition ratios Hit@k index

Split Ratio indicator	0.1	0.3	0.5	0.7	0.9	Datasets
Hit@1	36.07	36.26	37.46	38.23	39.27	DBP_{JP-EN}
Hit@5	62.18	62.96	63.77	65.21	65.78	
Hit@10	76.85	77.54	78.36	79.14	79.81	
Hit@1	36.97	37.35	39.14	39.91	40.02	DBP_{ZH-EN}
Hit@5	63.12	63.33	64.30	65.39	65.71	
Hit@10	76.35	76.95	78.57	79.14	79.81	

5 Conclusions

This paper introduces a cross-knowledge graph entity alignment model based on neural tensor network proposed in this paper. The model is mainly divided into two parts: joint knowledge representation learning and neural tensor network similarity evaluation. The entity alignment method based on neural tensor network is verified experimentally. The experimental results show that the method based on neural tensor network has good entity alignment performance under given experimental conditions. Compared with previous algorithms, the indexes HIT@5 and HIT@10 have been improved, but the improvement effect on HIT@1 is not obvious, which means that the method has short board in alignment accuracy.

Acknowledgments. The authors would like to thank the anonymous referees for their valuable comments and helpful suggestions. The work is supported by Science and Technology Project of the Headquarters of State Grid Corporation of China, "The research and technology for collaborative defense and linkage disposal in network security devices" (5700-202152186A-0-0-00).

References

1. Bordes, A., Glorot, X., Weston, J., et al.: Joint learning of words and meaning representations for open-text semantic parsing. In: International Conference on Artificial Intelligence and Statistics, pp. 127–135 (2012)
2. Nickel, M., Rosasco, L., Poggio, T.: Holographic embeddings of knowledge graphs (2015)
3. Sun, Z., Hu, W., Zhang, Q., et al.: Bootstrapping entity alignment with knowledge graph embedding. International Joint Conference on Artificial Intelligence, pp. 4396–4402 (2018)
4. Lasmar, N., Baussard, A., Chenadec, G.L.: Asymmetric power distribution model of wavelet subbands for texture classification. Pattern Recogn. Lett. **52**, 1–8
5. Schoenharl, T.W., Madey, G.: Evaluation of measurement techniques for the validation of agent-based simulations against streaming data. In: Proceedings of the 8th International Conference on Computational Science, Part III (2008)
6. Xia, P., Zhang, L., Li, F.: Learning similarity with cosine similarity ensemble. Inf. Sci. **307**, 39–52
7. Sun, Z., Hu, W., Li, C., et al.: Cross-lingual entity alignment via joint attribute-preserving embedding. In: International Semantic Web Conference, pp. 628–644 (2017)
8. Ye, R., Li, X., Fang, Y., Zang, et al.: A vectorized relational graph convolutional network for multi-relational network alignment. In: International Joint Conferences on Artificial Intelligence, pp. 4135–4141 (2019)

Fusion of Traffic Data and Alert Log Based on Sensitive Information

Jie Cheng[1], Ru Zhang[2(✉)], Siyuan Tian[2], Bingjie Lin[1], Jiahui Wei[1], and Shulin Zhang[1]

[1] State Grid Information and Telecommunication Branch, Beijing, China
[2] Beijing University of Posts and Telecommunications, Beijing, China
`liujy@bupt.edu.cn`

Abstract. At present, the attack behavior that occurs in the network has gradually developed from a single-step, simple attack method to a complex multi-step attack method. Therefore, the researchers conducted a series of studies on this multi-step attack. Common methods usually use IDS to obtain network alert data as the data source, and then match a multi-step attack based on the correlation nature of the data. However, the false positives and omissions of the alert data based on IDS will lead to the failure of the resulting multi-step attack. Multi-source data is the basis of analysis and prediction in the field of network security, and fusion analysis technology is an important means of processing multi-source data. In response to this problem, this paper studies how to use sensitive information traffic as data to assist IDS alert data, and proposes a method for fusion of traffic and log data based on sensitive information. This article analyzes the purpose of each stage of the kill chain, and relies on the purpose to divide the multi-step attack behavior in stages, which is used to filter the source data. And according to the purpose of the multi-step attack, the kill chain model is used to define the multi-step attack model.

Keywords: Sensitive information · Multi-setp attack · Alert log

1 Introduction

Since the birth of the Internet, cyber attacks have been threatening users and organizations. They also become more complex as computer networks become more complex. Currently, an attacker needs to perform multiple intrusion steps to achieve the ultimate goal. In order to detect network attacks, security researchers rely heavily on intrusion detection systems (IDS). However, due to the underreporting of IDS alert data and The nature of false positives. Multi-step attacks based only on alert logs are incomplete or incorrect.

In response to this problem, this paper studies and designs a flow and log data fusion method based on sensitive information. Based on the Spark framework, sensitive traffic is screened out from huge traffic information, the sensitive traffic is preprocessed, and merged with the alert log, and finally normalized data is obtained as the data source. The

© The Author(s) 2022
Z. Qian et al. (Eds.): WCNA 2021, LNEE 942, pp. 75–83, 2022.
https://doi.org/10.1007/978-981-19-2456-9_9

normalized data is preliminarily clustered based on the single feature of the IP address, combined with the kill chain model to filter within and between clusters, and finally a highly complete attack cluster that meets the kill chain attack stage is obtained.

2 Related Work

Multi-step attacks are the current mainstream attack method. So far, the correlation analysis methods of multi-step attacks can be divided into five categories: similarity correlation, causal correlation, model-based, case-based, and hybrid.

Similarity correlation is based on the idea that similar alerts have the same root cause and therefore belong to the same attack scenario. With the correct selection of similarity features, a more accurate attack scenario can be reconstructed, but it depends on the similarity of a small number of data segments.

The causal association method is based on a priori knowledge or a list of prerequisites and results of alerts determined under big data statistics. This method can correlate common attack scenarios more accurately, but the causal association based on prior knowledge lacks in reconstructing rare attacks Scenario means, due to the randomness of the attack process, the results of big data statistics lack confidence.

Model-based methods use existing or improved attack models for pattern matching, such as attack graphs, Petri nets, network kill chains, etc., which can match and reconstruct attacks that conform to the model, but lack detection methods for new attacks or APT attacks. Noel et al. [1] was the first to use the attack graph to match IDS alerts, which relies on prior knowledge such as the integrity of the attack graph and cannot detect unknown attacks. Chien and Ho. [2] proposed a color Petri net-based approach. Associated system, the attack types are divided in more detail. Yanyu Huo et al. [3] used the network kill chain model for correlation analysis.

Case-based methods can only target a certain type of attack. Vasilomanolakis et al. [4] collected real multi-step attacks through honeypots, etc., and developed case-based signatures. Salah et al. [5] modeled through reasoning or human analysis and added it to the attack database.

The hybrid method can combine the advantages and disadvantages of several methods and is the most commonly used method in recent years. Farhadi et al. [6] combined the attribute association and statistical relationship methods in the ASEA system, and used HMMs for plan identification. Shittu [7] combines Bayesian inference with attribute association.

3 Algorithm Design

3.1 Meaning of Sensitive Information

Researchers rarely use traffic data as the analysis data source, mainly due to the huge amount of traffic data and poor data readability. In order to solve these two problems, this paper proposes the meaning of sensitive information and a method of filtering sensitive information traffic based on the Spark framework.

Table 1. Sensitive information.

	Database information	Administrator account password, user profile information
Sensitive information	Site Information	Website script files, website front-end files
	system message	Registry file, domain name resolution file, passwd, shadow, source.list file
	company information	Confidential documents, personnel files
	Linux	/usr/bin、/usr/src、/proc/cpuinfo、/proc/devices、/etc/xinetd、/etc/rc.d
Sensitive path	Window	windows startup directory entry、windows registry directory
	web service	Web service system directory, Web background network path, etc.

The ultimate goal of the attack is defined as modifying, adding, stealing system data or destroying system behavior. Therefore, this article has obtained the sensitive information that may be contacted during the attack through a questionnaire survey by security personnel and a statistical analysis of multi-step attack behavior. Table 1 shows.

3.2 Sensitive Information Flow Screening Method Based on Spark Framework

The initially extracted traffic data contains basic information fields: time, IP information, port information, and the transmitted content body msg. In this paper, through distributed calculation of the content main body msg, the sensitive information flow is filtered out from the mass flow data according to the sensitive information list Sl (Fig. 1).

time		Source Ip	Source port	Destination ip	Destination port	protocol		type	name	
	2019/3/24 15:35	192.168.244.1	56934	192.168.244.136		80	http	Exploit	SQL injection	
	2019/3/24 15:35	192.168.244.1	56934	192.168.244.136		80	http	Exploit	SQL injection	
	2019/3/24 15:35	192.168.244.1	56934	192.168.244.136		80	http	Exploit	SQL injection	Alarm data
	2019/3/24 15:35	192.168.244.1	56934	192.168.244.136		80	http	Exploit	SQL injection	
	2019/3/24 15:37	192.168.244.1	56934	192.168.244.136		80	http	Trojan ccweb		
time		Source Ip	Source port	Destination ip	Destination port	Sensitive information				
	2019/3/24 15:32	192.168.244.1	56934	192.168.244.136		80	Website backend			
	2019/3/24 15:32	192.168.244.1	56934	192.168.244.136		80	Website backend		Sensitive information	
	2019/3/24 15:32	192.168.244.1	56934	192.168.244.136		80	Website backend			
	2019/3/24 15:36	192.168.244.1	56934	192.168.244.136		80	Server root directory			

Fig. 1. Alert data and traffic data extracted for the first time.

3.3 Data Normalization

The methods of multi-step attacks are ever-changing, but their essence is to rely on a combination of many single-step attacks to achieve the ultimate goal. For most of the multi-step attack processes, they are in line with the characteristics of the kill chain model. The kill chain model defines the attack stage as: reconnaissance and tracking, weapon construction, load delivery, vulnerability exploitation, installation and implantation, command and control, and goal achievement. This article is based on the above division scheme, according to The purpose of different stages of attack, the multi-step attack stage is divided into: information collection stage (reconnaissance tracking, weapon construction), vulnerability exploitation stage (load delivery, vulnerability exploitation), upload Trojan remote command execution stage (installation and implantation), remote connection The Trojan connects to the seven stages of privilege escalation stage (command and control), horizontal transmission stage, destruction, stealing and modifying information (achieving the goal), and the stage of eliminating intrusion evidence. Under the original kill chain model, the attack behavior is divided in more detail. Considering that the current multi-step attack behavior may have the nature of worm propagation (such as Wannacry, etc.), this article adds a horizontal propagation stage; in addition, it adds sensitive information flow data. The host information process that cannot be detected only with IDS alert data can be detected, so the stage of eliminating intrusion evidence is added.

In summary, the kill chain model used in this article is shown in Fig. 2.

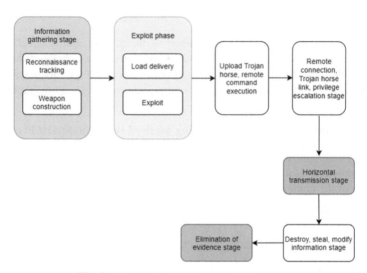

Fig. 2. This article kill chain model diagram.

The normalization process of data mainly depends on the selection of feature fields. The selection of feature fields mainly needs to consider the following three aspects: (1) The similarity of feature fields can indicate the similarity of attacks to a certain extent; (2) Feature fields can clearly contain this important piece of data; (3) Feature fields exist in all data sets. Based on the above considerations, this article selects the source IP address (src_ip), destination IP address (dst_ip), source port (src_port), destination port (dst_port), time (time), kill chain stage (killstep) and distinguishing flag (datatype). Finally get the normalized data set:

$$\text{data} = \{d_1, d_2, \ldots, d_n\}, \ d_i \text{ is a } 7 - \text{tuple data,}$$
$$d_i = \left[\text{src\_ip, dst\_ip, src\_port, dst\_port, time, killstep, datatype}\right]$$

3.4 Alert Log and Sensitive Information Flow Fusion Algorithm

Definition 1: Attack cluster collection:

attclusters = {attcluster_1, attcluster_2, attcluster_3, . . . , attcluster_n},

Where attcluster_i represents an attack cluster: $\text{attcluster}_i = \{d_a, d_b, \cdots, d_c\}d_x \in$ data

(A) IP similarity clustering

At present, the feature selection of network attack classification using similarity method mainly includes two types: one is to use multiple features such as IP, port, time, etc. to perform fuzzy clustering according to different weights; the other is to use a single feature for strong similarity Sexual clustering. This article considers that the subsequent multi-step attack model generation algorithm can supplement the missed multi-step attack behavior to a certain extent. Therefore, this article uses the similarity of single feature IP addresses to cluster, the formula is shown in 1:

IP address similarity formula (a):

$$F_{ip}(ip_1, ip_2) = \begin{cases} 1, & \text{if } Similar(src_{ip1}, src_{ip2}) \text{ and } Similar(dst_{ip1}, dst_{ip2}) \\ & \text{or } dst_{ip1} = src_{ip2} \\ 0, & \text{otherwise} \end{cases} \tag{1}$$

Among them, src_{ip}, dst_ip indicates the source and destination IP addresses of the data respectively. If the source IP addresses of two pieces of data are in the same network segment and the destination IP addresses are also in the same network segment, then the similarity value is 1, and the two pieces of data can be considered to belong to the same Attack process. For example: there are two IPs, IP1 = A1.A2.A3.A4, IP2 = B1.B2.B3.B4, then the formula is as shown in 2:

IP address similarity formula (b):

$$Similar(IP1, IP2) = \begin{cases} True, & A1 == B1 \text{ and } A2 == B2 \\ False, & otherwise \end{cases} \tag{2}$$

(B) Combine and filter within the attack cluster (Sim_in, CFD_in)

According to the analysis of normal attack behavior, there will usually be a large number of similar attack behaviors in a short period of time. Therefore, in this paper, each attack cluster is internally merged and filtered. The similarity formula within the attack cluster is shown in3, and the confidence formula is shown in 3:

(1) Similarity within the attack cluster:

$$Sim\_in(d_1, d_2) = \begin{cases} 1 & \text{if sametime and ip}(d_1, d_2) \\ \text{or neartime}(d_1, d_2) \text{ and same msg and ip}(d_1, d_2) \\ 0 & \text{otherwise} \end{cases} \tag{3}$$

(2) The built-in reliability of the attack cluster:

$$CFD\_in(d_1) = \begin{cases} 0 & \text{if killstep}(d_i) > 3 \text{ and killstep}(d_1) < maxkillstep \\ 1 & \text{otherwise} \end{cases} \tag{4}$$

If the time and IP address of the two pieces of data are the same, the similarity is 1, which is the same piece of data generated by sensitive information traffic and alert logs; the similarity of data with the same attack name and IP address within similar time is also 1, Which means the same attack in a short period of time. In this paper, a merge operation is adopted for the data whose similarity is 1 value. For each piece of data, if its kill chain stage is greater than 3 and smaller than the maximum kill chain stage of the attack cluster to this data, the confidence is 0. This paper removes the data with confidence of 0 from the attack cluster.

(C) Filter between attack clusters (CFD_out)

Due to the rule-based rather than result-based detection nature of the IDS system, there will be a large amount of attack failure data in the actual acquired attack data. Therefore, the attack cluster that only depends on the classification of IP addresses must contain a large number of attacks. The unsuccessful attack behavior, the attack to a certain extent due to the change of the attacker's target or the unsuccessful attack caused the cluster set to abandon, etc., these incomplete attack behaviors will lead to the incompleteness of the subsequent multi-step attack model; therefore In order to filter incomplete and incorrect attack clusters, this paper gives the confidence formula between attack clusters as shown in formula 5:

$$CFD\_in = \sum_{i=1}^{N} killstep(d_i) * typeCFD(d_i) \qquad (5)$$

where N represents the number of attack data of the attack cluster, and for each piece of data, its kill chain stage killstep is used as the product of authority and type confidence typeCFD to represent the confidence value of the corresponding data.

4 Experimental Design and Analysis

4.1 Dataset

(1) Simulation data D1
    This article uses the website management system CMS to build a Web site that contains a SQL injection backdoor, and sequentially uses Yujian to scan the website background, SQL injection to obtain the administrator account password, log in to the background, upload a sentence Trojan horse, and Chinese kitchen knife connection operations. Traffic data for this series of attacks. The attack process is shown in Fig. 3:

Fig. 3. Simulation experiment attack process.

(2) Campus network data D2
    In this paper, a traffic monitoring system is arranged on the three subnet nodes of the campus network. One of the subnets includes the CTF competition environment in the school. Accumulatively collected 2G traffic data in the network, and passed the IDS system and sensitive information screening., 10870 pieces of alert data and 205,408 pieces of sensitive information traffic were obtained.

(3) LLDDos 1.0 D3 of Darpa2000

This data set is widely used by researchers in the construction of multi-step attack scenarios. This article is based on its five attack steps: the attacker IPsweep scans all hosts in the network, detects the surviving hosts obtained in the previous stage, and determines which ones are The host is running the sadmind remote management tool on the Solaris operating system, the attacker enters the target host through a remote buffer overflow attack, the attacker establishes a telnet connection through the attack script, installs the Trojan horse mstream ddos software using rcp, and the attacker logs in to the target host to initiate a DDOS attack Launch attacks on other hosts in the LAN. An attack cluster is obtained through aggregation and screening, which contains 18-tone alert information.

4.2 Experimental Results

(1) The feasibility of the fusion algorithm of alert log and sensitive information flow.

First, the collected traffic data is passed through the IDS system to obtain the alert data. The pyspark module of python uses the Spark framework to extract the sensitive information flow from the flow. After the sensitive information flow and the alert log fusion algorithm, the detection accuracy and detection integrity are compared.

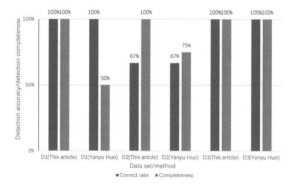

Fig. 4. Comparison of detection accuracy and detection completeness.

Figure 4 shows the experimental results of the three data sets and the comparison results of Yanyu Huo et al. [6] in detection accuracy and detection integrity. It can be seen that after the sensitive information traffic data is added, the multi-step attack is more effective. The detection integrity has been improved to a certain extent, and the detection accuracy is equivalent to the method of Yanyu Huo et al. [6], but the method in this paper does not need to be classified by a preset threshold, so the sensitive information flow and alert log fusion algorithm proposed in this paper It is feasible in practice. The D3 data set has no difference in detection accuracy and detection integrity because the alert data covers all the attack steps.

5   Conclusion

Figure 4 shows the results of detection accuracy and detection completeness of the three data sets. The conclusion that can be drawn is that, compared with only using IDS alert logs as source data, the alert log and sensitive information flow fusion algorithm proposed in this paper can indeed be used to a certain extent. In order to compensate for the false positives and false negatives of the alert data, and based on the integrity of the attack process in the traffic data, the attack behavior can be more deeply and completely identified. Combined with the kill chain model proposed in this paper, the horizontal transmission stage is added and the evidence of intrusion is eliminated. An attack cluster with higher correlation, higher attack success rate and a certain attack stage sequence can be obtained, and then a more complete multi-step attack behavior can be obtained when the subsequent multi-step attack prediction is performed.

Acknowledgement. The authors would like to thank the anonymous referees for their valuable comments and helpful suggestions. The work is supported by Science and Technology Project of the Headquarters of State Grid Corporation of China ,"The research and technology for collaborative defense and linkage disposal in network security devices" (5700-202152186A-0-0-00).

References

1. Noel, S., Robertson, E., Jajodia, S.: Correlating intrusion events and building attack scenarios through attack graph distances. In: 20th Annual Computer Security Applications Conference, pp. 350–359. IEEE (2004)
2. Chien, S.-H., Ho, C.-S.: A novel threat prediction framework for network security. In: Advances in Information Technology and Industry Applications, pp. 1–9. Springer (2012)https://doi.org/10.1007/978-3-642-26001-8_1
3. Zhang, R., Huo, Y., Liu, J., et al.: Constructing APT attack scenarios based on intrusion kill chain and fuzzy clustering. Secur. Commun. Networks (2017)
4. Vasilomanolakis, E., Srinivasa, S., García Cordero, C., Mühlhäuser, M.: Multi-stage attack detection and signature generation with ICS honeypots. In: 2016 IEEE/IFIP Network Operations and Management Symposium, NOMS 2016, pp. 1227–1232. https://doi.org/10.1109/NOMS.2016.7502992.2016
5. Salah, S., Maciá-Fernández, G., Díaz-Verdejo, J.E.: A model-based survey of alert correlation techniques. Comput. Netw. **57**(5), 1289–1317 (2013)
6. Farhadi, H., AmirHaeri, M., Khansari, M.: Alert correlation and prediction using data mining and HMM. ISC Int. J. Inf. Secur. 3(2) (2011)
7. Shittu, R.O.: Mining intrusion detection alert logs to minimise false positives & gain attack insight. City University London. Thesis (2016)

Mixed Communication Design of Phasor Data Concentrator in Distribution Network

Yan Wu[(⊠)], Weiqing Tao, Yingjie Zhang, and Xueting Li

School of Electrical Engineering and Automation, Hefei University of Technology, Hefei, China
1302881934@qq.com, wqtao@hfut.edu.cn

Abstract. Phase Data Concentrator (PDC) is an important part of Wide Area Measurement System (WAMS) and is widely used in transmission systems. WAMS technology will also be applied in smart distribution network, which has many nodes, complex architecture and various types of data transmission services, and a single communication mode cannot meet its needs. In order to solve this problem, this paper first introduces the composition of WAMS system, communication network mode, and discusses the access layer communication network mode. According to the main station, sub-station interaction process design a synchronous phase data set device that can carry out up-down communication and mix network by various means of communication. Finally, the experimental environment of Power Line Carrier (PLC) and twisted pair network communication is set up to verify.

Keywords: WAMS · Phasor data concentrator · Mixed communication · Upstream and downstream communication

1 Introduction

With the establishment of the goal of "double carbon", the country for the first time put forward the new concept of "new power system with new energy as the main body" of the future grid blueprint [1]. The wide area measurement system can monitor the distribution network status in real time by using synchronous phase measurement technology, which provides a new scheme for the safe operation and stable control of the high proportion of new energy distribution network in the future [2-3]. The data measured by WAMS has three characteristics: time synchronization, spatial wide area and direct measurement of phase angle data, which provides data for the good control of power system [4]. Reference [5] analyzes the development of synchronous measurement technology at home and abroad and the future development direction of distribution network. In Reference [6], a new PDC with blade structure is designed to make it extensible. For Phasor Measurement Unit (PMU), intelligent substation platforms have applicability, low energy consumption, strong storage capacity, strong communication makes WAMS system more reliable. Reference [7] analyzes the communication mode and existing problems of the existing distribution network communication network, and proposes a communication scheme of hybrid optical fiber and power line carrier network. This paper will discuss WAMS communication network and access layer communication mode, and design a PDC that can process data from multiple channels. Finally, the PDC hybrid network experimental environment was built for verification.

© The Author(s) 2022
Z. Qian et al. (Eds.): WCNA 2021, LNEE 942, pp. 84–92, 2022.
https://doi.org/10.1007/978-981-19-2456-9_10

2   WAMS Network

WAMS system is mainly composed of communication network, PMU, GPS, PDC and data center station [8]. WAMS collects phasor data through GPS and aggregates data from the entire power system through a communication network. In this way, the dynamic information of the power grid can be obtained to achieve the role of the monitoring system and improve the security and stability of the power grid. GPS synchronous clock provides a unified high precision clock signal for power system. PMU can unify the state quantity of different nodes and lines, and establish a connection with the dispatch center through the communication network, and save and transmit data in real time to ensure the synchronization of data of the whole network.

Distribution network WAMS communication network generally includes access layer and backbone layer communication. The backbone layer communication is the communication between the main station and the PDC, and the communication mode is mainly Synchronous Digital Hierarchy (SDH) fiber. Access layer communication is PDC to multiple PMUs of communication, there are fiber optic, PLC, wireless network and other communication methods mixed [9]. Most of the PMUs in the distribution network are installed on the lines and important nodes, a distribution network main station will connect a large number of PMUs, a single main station cannot process a large number of communication messages in a timely manner, will make the sent message conflict. The double-layer communication structure of master station connecting PDC and PDC connecting PMU can greatly reduce the communication pressure of master station and ensure the stability and reliability of data transmission.

3   Access Layer Communication Network Analysis

Compared with the backbone layer communication network, the coverage of access layer communication network is obviously insufficient. This is due to the restriction of economic and technical level, the degree of distribution network construction in different places is very different. Access layer communication mode can be divided into wired and wireless mode, wired communication mainly includes power line carrier, optical fiber, field bus. Wireless communication mainly includes 230 MHz wireless private network, wireless public network, 4G, 5G. Optical fiber communication is suitable for distribution network backbone communication or pre-buried lines, high transmission bandwidth, simple network is less affected by the environment, high reliability. However, the cost of fiber optic construction is large, and the construction and installation of old urban areas and economically backward areas is difficult. PLC communications can be transmitted using existing power lines without laying additional lines, and the installation is convenient and secure, saving costs, but real-time, reliability is not high. 230 MHz wireless network communication can save line investment, construction facilities and a wide range of applications, but low bandwidth coverage is small, real-time cannot be guaranteed. Therefore, a single means of communication cannot meet the existing distribution network communication needs. Only in the access network using a hybrid network, a variety of communication methods complement each other, and further improve the quality of communication.

4 Distribution Network PDC Software Design

The PDC needs to have up-and-down communication as an intermediate device between the primary and PMUs. PDC communication needs to meet the main and sub-station interaction processes specified in G.BT 26865.2-2011.There are two kinds of communication between master station and sub-station: real-time communication and offline communication. There are four data formats for real-time distribution network communication: data frame, head frame, configuration frame, and command frame [10]. The data frame contains information such as switching quantity, analog quantity, amplitude and phase angle. The head frame uses the ASCII code to represent information such as synchronous phase measurement devices, data sources, etc. The configuration frames are divided into CFG-1 and CFG-2, representing the output and configuration of the sub-stations respectively. The command frame is responsible for transmitting the instructions sent.

PDC devices should meet the functions of distribution network, dynamic data collection and storage, fault recording data storage, time-to-time and so on. In WAMS system, PDC mainly takes the role of PMU networking, PMU vector data collection and sending to the master station. The data aggregated by PDC mainly includes the configuration information of the underlying PMU, real-time data information and historical data information. Configuration information is generally used only before the PDC aggregates data, and the amount of data is small. Real-time data is continuously uploaded to the PDC at a fixed number of frames per second, data is sent frequently, the amount of data per PMU is small but the real-time requirements of uploading PDC are high. Historical data information is a historical event that records the PMU, is saved as a file, and the amount of data information is large but the upload time is long. Based on LINUX system, this paper uses libuv function based on event-driven asynchronous IO library to implement PDC software operation.

4.1 PDC Up and Down Communication Design

PDC communication is divided into upstream and downstream communication, upstream communication with the dispatching center master station, downstream communication with multiple PMU. PDC needs to build data channels, file channels, and command channels when communicating up and down the line. When communicating upstream, the PDC, as a server, needs to respond to a command request sent by the master and accept the configuration frames sent by the master. The communication flow of the PDC connecting multiple master stations when communicating upstream is shown in Fig. 1. When the PDC communicates uplink with multiple master stations, each master needs to be connected in turn. In the figure, n is the number of connected master stations. The IP and port number parameters are first configured for each master station to be connected to by the PDC through the for loop. The listening is then bound based on the IP and port number of the PDC. When a request for a connection is received and commands, data, and file connections are established, the PDC can communicate with each master.

When communicating downstream, the PDC, as a client, is required to accept real-time data uploaded by multiple PMUs, offline data, and command requests to the PMU.

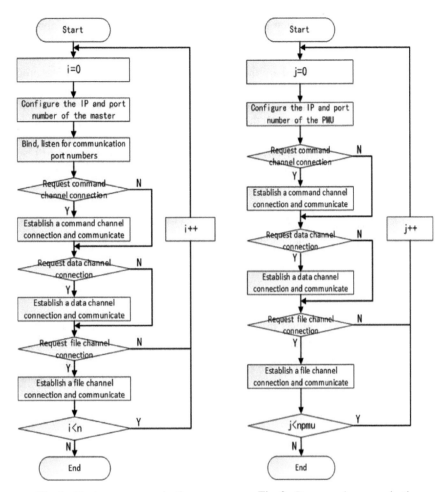

Fig. 1. Upstream communication.

Fig. 2. Downward communication.

The downstream communication process is shown in Fig. 2. In the figure, npmu is the number of PMUS connected to the PDC. When communicating downstream, the IP, command port number, data port number, file port number, and so on of each PMU to which the PDC is connected are first configured through the for loop. The program connects data, commands, and file channels based on the parameter configuration of each PMU. After the connection is established, the PDC will send command requests to each downstream PMU through the command channel to realize the real-time data upload of each PMU.

For the aggregation of real-time vector data, the libuv network interface API is used to implement. The libuv function used for PDC up-and-down communication is shown in Table 1.

Table 1. Libuv function table.

Connect the PMU		Listen to the main station	
Function	Instructions	Function	Instructions
uv_tcp_init()	establish a TCP handle	uv_tcp_init()	initialize the TCP server object
uv_ip4_addr()	fill the PMU's IP address and port number	uv_ip4_addr()	fill the PDC's IP address and port number
uv_tcp_connect()	apply for connection	uv_tcp_bind()	bind the server to the local IP address and port number
uv_read_start()	read vector data uploaded by PMU	uv_listen()	establish TCP server monitoring

4.2 Software Running Script

When the PDC program stops unexpectedly, it disconnects upstream and downstream traffic, making it impossible for PMU data to be uploaded in real time. The detection of PDC program is very important, and the detection function of the program needs to be realized through the script file. The script is primarily implemented by the ps-ef command in linux, which can view related activity processes. The specific script code is shown in Fig. 3.

Diagram #! is a special representation, /bin/sh is the shell path to interpret the script, while loop means that the script keeps running. The fourth line in the figure indicates that the number of processes containing 'pdc' is viewed and assigned to procnum through the ps-ef command. The fifth line says if pronum equals zero, then proceed down, otherwise re-enter the path of the PDC and run the program. Set to check whether the PDC program is in running state every 10 s. The PDC program is not interrupted and the data is uploaded in real time.

```
#!/bin/sh
while true
do
procnum=`ps -ef |grep "pdc" |grep '/home/csg/pdc/PDC-7-8/pdc' |grep -v grep |wc -l`
if [ $procnum -eq 0 ];then
cd //home/csg/pdc/PDC-7-8/pdc;
./pdc

fi
sleep 10
done
```

Fig. 3. PDC run script.

5 PDC Mixed Networking Testing

Build the test environment shown in Fig. 4. Figure 4 synchronous clock device to PMU1, PMU2 to provide time-to-time function, PDC uplink through the network cable connection analog main station. The PDC downlink connects PMU1 and PMU2 via twisted pair cable and PLC. The test begins by simulating commands from the main station, summoning real-time data, and observing the frame rate of data transmission.

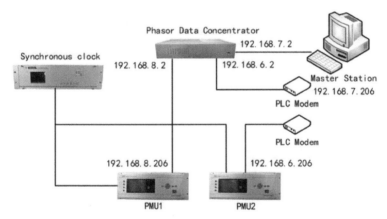

Fig. 4. Experimental environment.

The communication parameters that simulate the master, PDC, and PMU in the test are shown in Table 2.

Table 2. Device communication parameters.

Equipment	IP	Command port number	Data port number	File port number
PMU1	192.168.8.206	9000	9100	9600
PMU2	192.168.6.206	9001	9101	9601
master station	192.168.7.206	any port	any port	any port
PDC eth1	192.168.8.2	8001	8000	8600
PDC eth2	192.168.6.2			
PDC eth3	192.168.7.2			

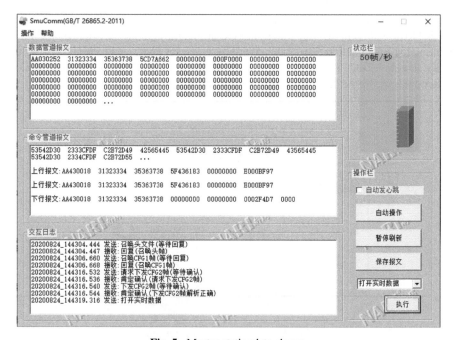

Fig. 5. Master stationdata shows.

The test results are shown in Fig. 5. When the master station sends the command correctly, the data channel connection is established to open the real-time data. From the figure, it can be seen that the data of the two PMUs converges in the PDC and is transmitted steadily to the analog master station at 50 frames/s. It is proved that PDC can mix network and carry out stable communication by PLC and twisted pair communication.

6 Conclusion

Based on the data transmission protocol of real-time dynamic monitoring system, this paper introduces the form of WAMS communication network, discusses the feasibility of the access layer hybrid network communication mode. Based on the libuv function, PDC software is developed to realize PDC up and down communication, and the data of multi-channel PMU is pooled and sent to the analog master station in real time, so as to ensure that the operation of the PDC program is not interrupted by script files. The up-and-down communication, twisted pair network cable and PLC networking function of PDC are verified by setting up the test environment of analog main station, PDC and multi-PMU.

References

1. Khodabakhsh, J., Moschopoulos, G.: Uncertainty reduction for data centers in energy internet by a compact AC-DC energy router and coordinated energy management strategy. In: Proceedings of the IEEE Energy Conversion Congress and Exposition (ECCE), pp. 4668–4673 (2020)
2. Gang, D., Yaqin, Y., Xiaodong, X., et al.: Development status and prospect of wide-area phasor measurement technology. Autom. Electr. Power Syst. **39**, 73–80 (2015)
3. Hao, L., Tianshu, B., Quan, X., et al.: Technical scheme and prospect of high precision synchronous phasor measurement for distribution network. Autom. Electr. Power Syst. **44**, 23–29 (2020)
4. Aminifar, F., Fotuhi-Firuzabad, M., Safdarian, A., Davoudi, A., Shahidehpour, M.: Synchrophasor measurement technology in power systems: Panorama and state-of-the-art. IEEE Access. **2**, 1607–1628 (2014)
5. Kasembe, A.G., Muller, Z., Svec, J., Tlusty, J., Valouch, V.: Synchronous phasors monitoring system application possibilities. In: Proceedings of the IEEE 27th Convention of Electrical and Electronics Engineers, Israel, pp.1–3 (2012)
6. Wei, L., Liang, W., Yulin, C., et al.: Design and implementation of phasor data concentrator with blade frame in wide area measurement system. Autom. Electr. Power Syst. **36**, 61–65 (2012)
7. Jun, Z., Shiqi, G., Yang, H., Li Jin, L., Wansheng, C., Lijuan, S.: Research on hybrid communication network in power distribution communication access network. Power Syst. Commun. **32**, 36–41 (2016)
8. Beg Mohammadi, M., Hooshmand, R., Haghighatdar Fesharaki, F.: A new approach for optimal placement of PMUs and their required communication infrastructure in order to minimize the cost of the WAMS. IEEE Trans. Smart Grid. **7**, 84–93 (2016)
9. Wenxia, L., Hong, L., Jianhua, Z.: System effectiveness modeling and simulation of WAMS communication service. Proc. CSEE. **32**, 144–150 (2012)
10. Yingtao, W., Daonong, Z., Xiaodong, X., Jiang, H., Yuehai, Y., Zhaojia, W.: Power system real-time dynamic monitoring system transmission protocol. Power Syst. Technol. 81–85 (2007)

Devices, Tools, and Techniques for WSN and Other Wireless Networks

Research on Universities' Control of Online Discourse Power in the Period of COVID-19: A Case Study of Shanghai Universities

Lei Sun[1] and Zhuojing Fu[2(✉)]

[1] Department of Cultural Management, Shanghai Publishing and Printing College, Shanghai, China
[2] Schools of Marxism, Shanghai University of Medicine and Health Sciences, Shanghai, China
simple37@163.com

Abstract. Under the situation of the normalization of the prevention and control of COVID-19, related online public opinion occurs from time to time. University administrators must grasp the right of online discourse to guide the direction of online public opinion and ensure the stability of campus order. This paper analyzes the necessity and feasibility of university administrators to grasp the right of online discourse from the basis of reality, compares two kinds of measures and their combinations through questionnaires and computer simulation experiments: publishing authoritative information and focusing on opinion leaders, argues the effectiveness of these two types of measures, and puts forward specific countermeasure suggestions on this basis.

Keywords: COVID-19 · The right of online discourse · Online public opinion

1 Introduction

Under the normalized situation of the prevention and control of COVID-19, news about the epidemic often occupies the hot search list of major Chinese websites. As the main force of the network, the self-expression of university students in the network is very likely to trigger the university network public opinion. In this context, it is important for university administrators to grasp the right of online discourse to guide the direction of online public opinion and maintain social stability.

Related scholars in China have conducted research in terms of opinion leaders and controllers of online discourse, and formed a map of online discourse control, in which algorithms are studied and aided by simulation experiments for verification. Fang Wei et al. [1], Wang Ping [2] and Liu Xiaobo [3] conducted theoretical and simulation simulation experimental research on the formation and evolution mechanism of online public opinion. Jiang Kan et al. [4], CHEN Yuan et al. [5], and Wang Zheng [6] conducted studies on the influence exerted by opinion leaders in online public opinion. Zeng Runxi [7] did studies on how opinion managers conduct online opinion guidance. Fu Zhuojing

Z. Qian et al. (Eds.): WCNA 2021, LNEE 942, pp. 95–103, 2022.
https://doi.org/10.1007/978-981-19-2456-9_11

et al. [8, 9] and Wang Huancheng [10] made studies on improving the monitoring mechanism of online public opinion and grasping the right to master the discourse of public opinion guidance in universities.

Different studies have recognized the role that administrators play in online public opinion, so how specifically can we, as university administrators, master online discourse in the new situation where epidemics are normalized? In this paper, we will conduct simulation experiments based on survey data and previous studies to come up with targeted countermeasures.

2 The Questionnaire Survey

In mid-December 2020, we conducted a survey for college students in six universities in Shanghai. The survey focused on understanding the impact of the Internet on students' study and life on campus during the epidemic. 351 people participated in the survey, with education levels involving senior, college, bachelor, master and doctoral degrees, and majors covering science and technology, arts, economics, management, law and medicine. The survey shows that as high as 89.17% of students choose to go online, and the Internet is more closely connected with the study and life of college students.

2.1 Mainstream Media Show Authority

The survey showed that at the beginning of the emergence of COVID-19, students were easily confused by the Internet rumors related to the epidemic, and only 35.5% of students did not have the experience of being confused. When there were more online rumors, 54.2% of students chose to actively search for relevant information, as many as 96.64% of students chose to clarify online rumors through official releases, 25.21% of students chose to clarify through online celebrities on social media platforms, 23.11% of students chose to clarify through teachers and parents, and 19.33% learned the truth through classroom learning. When the epidemic was more serious, 81.3% of students actively searched for relevant information, a figure that declined after the state released real-time developments of the epidemic. After the official release of the real-time news of the epidemic and the provision of a small platform for disinformation, up to 56.64% of students chose to stop believing the unofficial news forwarded by their friends and replaced it with the official news. As many as 72.9% of students trust the official information about the Newcastle Pneumonia outbreak, while only 0.27% of students do not trust it at all.

A whopping 79.67% of the respondents said that they browse social networking platforms multiple times a day. The main channel for students to get information about COVID-19 (multiple choices) was Weibo in the first place, accounting for 67.21%, followed by WeChat friend circle 57.72%, mainstream media public number 55.83% in the third place, mainstream media microblog 49.05% in the fourth place, and only 16.26% got the information through classroom. Mainstream media public numbers and mainstream media microblogs are the best channels for students to get authoritative information related to the epidemic.

2.2   Proactive Screening and Careful Forwarding

The survey showed that 69.65% of students had half-confidence in the authenticity and credibility of the unofficial information about the Newcastle pneumonia outbreak. Only 5.96% of students believe it completely, and even if they believe it completely or partially, the proportion of students who would forward it is only 38.35%. Up to 74.07% of students would choose to use online engines to search authoritative websites to get authoritative information; followed by finding answers from the news, accounting for 59.6%; at the bottom of the list is communicating with teachers of professional courses, accounting for only 12.12%, with more specialist, undergraduate and doctoral students choosing to communicate with their teachers. If university administrators can forward authoritative information immediately can control online rumors from the source of information, which is more helpful to prevent online public opinion.

A whopping 39.92% of students said that the school's interpretation of relevant policies could ease their anxiety about the epidemic, and another whopping 47.29% said they would actively open news about the epidemic shared by their teachers in their class groups, a percentage second only to students who would actively view news with authoritative experts expressing their professional opinions (62.96%) and news that made it to the top of the list (58.69%), and is higher than WeChat's precisely placed public service videos (30.77%).

3   Simulation Experiments

The experiment is based on the Netlogo platform [11], combined with the Language Change model [12], and is built on the basis of the communication model proposed by Zhuojing Fu et al. [8, 9], adapted to test the effectiveness of different measures taken by university administrators to grasp online discourse and influence online public opinion.

3.1   Model Design

It is assumed that the online information dissemination space is a 99×99 square and that students are in this space forming a social network with some linking hubs in the network. The dots represent a student and the links represent the connections and communication channels between them. White dots (0) represent students who are able to transmit positive energy in their online participation, black dots (1) represent students with more negative online feelings, and grey dots (0.5) represent students in a neutral state. Nodes with connection lines greater than or equal to 5 are shown as larger key dots, and the network participants represented by these dots are network opinion leaders or special network connectors in an active position, such as moderators, followers of comments, etc.

The parameters of the experiment were set according to the survey results; 46.72% of the students feel anxious and upset about the epidemic, which can be interpreted as a corresponding percentage of nodes with a black negative state in the initial state. In each system operation cycle, 38.35% of the nodes will disseminate their state to their neighbors, 5.96% of the nodes fully receive and adjust to the incoming state; 69.65% of the nodes will half believe the received message, of which 74.07% choose to corroborate their judgment by searching for authoritative information; if there is no valid authoritative information released at this time, the experiment shows that there will be 46.72% of the nodes would choose to receive messages that they believed half-heartedly before.

Judging from surveys and past experience, there are two basic measures that can help college and university administrators capture online discourse.

Measure 1 (C1): by publishing official authoritative information across the network, it makes a lot of positive information available on mainstream media, and most (72.9%) of the nodes will accept the positive information after querying, and another 0.27% of students will not accept it at all. The variable C1 is set in this model, taking the value range 0–100%, and the proportion of positive information coverage on the network can reach the level of C1 after taking this measure C1 (assuming that the rest is invalid information).

Measure 2 (C2): focus on network opinion leaders (key nodes), targeted push, and timely push messages to other nodes. The switch C2 is set in this model and turning on C2 means starting to implement measure 2. The experiment is set to select the larger dot after every 5 system times, assign a positive status to that dot, and propagate the positive message to its neighbors.

3.2   Initial Experiments

Simulates the initial state without any measures, with C1 at 0% and C2 off.

The experimental run was started and after 45 system times (T), the negative messages covered all network nodes. Figure 1 shows the results of the experiment without any measures: the world view window shows all dots as black and the statistical curve shows that the node state mean reaches 0 at T = 45 (0 is black, 0.5 is gray, 1 is white).

The initial experimental results show that if university administrators do not take measures to intervene during the outbreak of online public opinion, it will lead to the rapid spread of negative information such as online rumors, and the online public opinion will be out of control in a short period of time.

3.3   Comparative Experiments

Comparative Experiment 1. This experiment tests the effect of publishing authoritative information across the network. The other settings are the same as the initial experiment, and the C1 ratio is turned up to 10%, 20%, 50%, and 100% in that order and run for observation. Figure 2 shows the results of the experiment with measure 1. The results show that only measure 1 makes all the dots white, and the rate of change increases in tandem with the percentage of positive messages in C1, but the increase slows down after C1 exceeds 50%.

The results of Comparative Experiment 1 shows that if measures 1 are taken alone, university administrators can improve the psychological state of the student group in a short time by publishing official authoritative information and making students search for authoritative information on mainstream media (coverage does not have to be high) as soon as possible, thus effectively guiding the direction of online public opinion until positive information dominates the Internet.

Comparative Experiment 2. This experiment tests the directed push of authority information to key nodes. The other settings are the same as the initial experiment, and the C2 switch is turned on and run for observation. Figure 3 shows the results of the experiment for Measure 2. After several effective runs, when the system time reaches above 200–300, most of the nodes show white; while when the system time reaches around 400 interval, only individual end small groups are left black, and sometimes the dots can all be converted to white.

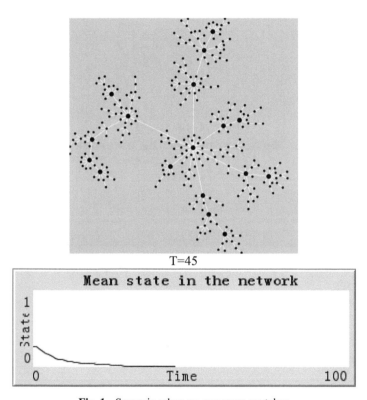

Fig. 1. Scenario when no measures are taken

The results of Comparative Experiment 2 shows that if measure 2 is taken alone, university administrators directed to influence key nodes to ensure that the information they disseminate to surrounding nodes is positive and timely, and can also positively guide the direction of online public opinion, however, measure 2 is not as efficient as measure 1, as reflected by the long time spent and the small range of groups covered.

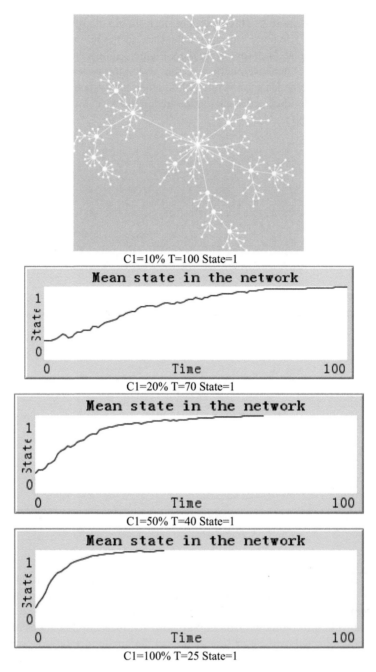

Fig. 2. Results of a typical run of Comparative Experiment 1

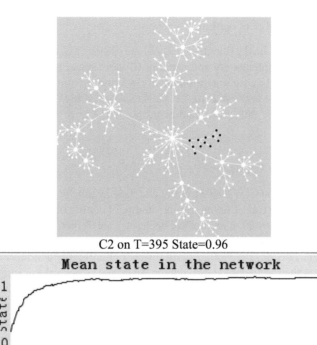

Fig. 3. Results of a typical run of Comparative Experiment 2

3.4 Conclusions of the Experiments

The above experimental situation shows that if university administrators do not take any measures, online public opinion will quickly get out of control; whereas, if conditions permit, prioritizing measure 1 to popularize authoritative information among students in general will quickly control the direction of online public opinion. In the stage when authoritative information is not yet available and online public opinion begins to emerge, adopting Measure 2 to target and influence online opinion leaders or relevant online participants in an active position can be an effective supplement when Measure 1 cannot be taken.

4 Countermeasures and Suggestions

In the context of normalized epidemic prevention and control, the authority trusted by Chinese college students is the mainstream media, and students pay attention to the information about the epidemic and the interpretation of relevant policies forwarded by their schools. In the network public opinion that may break out at any time, university administrators should take this opportunity to grasp the guidance of public opinion and build a mechanism to prevent university network public opinion.

4.1   Leverage the Power of Authority

In the COVID-19 outbeak, the scientific study of the epidemic by the authoritative expert group greatly relieved the anxiety and panic of Chinese social groups; the mainstream media's notification of the case situation shattered all kinds of rumors about the epidemic, and the opinion leaders and authoritative views showed a high degree of integration. Leveraging authority by university administrators is the most effective way to guide online public opinion.

4.2   Focus on the Key Points

Online public opinion on COVID-19 usually matches the time of case confirmation, and is the stage of rapid spread of online rumors and the budding of online public opinion when authoritative information has not yet been released. Experiments have shown that when authoritative information is not yet in play, voices can be raised with the help of online opinion leaders or active online participants. For university administrators, firstly, they should establish a network management team and occupy the position of active network participants; secondly, they should screen out negative emotion groups and lock the key pushing targets; thirdly, they should carry out accurate pushing of network information, including pushing network information that conveys positive energy and publishing positive comments in the comment section.

Acknowledgements. This paper was supported by the 2022 Shanghai Education Science Research Project "Research on University administrators' Control of Online Discourse Power in Emergent Hot Events".

References

1. Fang, W., He, L., Sun, K.: A study of online opinion dissemination model using metacellular automata. Comput. Appl. (3) (2010) (in Chinese)
2. Wang, P., Xie, C.: Research on the formation and evolution mechanism of online public opinion on sudden public events. Modern Commun. (J. Commun. Univ. China) (3) (2013) (in Chinese)
3. Liu, X.: Implementation of an opinion evolution model based on the NetLogo platform. Intelligence Data Work **1** (2012) (in Chinese)
4. Jiang, K., Tang, Z.: Identification of key nodes and analysis of diffusion patterns of online public opinion in microblogging context. Library and Intelligence Work (2015) (in Chinese)
5. Chen, Y., Liu, X.: A study on the identification of opinion leaders based on social network analysis. Intelligence Sci. (4) (2015) (in Chinese)
6. Wang, Z.: A study of micro-evolutionary prediction algorithms for control mapping of final discourse on the Internet. Intelligence Theory Practice **7** (2019) (in Chinese)
7. Zeng, R.X.: A comparative study of the dynamics of online public opinion information dissemination mechanisms. Library and Intelligence Work (2018) (in Chinese)
8. Fu, Z.J., Sun, L.: A study on the balance between legal protection of students' discourse rights and public opinion guidance in universities under the Internet space. China Telegraphic Education (4) (2015) (in Chinese)

9. Fu, Z.J., Xu, Y., Sun, C., Sun, L.: Innovation of Ideological and Political Education Mode with "Internet Onlookers" as an Entry Point. J. Comput. Inform. Syst. 1–8 (2013)
10. Wang, H.C.: Analysis of online discourse and public opinion monitoring in universities. Manag. Observer (2017) (in Chinese)
11. Wilensky, U.: Center for Connected Learning and Computer-Based Modeling. Northwestern University, Evanston, IL (1999) http://ccl.northwestern.edu/netlogo/
12. Troutman, C., Wilensky, U.: Center for Connected Learning and Computer-Based Modeling. Northwestern Institute on Complex Systems, Northwestern University, Evanston, IL (2007) http://ccl.northwestern.edu/netlogo/models/LanguageChange

Multivariate Passenger Flow Forecast Based on ACLB Model

Lin Zheng, Chaowei Qi, and Shibo Zhao[✉]

School of Computer and Network Security, Chengdu University
of Technology, Chengdu 610059, China
12523177@qq.com

Abstract. With the rapid increase in urban population, urban traffic problems are becoming severe. Passenger flow forecasting is critical to improving the ability of urban buses to meet the travel needs of urban residents and alleviating urban traffic pressure. However, the factors affecting passenger flow have complex non-linear characteristics, which creates a bottleneck in passenger flow prediction. Deep learning models CNN, LSTM, BISTM and the gradually emerging attention mechanism are the key points to solve the above problems. Based on summarizing the characteristics of various models, this paper proposes a multivariate prediction model ACLB to extract the nonlinear spatio-temporal characteristics of passenger flow data. We compare the performance of ACLB model with CNN, LSTM, BILSTM, CNN-LSTM, FCN-ALSTM through experiments. ACLB performance is better than other models.

Keywords: CNN · Attention · LSTM · BILSTM · Passenger flow

1 Introduction

Due to the rapid growth of urban population, the pressure of urban traffic load is increasing. City buses are the most important and popular transportation for most urban residents. Accurate prediction of passenger flow in various periods has important significance for allocating buses according to passenger travel rules and improving the utilization of vehicles to meet the needs of passengers. However, the passenger flow has non-linear dynamics, affected by time and external factors, and has complex temporal and spatial characteristics. Therefore, it is crucial to develop a multi-variable prediction model that integrates multiple influencing factors to predict the passenger flow.

There are two ways to develop the passenger flow prediction model. On the one hand, the passenger flow forecasting is regarded as a regression problem, and the data of time and other external factors are used to construct the feature space. Use Linear Regression, Support Vector Regression (SVR) and other machine learning algorithms to establish a prediction model. In addition, bus passenger flow data has time series characteristics and is typical time series data. Therefore, bus passenger flow forecasting can be regarded as a time series forecasting problem. Time series forecasting needs to examine the data mining time series information of passenger flow in a time segment, and establish a time

© The Author(s) 2022
Z. Qian et al. (Eds.): WCNA 2021, LNEE 942, pp. 104–113, 2022.
https://doi.org/10.1007/978-981-19-2456-9_12

series prediction model based on the overall time series characteristics of the data. This method takes into account the time series characteristics of the data and is widely used in the prediction of passenger flow and traffic flow. In recent years, the application of deep learning in various fields has made breakthrough progress. Therefore, researchers at home and abroad have also begun to pay attention to the application of deep learning in time series prediction tasks. Convolutional neural networks (CNN) can extract local features of time series data and Recurrent Neural Network (RNN) and improved long short-term memory (LSTM) and bi-directional long short-term memory (BILSTM) can capture the time series characteristics of data. In addition, the attention mechanism (Attention) is applied in the recurrent neural network. It can improve the processing performance of RNN for ultra-long sequences. On the basis of these research results, this paper proposes a neural network model ACLB that combines attention mechanism, CNN, LSTM, and BISLTM based on the characteristics of multivariate bus passenger flow sequence data.

2   Related Work

Traditional time series forecasting models are Smoothing Methods and autoregressive methods, including ARIMA and SARIMA. etc. Li Jie, Peng Qiyuan [1] have used the SARIMA model to predict the flow of people on the Guangzhou-Zhuhai Intercity Railway and achieved good results. Many researchers have begun to apply Deep Learning to solve time series related problems [2–5]. Yun Liu et al. combined CNN and LSTM to propose the DeepConvLSTM [7] model to be applied to the field of human activity recognition (HAR). This model can automatically extract human behavior characteristics and time feature. Fazle Karim [8] used Fully Convolutional Network (FCN) to replace the pooling layer and fully connected layer of CNN in the task of time series classification, and then combined with LSTM to establish the LSTM-FCN model and ALSTM-FCN. Xie Guicai [4] et al. proposed a multi-scale fusion timing mode convolutional network based on CNN. The model designed short-term mode components and long-term mode components to extract the short-period and long-period spatiotemporal features of the time series, and then obtained Feature fusion recalibration of the final output prediction value comparison, but the model does not consider the influence of external factors other than the flow of people.

3   Model:ACLB

Bus passenger flow prediction should consider the complex non-linear relationship between urban bus passenger flow and time and space factors. The passenger flow of a certain time period is not only affected by the adjacent time period, but also related to various current external factors. For example, the passenger flow of weekdays has obvious morning peak and evening peak, and the peak passenger flow of holidays will be postponed later. Temporary rainfall may lead to a sharp drop in the number of people taking public transportation. And each feature of the data is of different importance to the final prediction result. Therefore, the prediction model should not only consider the temporal and spatial characteristics of the time series data, but also consider reducing the interference of the less correlated data on the prediction result. In order to overcome these problems, this paper proposes a new neural network model ACLB. The structure of the ACLB model is shown in Fig. 1:

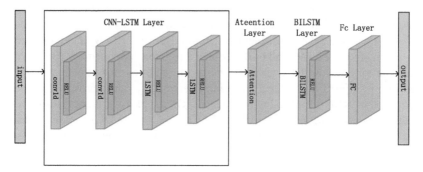

Fig. 1. The structure of the ACLB mode

The ACLB model consists of a CNN-LSTM layer, a BILSTM layer, an attention layer, a fully connected layer, and an output layer. The ACLB model incorporates an attention mechanism on the basis of CNN-LSTM, so that the model can extract the spatiotemporal features of the data and focus the model's attention on key features, and the BILSTM layer is added to extract the bidirectional time dependence of time series data.

3.1 CNN-LSTM Layer

The CNN is used as a feature extractor, and then the sequence output from the CNN is input to the LSTM for training. This CNN-LSTM structure model is mainly used for image caption generation [4], but in research, it is found that CNN-LSTM can also be applied to Time series forecasting [2, 9–11], such as electricity forecasting [12, 13], stock closing price forecasting and other fields. The CNN-LSTM layer in the ACLB model uses the combined structure of CNN and LSTM to extract the local features and timing features of the data. CNN-LSTM Layer is shown in Fig. 2:

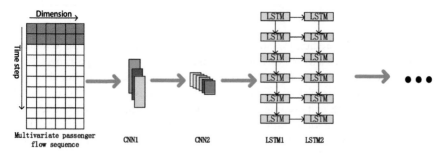

Fig. 2. The structure of the CNN-LSTM layer

Convolutional Neural Networks. In the task of machine learning, feature extraction is a very critical step. For time series prediction, extracting data features can also significantly improve the performance of the model. CNN consists of a convolutional layer, a pooling layer, a fully connected layer and an output layer. It is generally used for feature extraction in image processing, text processing and other fields. At the same time, CNN also has a good effect on time series data. The core part of the CNN convolutional layer is an automatic feature extractor and reduces the overall computational consumption of the model.

Long Short-term Memory. CNN can effectively extract local features of time series data, but CNN cannot capture the time dependence of time series. Therefore, after CNN extracts spatiotemporal features, the LSTM [14, 15] is used to extract the time dependence of time series. LSTM is an improvement of RNN. It adds forget gate, update gate, output gate, memory cell C on the basis of RNN, alleviating the problem of RNN gradient explosion so that the LSTM can capture long-term dependencies. The structure of an LSTM node is shown in Fig. 3:

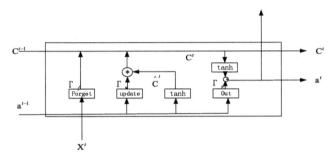

Fig. 3. The structure of an LSTM node

$$\hat{C}^t = tanh(w_c[a^{t-1}, X^t]) + b_c \tag{1}$$

$$\Gamma_u = \sigma(w_u[a^{t-1}, X^t]) + b_u \tag{2}$$

$$\Gamma_f = \sigma(w_f[a^{t-1}, X^t]) + b_f \tag{3}$$

$$\Gamma_o = \sigma(w_o[a^{t-1}, X^t]) + b_o \tag{4}$$

$$C^t = \Gamma_u * \hat{C}^t + \Gamma_f * C^{t-1} \tag{5}$$

$$a^t = \Gamma_o * tanh(c^t) \tag{6}$$

$\hat{C}^{(t)}$ is the memory cell value to be refreshed, a^t is the activation value of the previous LSTM node, X^t is the input value of the current node, C^t is the memory cell value, Γ_u

is the update gate, $\boldsymbol{\Gamma}_f$ is the forget gate, $\boldsymbol{\Gamma}_o$ is the output gate, partial is the range of the activation function from 0 to 1, $\boldsymbol{a}^{t-1}$ is the hidden state of tht node, $\boldsymbol{b}_c, \boldsymbol{b}_u, \boldsymbol{b}_f, \boldsymbol{b}_o$ are all offset values. Memory cell C is the key structure in STM. It transmits information on the entire LSTM, so that key sequence information is retained or discarded, and the problems of gradient explosion and gradient disappearance are alleviated. From Fig. 3 and formula (1)–(6), it can be found that when the memory cell value is passed from the previous node to the current node, its value is controlled by the current node's forgetting gate, the update gate and the input value X of the current node.

3.2 Attention Layer

The attention [14, 16, 17] mechanism is inspired by the cognitive mechanism of the human brain. The human brain can grasp the key information from the complex information and ignore the meaningless information. The attention mechanism assigns weights to the input data to make the model focus on the important features of the data. The structure of the attention mechanism is shown in Fig. 4:

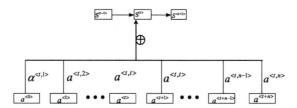

Fig. 4. The structure of attention layer

$$\alpha^{t,i} = \frac{\exp(e^{t,i})}{\sum_{i=0}^{t+n} \exp(e^{t,i})} \tag{7}$$

$$e^{t,i} = S^{t-1} \cdot a^i \tag{8}$$

$[a^0, a^1, \ldots a^n]$ is the hidden state from the CNN-LSTM layer. $\alpha^{t,i}$ represents the ratio of the model's attention to a^i in the input sequence when the attention layer outputs the value S^t. The attention mechanism makes the model always focus on the most critical information.

3.3 BILSTM Layer

BILSTM [18, 19] consists of two LSTMs with opposite information propagation directions. This structure enables BILSTM to capture the forward and backward information of the sequence.

$[S^1, S^2, \ldots, S^t, \ldots, S^{n-1}, S^n]$ is from Attention Layer, It is input into BILSTM to get $[H^1, H^2, \ldots, H^t, \ldots, H^{n-1}, H^n]$. The formula is as follows

$$\vec{H}^t = \overrightarrow{LSTM}(\vec{C}, S^t, \vec{h}^t) \tag{9}$$

$$\overleftarrow{H}^t = \overleftarrow{LSTM}(\overleftarrow{C}, S^t, \overleftarrow{h}^t) \tag{10}$$

$$H^t = w_1 \vec{H}^t \cdot w_2 \overleftarrow{H}^t \tag{11}$$

In the formula (9), (10) and (11), C is the memory cell value, S is the current input, h is the hidden state of the previous node. The ($\leftarrow$, $\rightarrow$) in the formula represents the direction of information flow. $\vec{H}^t$, $\overleftarrow{H}^t$ is the output of the LSTM in the opposite direction. H^t is the output of BILSTM.

4 Experiment

4.1 Construct Training Set

The data set is historical bus card data and weather information data from aity in Guangdong from August 1, 2014 to December 31, 2014. Count the number of passengers in different time periods at one-hour intervals, remove useless fields, and insert weather information corresponding to each time period. $x_i = $ [passenger flow, temperature, $rainfall$, ...] represents passenger flow and external factor data in the i period of the day, $X_i = (x_{i-k}, x_{i-k+1}, \ldots x_i)$ represents a time series from $i - k$ to i. The passenger flow forecast problem is defined as (12)

$$Y_{i+h} = f(X_i) \tag{12}$$

Y_{i+h} is the passenger flow predicted by model at $i + h$. In the following experiment, h is set to 1, which is to predict the passenger flow 1 h away from the current moment. we uses the original data to construct a training set $Z = (X_1, X_2, X_3, \ldots, X_n)$. Among them, the data from August 1 to November 30, 2014 is the training set, December 1 to December 15 is the test set, and December 16 to December 31 is the verification set.

4.2 Model Details

The CNN-LSTM layer in the ACBL model has 2 CNN, 2 pool, and 2 LSTM layers, and the convolution kernels are all set to 3×1. The LSTM has 100 hidden neurons, dropout $= 0.5$ and the BILSTM layer has 100 hidden neurons. During the training, the learning rate is 0.001 and the bachsize is 10. In order to reflect that the improvement of the ACLB model is effective, the performance of the ACLB model is compared with CNN, LSTM, BILSTM, CNN-LSTM and FCN-ALSTM.

4.3 Result

The evaluation indicators adopt RSME and MAPE. In order to avoid the influence of different dimensions on the model, the passenger flow data have been normalized. From the data in Table 1, compared with the single models CNN, LSTM, and BILSTM, the RMSE of CNN-LSTM is reduced by 0.188, 0.159, 0.003, respectively, and the MAPE is reduced by 12.6%, 11.6%, and 2.6%, respectively. Compared with the CNN-LSTM and FCN-ALSTM models, the ACLB model has reduced RMSE by 0.024 and 0.022, and MAPE reduced by 1.3% and 1.5%, respectively.

Table 1. Model performance evaluation (passenger flow prediction result when h = 1)

Model	RMSE	MAPE
LSTM	0.201	20%
CNN	0.230	21%
BILSTM	0.045	9%
CNN-LSTM	0.047	8.4%
FCN_ALSTM	0.045	8.6%
ACLB	0.023	7.1%

Therefore, the ACLB model effectively reduces reduce the error of passenger flow forecast and improves the accuracy.

Figure 5(a)–(e) is the RMSE comparison chart of ACLB and all models for each period from December 29 to 31.

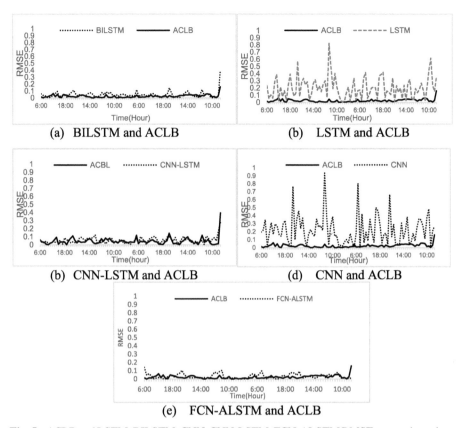

Fig. 5. ACLB and LSTM, BILSTM, CNN, CNN-LSTM, FCN-ALSTM RMSE comparison chart. Passenger flow data has been normalized, so RMSE has no unit

5 Conclusion

In this article, we propose a new model ACLB for passenger flow prediction. In order to evaluate the performance of the ACLB model, in the experiment we used the ACLB model and other models to predict the passenger flow in the next hour. The experimental results show that the ACLB model works well. However, the data set in this article is only a small sample of data. In the next step, we will verify the performance of the ACLB model on a larger range of data sets.

References

1. Jie, L., Qiyuan, P., Yuxiang, Y.: Guangzhou-Zhuhai intercity railway passenger flow forecast based on SARIMA model J. J. Southwest Jiaotong Univ. **55**(1), 51 (2020)
2. Elmaz, F., Eyckerman, R., Casteels, W., Latré, S., Hellinckx, P.: CNN-LSTM architecture for predictive indoor temperature modeling. J. Build. Env. **206**, 108327 (2021)

3. Donahue, J., Anne Hendricks, L., Rohrbach, M., Venugopalan, S., Guadarrama, S., Saenko, K.: Long-term recurrent convolutional networks for visual recognition and description. In: Proceedings of the IEEE conference on computer vision and pattern recognition, pp. 2625–2634 (2015)
4. Xie, G., Duan, L., Jiang, W., Xiao, S., Xu, Y.: Multi-scale time-dependent prediction of pedestrian flow in campus public areas. J. Softw. **32**(3), 831–844 (2021)
5. Vinyals, O., Toshev, A., Bengio, S., Erha, D.: Show and tell: a neural image caption generator. In: Proceedings of the IEEE conference on computer vision and pattern recognition, pp. 3156–3164. (2015)
6. Qu, W., et al.: Short-term intersection traffic flow forecasting. J. Sustain. **12**(19), 8158 (2020)
7. Ordóñez, F.J., Roggen, D.: Deep convolutional and lstm recurrent neural networks for multimodal wearable activity recognition. J. Sensors **16**(1), 115 (2016)
8. Karim, F., Majumdar, S., Darabi, H., Chen, S.: LSTM fully convolutional networks for time series classification. J. IEEE Access **6**, 1662–1669 (2017)
9. Abbas, G., Nawaz, M., Kamran, F.: Performance comparison of NARX & RNN-LSTM neural networks for lifepo4 battery state of charge estimation. In: 2019 16th International Bhurban Conference on Applied Sciences and Technology (IBCAST), IEEE, pp. 463–468 (2019)
10. Yoshida, K., Minoguchi, M., Wani, K., Nakamura, A., Kataoka, H.: Neural joking machine: Humorous image captioning, arXiv preprint arXiv:1805.11850 (2018)
11. Alayba, A.M., Palade, V., England, M., Iqbal, R.: A combined CNN and LSTM model for arabic sentiment analysis. In: Holzinger, A., Peter Kieseberg, A., Tjoa, M., Weippl, E. (eds.) CD-MAKE 2018. LNCS, vol. 11015, pp. 179–191. Springer, Cham (2018). https://doi.org/10.1007/978-3-319-99740-7_12
12. Jia, R., Yang, G., Zheng, H., Zhang, H., Liu, X., Yu, H.: Based on adaptive weights CNN-LSTM&GRU combined wind power prediction method. ChinaPower. https://kns.cnki.net/kcms/detail/11.3265.TM.20211001.1133.002.html
13. Taylor, J.W., McSharry, P.E., Buizza, R.: Wind power density forecasting using ensemble predictions and time series models. J. IEEE Trans. Energy Convers. **24**(3), 775–782 (2009)
14. Tang, F., Kusiak, A., Wei, X.: Modeling and short-term prediction of HVAC system with a clustering algorithm. Energy Build. **82**, 310–321 (2014)
15. Hochreiter, S., Schmidhuber, J.: Long short-term memory. J. Neural Comput. **9**(8), 1735–1780 (1997)
16. Vaswani, A., et al.: Attention is all you need. In: Advances in neural information processing systems, pp. 5998–6008. (2017)
17. Zhang, X., Qiu, X., Pang, J., Liu, F., Li, X.W.: Dual-axial self-attention network for text classification. J. Sci. China Inform. Sci. **64**, 222102 (2021)
18. Schuster, M., Paliwal, K.K.: Bidirectional recurrent neural networks. IEEE Trans. Signal Process. **45**(11), 2673–2681 (1997). https://doi.org/10.1109/78.650093
19. Tianyu, H., Li, K., Ma, H., Sun, H., Liu, K.: Quantile forecast of renewable energy generation based on indicator gradient descent and deep residual BiLSTM. Control Eng. Pract. **114**, 104863 (2021)

Resource Scheduling Strategy for Spark in Co-allocated Data Centers

Yi Liang[(✉)] and Chaohui Zhang

Faculty of Information Technology, Beijing University of Technology, Beijing 100124, China
yliang@bjut.edu.cn

Abstract. The co-allocated data centers are to deploy online services and offline workloads in the same cluster to improve the utilization of resources. Spark application is a typical offline batch workload. At present, the resource scheduling strategy for co-allocated data centers mainly focuses on online services. Spark applications still use the original resource scheduling, which can't solve the data dependency and deadline problems between spark applications and online services. This paper proposes a data-aware resource-scheduling model to meet the deadline requirement of Spark application and optimize the throughput of data processing on the premise of ensuring the quality of service of online services.

Keywords: Co-allocated data centers · Resource scheduling · Deadline

1 Introduction

With the rapid development of the Internet [1], the data scale of the data center has developed rapidly. When the amount of data in the data center is increasing rapidly, the utilization of resources has become an issue of widespread concern in the industry [2]. To improve the utilization of resources, Co-allocated data centers have become an option for many companies. It is to deploy online services and offline workloads on the same cluster and share the data resources of the cluster to improve resource utilization.

There are new deadline requirements in offline applications in many enterprises [3]. For example, a shopping platform recommendation system has a data dependency relationship between offline workloads and online services. Offline workloads need to process intermediate data generated in real-time and provide timely feedback to users, guaranteeing the timeliness of the result data. Spark application is a typical offline batch workload, in traditional resource scheduling; it can't solve the problems encountered in this scenario. In the current scenario, the input data of Spark applications, which is generated from online services can be partitioned and processed in a few phases on demand. The goal of Spark applications is to improve the throughput of data processing while ensuring the deadline requirement. Multiple Spark applications are executed at the same time in the co-allocated data center. How to partition the data and allocate resources among multiple applications has become a big challenge. This paper proposes a resource-scheduling model for Spark in co-allocated data centers, which can reasonably provide

Z. Qian et al. (Eds.): WCNA 2021, LNEE 942, pp. 114–122, 2022.
https://doi.org/10.1007/978-981-19-2456-9_13

data-resource allocation for Spark applications and process more data while meeting the deadline requirement.

The rest of the paper is organized as follows. Section 2 introduces the related work of this article. Section 3 introduces the detailed design of time prediction modeling and the data-away resource scheduling strategy of the Spark application. Section 4 conducts experimental evaluation and analysis. Section 5 summarizes the main contributions of this paper.

2   Related Work

Resource scheduling of applications has been a major research direction in recent years. In the previous resource scheduling research, Kewen Wang and Mohammad Khan Divide a single application into multiple intervals to dynamically, allocate resources to save more resources and improve the utilization of resources [4]. Zhiyao Hu et al. optimized the Shortest Job First Scheduling, by fine-tuning the resources of one job for another job, until the predicted completion time of the job stops decreasing, reducing the overall running time [5].

However, more and more applications have new requirements for the deadline, which has not been considered in previous studies; Guolu Wang et al. proposed a hard real-time algorithm DVDA [6]. Compared with the traditional EDF algorithm, it not only considers the deadline of the application, but also considers the value density, resets the value weight function, and allocates resources to the highest weighted application by priority. With the advent of the data center, there is a dynamic change of available resources, Dazhao Cheng et al. propose a resource and deadline-aware Hadoop job scheduler RDS [7]. The resource allocation is adjusted in time through time prediction, Each job is divided into ten intervals, the resource allocation is adjusted through the execution time and forecast time of each interval, and a simple and effective model is also proposed to predict future resource availability through the recent historical available resources.

With the rapid increase of job scale, many parallel jobs are limited by the network that the cluster is difficult to expand. It is necessary to reduce the cross-rack network traffic by improving the locality of rack data. Faraz and Srimat proposed that ShufflerWatcher [8] tried to arrange the Reducer on the same rack as most M apers to localize the Shuffle stage, but only considering the situation of a single job for independent scheduling, Shaoqi Wang et al. found that there are data dependencies between many jobs in reality [9], and proposed Dawn composed of the online plan and network adaptive scheduler. The online plan determines the preferred rack according to the input data position of the task and the task relevance. After the network adaptive scheduler finds the idle resources of the rack, it selects the appropriate job to schedule on the rack according to the current network status.

3   Model Design

This chapter first introduces the framework overview, then it introduces the design scheme of the time prediction modeling and the data-aware resource scheduling.

The scheduling goal of this paper is the proportion of meeting deadline requirements and the throughput of data processing. The expression is as follows:

$$\text{DAR} = \frac{1}{n} \sum_{i=1}^{n} f(y_i, y_i^\wedge), f(y_i, y_i^\wedge) = \begin{cases} 0, & y_i > y_i^\wedge \\ 1, & y_i \leq y_i^\wedge \end{cases}, \tag{1}$$

$$\text{DTR} = \sum_{i=1}^{n} D_i. \tag{2}$$

y_i and $y_i^\wedge$ represent the actual execution time and deadline time of application i respectively, and the function $f(y_i, y_i^\wedge)$ represents whether application i is completed before the deadline. D_i represents the throughput of data processing for application i.

3.1 Framework Overview

This paper proposes a data-away resource scheduling model based on time prediction. The model is mainly divided into two parts, the first part is to perform time prediction modeling for each Spark application separately to ensure that it can be completed while meeting deadline requirements. The second part is the resource scheduling optimization algorithm, which uses the heuristic algorithm to select the best data-resource allocation plan to ensure that each application is completed while meeting deadline requirements and maximizing the data processing capacity of the spark application. The overall design framework is as follows (Fig. 1):

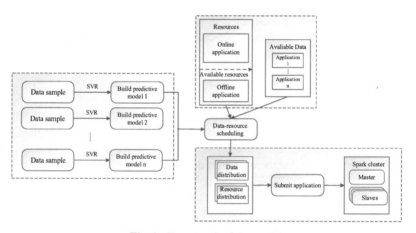

Fig. 1. Framework of the model

3.2 Prediction of Spark Application Execution

This paper selects SVM as a time predictive modeling tool [10, 11]. SVM is a machine learning method developed in the mid-1990s, mainly to minimize the experience risk and confidence range to improve the generalization ability of the learning machine so that better statistics can be obtained in a small sample. Our goal is to predict the execution time of the Spark application, so we need to select the key factors that affect the time prediction. Since this paper models each application separately, internal factors such as the number relationship between the action operator and the transformation operator of the application, the number of shuffles, etc. are not included in the influencing factors. The main influencing factors selected in this paper are input data scale, core and memory resources.

Support vector regression is to transform the original input data x through a non-linear mapping into the corresponding high-dimensional feature space. The linear representation is $\varphi(x)$, and the linear regression is completed. SVR is the method of regression prediction [12, 13]. Through the Lagrangian multiplier method and KKT condition, the SVR can be expressed as:

$$f(x) = \sum_{i=1}^{m} (\alpha_i' - \alpha_i) k(x, x_i) + b. \tag{3}$$

where $k(x, x_i) = \emptyset(x_i)^T \emptyset(x_j)$ is the kernel function. Commonly used kernel functions are linear kernel function, polynomial kernel function, and radial basis kernel function. Select 75% of the samples as the training data, and select the best kernel function through experiments to construct the prediction model.

PSO-Based Resource Scheduling Strategy. The Particle Swarm Optimization (PSO) is a search optimization algorithm with simple operation and fast convergence speed [14, 15]. Each particle in PSO represents a feasible solution to the target problem; each particle mainly contains two attributes: position and velocity. The position represents a feasible solution, the velocity represents the moving speed and direction of the particle, and the movement process of the particle is called the search process of the particle. The update formula for the velocity and position of each particle is as follows:

$$V_i(t+1) = \omega * V_i(t) + C_1 * \text{rand} * (Pb_i(t) - X_i(t))$$
$$+ C_2 * \text{ran} * (gb(t) - X_i(t)), \tag{4}$$

$$X_i(t+1) = X_i(t) + V_i(t). \tag{5}$$

In Eq. (4), t represents the number of iterations. Pb_i and gb respectively represent the optimal position of the ith particle and the global optimal position. ω is the inertia factor, C_1 represents the cognitive ability of the particle, and C_2 represents the learning ability of the particle swarm. *rand* represents a uniform function in [0,1].

Definition of Particles. In the PSO, the definition of particle swarm P is expressed as follows:

$$P = \{P_q | 1 \leq q \leq \text{pNumber}\}. \tag{6}$$

pNumber represents the number of particle swarms, and P_q represents particles. The formula of P_q is as follows:

$$P_q = \{(d_i, c_i, m_i) | 1 \leq i \leq n\}. \tag{7}$$

In Eq. (7), n represents the number of spark applications, (d_i, c_i, m_i) represents a data-resource scheduling solution of the ith spark application, d_i represents the throughput of data processed by the ith spark application, c_i and m_i respectively represent the core and memory resources allocated by the cluster.

Definition of Particle Fitness. Each particle represents a data-resource scheduling solution between spark applications, and the fitness function of the particle represents the revenue that each particle can bring. The scheduling goal of this paper is that Spark applications can improve the throughput of data processing while ensuring the deadline requirement. Therefore, this paper sets the particle fitness as the sum of the data processed by each application, the fitness expression of particles is as follows:

$$E = d_1 + d_2 + \ldots + d_n, \tag{8}$$

$$s.t. y_i \leq deadline, \tag{9}$$

$$\sum_{i=1}^{n} c_i \leq C, \sum_{i=1}^{n} m_i \leq M, \sum_{i=1}^{n} d_i \leq D, \tag{10}$$

$$c_i \geq 0, m_i \geq 0, d_i \geq 0. \tag{11}$$

The constraints of Eqs. (10) and (11) respectively indicate that each application needs to be completed before the deadline, and the allocated data-resources are less than the currently available data-resources.

4   Experimental Results and Analysis

4.1   Experimental Setup

The selection of the experimental environment in this paper is a Spark cluster composed of 15 nodes, including 1 master node and 14 worker nodes. The detailed configuration of each Spark node is shown in Table 1.

The experiment in this paper is divided into two parts, one is the experiment of time prediction model accuracy, and the other is the experiment of resource scheduling strategy performance comparison. In the experiment, Wordcount, Sort, and Pagerank in Hibench are selected as the experimental workload.

Table 1. Experimental environment configuration

Resource type	Resource name	Resource allocation
Hardware	Cpu	Intel(R) Xeon(R) CPU E5–2660 0 @ 2.20 GHz × 32
	Memory	64 GB
	External memory	1 TB
	Operating system	Centos7.4
	Spark	3.0.0
Software	Scala	2.12.10
	JVM	jdk1.8.0
	Hadoop	2.7.3

4.2 Accuracy of Spark Application Execution Time Prediction

We use different kernel functions to model the time prediction of each application, and evaluate the accuracy of each model through RMSE and MAPE, and select the best time prediction model. The results of the prediction model for different workloads are shown in Fig. 2 and Fig. 3.

Through the comparison of time prediction accuracy under different kernel functions in Fig. 2, we can see that different applications can get a better prediction effect when using linear kernel function for time prediction. The time prediction results obtained by using the linear kernel function are more similar to the real execution time.

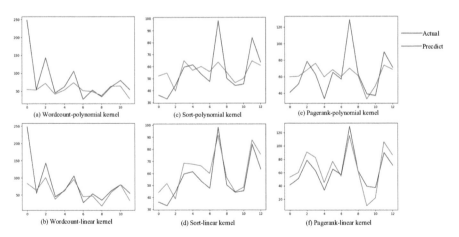

Fig. 2. Time prediction accuracy of different kernel functions

Figure 3 evaluates the large error and relative error in the prediction results by using the evaluation indexes RMSE and MAPE, it can be obtained that when the linear kernel function is used for time prediction, the RMSE is reduced by an average of 27%, and the MAPE is reduced by an average of 1.9%. Therefore, the linear kernel function is selected as the kernel function for time prediction modeling.

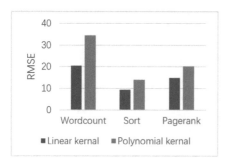
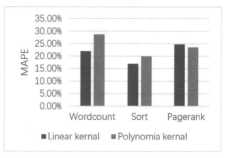

Fig. 3. Experimental evaluation of different kernel functions

4.3 Performance of The Resource Scheduling Strategy

Our resource scheduling strategy is compared with the conservative resource scheduling strategy and the radical resource scheduling strategy, using the DAR and TAR in Sect. 3 as the evaluation indicators of the experiment. The experiment is carried out in the cluster with variable resources, and the following experimental results are obtained.

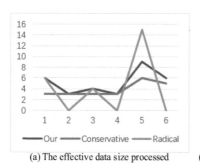
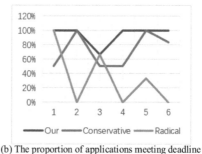

(a) The effective data size processed   (b) The proportion of applications meeting deadline

Fig. 4. Performance of different resource scheduling strategies

It can be seen from Fig. 4 that our method can bring about an increase in the throughput of data processing and the proportion of meeting deadline requirements. Compared with conservative and radical scheduling strategies, our resource scheduling strategy increases the throughput of application processing data by an average of 12% and 50%, respectively. The proportion of applications that meet deadline requirements has increased by 20% and 50%, respectively. Although the conservative resource scheduling strategy can ensure that there is output in the deadline demand, it cannot ensure that more data is processed before the deadline; Although the radical resource scheduling strategy can guarantee the processing of as much data as possible, it cannot guarantee the deadline requirements of Spark application, so there will be less effective data processing; the scheduling strategy in this paper takes into account both the demand for the deadline requirements and the demand for the throughput of data processing, and achieves a good result.

5 Conclusions

This paper proposes a resource-scheduling model for Spark in co-allocated data centers. This method is based on a time prediction model, which increases the throughput of data processing while meeting the deadline requirement, and it solves the new requirements of Spark applications. In the future, we intend to improve the performance of the data-away resource scheduling strategy by increasing the accuracy of time prediction and refining the conditions of scheduling policy.

References

1. Gantz, B.J., Reinsel, D., Shadows, B.D.: Big data, bigger digital shadows, and biggest growth in the far east executive summary: a universe of opportunities and challenges. Idc 1–16 (2007)
2. Delimitrou, C., Kozyrakis, C.: Quasar: resource-efficient and QoS-aware cluster management. ACM SIGPLAN Notices **49**(4), 127–144 (2014)
3. Tang, Z., Zhou, J., Li, K., Li, R.: MTSD: a task-scheduling algorithm for MapReduce base on deadline constraints. In: 2012 IEEE 26th International Parallel and Distributed Processing Symposium Workshops & PhD Forum. IEEE (2012)
4. Wang, K., Khan, M.M.H., Nguyen, N.: A dynamic resource allocation framework for apache spark applications. In: 2020 IEEE 44th Annual Computers, Software, and Applications Conference (COMPSAC), pp. 997-1004. IEEE (2020)
5. Hu, Z., Li, D., Guo, D.: Balance resource allocation for spark jobs based on prediction of the optimal resource. Tsinghua Sci. Technol. **25**(4), 487–497 (2020)
6. Wang, G., Xu, J., Liu, R., Huang, S.S.: A hard real-time scheduler for spark on YARN. In: 2018 18th IEEE/ACM International Symposium on Cluster, Cloud and Grid Computing (CCGRID), pp. 645-652. IEEE (2018)
7. Cheng, D., Zhou, X., Xu, Y., Liu, L., Jiang, C.: Deadline-aware MapReduce job scheduling with dynamic resource availability. IEEE Trans. Parallel Distrib. Syst. **30**(4), 814–826 (2018)
8. Ahamad, F., Chakradhar, S.T., Anand, R., Vijaykumar, T.N.: ShuffleWatcher: shuffle-aware scheduling in multi-tenant MapReduce clusters. In: 2014 USENIX conference on USENIX Annual Technical Conference, pp. 1–12. USENIX Association, USA (2014)
9. Wang, S., Chen, W., Zhou, X., Zhang, L., Wang, Y.: Dependency-aware network adaptive scheduling of data-intensive parallel jobs. IEEE Trans. Parallel Distrib. Syst. **30**(3), 515–529 (2018)
10. Yunmei, L., Yun, Z., Meng, H., Jing, (Selena) H., Yanqing, Z.: A survey of GPU accelerated SVM. In: Proceedings of the 2014 ACM Southeast Regional Conference (ACM SE '14), Article 15, pp. 1–7. Association for Computing Machinery, New York, NY, USA (2014)
11. Pandya, D.: Spam detection using clustering-based SVM. In: Proceedings of the 2019 2nd International Conference on Machine Learning and Machine Intelligence, pp. 12–15. (2019)
12. Ge, W., Cao, Y., Ding, Z., Guo, L.: Forecasting model of traffic flow prediction model based on multi-resolution SVR. In: Proceedings of the 2019 3rd International Conference on Innovation in Artificial Intelligence, pp. 1–5. (2019)

13. Qian, Z., Juan, D.C., Bogdan, P., Tsui, C.-Y., Marculescu, D., Marculescu, R.: Svr-noc: A performance analysis tool for network-on-chips using learning-based support vector regression model. In: 2013 Design, Automation & Test in Europe Conference & Exhibition (DATE), pp. 354-357. IEEE (2013)
14. Hu, M., Wu, T., Weir, J.D.: An adaptive particle swarm optimization with multiple adaptive methods. IEEE Trans. Evol. Comput. **17**(5), 705–720 (2012)
15. Guo, P., Xue, Z.: An adaptive PSO-based real-time workflow scheduling algorithm in cloud systems. In: 2017 IEEE 17th International Conference on Communication Technology (ICCT), pp. 1932–1936. (2017)

Measurement and Evaluation on China's Cargo Airlines Network Development

Chaofeng Wang[✉] and Jiaxin Li

Civil Aviation Flight University of China, Guanghan 618307, China
chaofengbrad@126.com

Abstract. In view of China's cargo airlines network, taking the airport of each city as the node and the number of flights between cities as the weight of the side, the network topology index and economic index are used to evaluate the current situation of the network and the development potential of the network. Then, the TOPSIS method is used to comprehensively evaluate China's cargo airlines network. The results show that the network ranking of each airline is: China Cargo Airlines, SF Airlines, China Post Airlines, Jinpeng Airlines, Longhao Airlines, Yuantong Airlines. Finally, considering the development stage of China's cargo airlines, the sensitivity analysis is conducted by resetting the weight to verify the effectiveness of TOPSIS method. At the same time, according to the different stages of the network of cargo airlines, some suggestions on the development of the network are given.

Keywords: Cargo airlines · Air transport network · Topology analysis · TOPSIS approach

1 Introduction

Compared with other modes of transportation, air transportation can fully meet the timeliness requirements of logistics services for medium and high value-added goods with its technical and economic advantages such as speed, mobility and flexibility. Civil aviation cargo transportation plays an irreplaceable role in medium and long haul distance and transnational transportation. The relevant research by the International Air Transport Association (IATA) suggests that a one-percentage-point increase in air cargo accessibility boosts trade by about six percentage points. With the rapid development of China's air cargo in recent years, its transportation volume has reached nearly 8 million tons in 2019, ranking second only to the United States in the world. Especially in the epidemic situation, air freight and logistics ensure the supply and stability of materials to a certain extent, and play a great role in epidemic prevention and fighting. As of the end of 2019, there were 13 airlines operating all-cargo aircraft in mainland China, with a total of 174 cargo aircraft. There are 8 main cargo airlines, SF Airlines, China Post Airlines, China International Air Cargo Company, China Southern Air cargo Company, Jinpeng Airlines, YuanTong Airlines, China Cargo Airlines and Longhao Airlines. As China's air cargo has been carrying cargo in the belly warehouse for a long time, the

Z. Qian et al. (Eds.): WCNA 2021, LNEE 942, pp. 123–138, 2022.
https://doi.org/10.1007/978-981-19-2456-9_14

number of cargo aircraft is insufficient and the cargo aviation network is not sound enough, the growth rate of air cargo is gradually slowing down. According to the data from Civil Aviation Administration of China, the average annual growth of cargo and mail transportation volume of the whole industry from 2014 to 2019 was 5.0%, and the year-on-year growth in 2019 was 2.1%. In this context, the research on the development status and trend of China's cargo aviation network is of great significance to promote the healthy development of aviation industry and improve the development efficiency and quality of national economy.

Complex network theory is a tool commonly used to analyze networks. The characteristics and main applications of complex networks in different practical fields are systematically compared and analyzed by Boccaletti et al. (2006) [1] and Costa et al. (2011) [2]. The use of complex network theory to study aviation network has also been a hot spot and focus in recent years, but the results of existing research on cargo airline network are still very limited. Starting from the air cargo routes, this paper studies the freight network relationship between cities and regions, and finds that China's air cargo network presents clear centralized characteristics (PAN Kunyou et al. 2007) [3]. XIE Fengjie and CUI Wentian (2014) analyzed the topological structure of specific enterprise's express route network and proposed that its network has the characteristics of a small-world network [4]. Dang Yaru (2012) concluded from the study: China's freight network is a scale-free network that has formed a relatively high agglomeration group, and the level of freight is very clear, but the network distribution is not balanced [5, 6]. Li Hongqi et al. (2017) studied the basic statistical characteristics and correlation of China's air cargo network from the perspective of complex network, obtained the statistical characteristics of China's air cargo network, and pointed out that China's air cargo network has scale-free and small world characteristics, large clustering coefficient and small average path length [7]. Mo Huihui et al. (2017) studied the cargo network of aviation enterprises from the perspective of Chinese cargo airlines, and concluded that Chinese cargo airlines are a hub-structured network with smaller scale and higher organizational efficiency, and maintained a stable network expansion trend [8].

Most of the existing researches on the network of China's cargo airlines are based on the passenger transport network, which is carried out in the manner of carrying cargo in the belly warehouse. Few people have discussed in depth the freight network composed of all-cargo aircraft. And most of the research is based on the network topology, the main indicators used are degree, strength, characteristic path length, clustering coefficient and so on, but less attention is paid to the economic characteristics of airlines. This can only evaluate the current status of the air cargo network, but cannot reflect the development and changes of the future network. Based on this, this paper comprehensively considers the existing network topology and economic benefit characteristics of cargo airlines to comprehensively evaluates their network development capabilities.

2   Chinese Cargo Airlines Network

China's cargo aviation network is mainly composed of 8 Airlines: SF Airlines, China Post Airlines, China International Air Cargo Company, China Southern Air cargo Company, Jinpeng Airlines, Yuantong Airlines, China Cargo Airlines and Longhao airlines. By the end of 2019, China had 236 civil airports in operation. Among them, there are two airports in Beijing and Shanghai, one airport in other areas. In order to facilitate analysis and statistics, we merged the data of Beijing Capital Airport and Beijing Daxing airport as one node, and so did Shanghai. The data in this paper includes the data volume from March 1 to 7, 2021. The total freight network contains 56 nodes and 324 edges (Figs. 1 and 2).

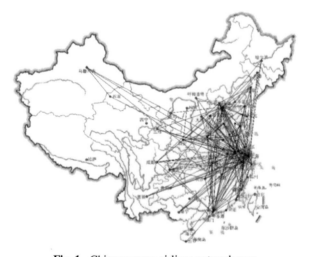

Fig. 1.   Chinese cargo airlines network map

Different cargo airlines have different networks. SF Airlines connects 27 airport nodes, China Post Airlines 41, Jinpeng Airlines 12, Yuantong Airlines 7, China Cargo Airlines 45 and Longhao Airlines 14. China International Air Cargo Company and China Southern Air cargo Company mainly operate international cargo routes, but this paper mainly studies the air cargo network in china, so we did not join these two companies when studying each airline network in detail.

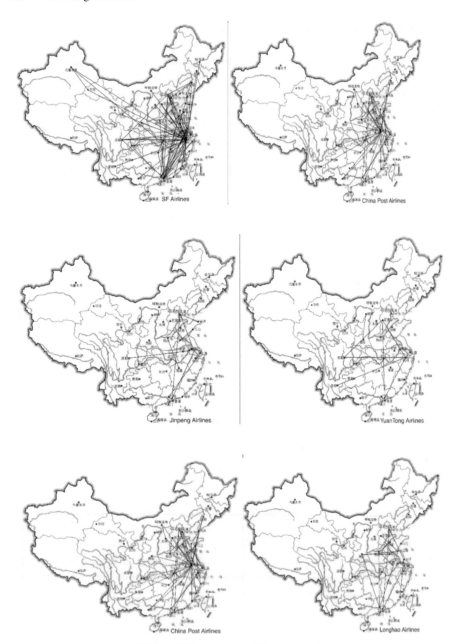

Fig. 2. Different cargo airlines network

3   Network Measurement Index System of Cargo Airlines

3.1   Index System

From the two aspects of network topology index and economic benefit index, among them, the network topology index reflects the current situation of the network, and the economic index reflects the development ability of the network. The evaluation index system is shown in Fig. 3.

3.2   Network Topology Index

Network topologies are widely exist in various social phenomena, basic transportation and biological systems. Different network topologies represent different network connections and dynamic processes (Hossain et al., 2013) [9]. Therefore, the analysis of network topology depends on specific indicators.

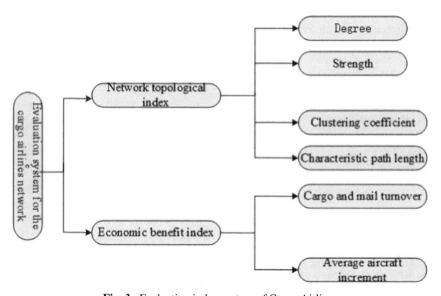

Fig. 3.   Evaluation index system of Cargo Airlines

Degree.   Degree is one of the important basic attributes of nodes in the network, and it is the embodiment of the most basic connection characteristics of nodes in the network. Degree k_i refers to the number of nodes directly connected to node i or the number of edges connected to node i.with that of node i defined as:

$$k_i = \sum_{j=1}^{n} a_{ij} \tag{1}$$

If node i is connected to node j, it is 1, otherwise it is 0. Generally speaking, the importance of degree is that the larger it is, the better the airport accessibility of the node corresponds, and the more important the node is. For the network, some very important indicators are formed, including the average degree k, which is a comprehensive index used to represent the average degree of all nodes. It can be written as:

$$k = \frac{1}{n} \sum_{i=1}^{n} k_i \tag{2}$$

Strength. Degree is the total number of nodes associated to a node. It only considers whether the nodes in the network are connected. However, cargo capacity, number of available seats and flight frequency can be used as weights to affect the connection between airport nodes. This paper selects the number of flights between node i and node j in a week as the weight w_{ij}, the introduction strength S_i can be expressed as:

$$S_i = \sum_{j=1}^{n} w_{ij} a_{ij} \tag{3}$$

The average strength S of all nodes is the average strength, which can be expressed as:

$$S = \frac{1}{n} \sum_{i=1}^{n} S_i \tag{4}$$

Clustering Coefficient. The clustering coefficient C_i is the ratio of the number of edges actually connected to node i and all nodes connected to it to the maximum possible number of connected edges. It describes the proportion of network nodes that are also connected to each other. It shows the closeness of the nodes in the small groups in the network. The larger the value, the higher the closeness. C_i can be written as:

$$C_i = \frac{1}{k_i(k_i - 1)} \sum_{i,j,k} a_{ij} a_{jk} a_{ik} \tag{5}$$

The average clustering coefficient C is the average value of the clustering coefficient of the whole network and can be expressed as:

$$C = \frac{1}{n} \sum_{i} C_i \tag{6}$$

where n is the total number of network nodes, $0 \leq C \leq 1$. The average clustering coefficient is used to describe the local properties of the whole network. If all nodes in the network are independent of each other, then $C = 0$; if all individual nodes in the network have edge connections with other nodes, then $C = 1$.

Characteristic Path Length. The characteristics path length L of the network is the average number of shortest paths for all node pairs. Node i and another node connected to i form a node pair. It can be written as:

$$L = \frac{1}{n(n - 1)} \sum_{i \in V} \sum_{j \neq i \in V} d_{ij} \tag{7}$$

where d_{ij} is the number of edges of the shortest path between node i and node j in the network, and n is the total number of nodes. The characteristic path length is usually used to measure the transmission efficiency of the network. The larger the characteristic path length value, the more edges the network passes through, and the lower the transmission efficiency.

3.3 Economic Benefit Index

Airlines obtain operating revenue and profits by transporting passengers and cargo. In order to further develop enterprises and meet the needs of the market, airlines will invest in opening up new routes. In the case of poor market conditions and poor business operation, the routes will be reduced, and the aviation network will be changed. Based on this, the cargo and mail turnover reflecting the market scale and the investment of aviation companies in aviation network are selected as important economic indicators.

1. Cargo and mail turnover is the total output produced by air cargo companies in a certain period of time. It is a composite index of transportation volume and transportation distance. It comprehensively reflects the total task and total scale of air transportation production. It is not only the most important index of civil aviation transportation companies, but also one of the main indicators for the state to assess air cargo companies.
2. Growth in the number of aircraft: The growth in the number of aircraft of airlines in recent years can reflect the economic situation and operation management of the company in recent years, and to a certain extent, it can also reflect the expansion speed of the company's network. Only when market conditions are good and economic operation management is good, airlines will increase flight density of routes or invest in new routes and purchase new aircraft.

3.4 Measurement Method

Based on the analysis of network topology index and economic benefit index, the entropy weight method is used to calculate the weight of each index, and then TOPSIS model is used to comprehensively evaluate the airport network of each cargo airlines.

Principle of Entropy Weight Method. Entropy weight method is an objective weighting method widely used in various fields. It weights different indicators according to the amount of information of different evaluation indicators, avoiding the differences between evaluation index data and reducing the difficulty of evaluation and analysis (Wang and Lee, 2009) [10]. The specific steps are as follows:

Step 1: According to relevant index data a_{ij} (i = 1, 2, ..., 6, j = 1, 2, ..., 6; i is the number of evaluation objectives; j is the number of indicators), in the future, the values of i and j are the same, and the original evaluation index system matrix A_{mn} is established.

$$A = \begin{bmatrix} a_{11} & a_{12} & \cdots & a_{1n} \\ a_{21} & a_{22} & \cdots & a_{2n} \\ \vdots & \vdots & \vdots & \vdots \\ a_{m1} & a_{m2} & \cdots & a_{mn} \end{bmatrix} \tag{8}$$

Step 2: The extreme value method is used to eliminate the errors caused by the possible differences in the properties, dimensions, orders of magnitude and other characteristics of each index, and then the data are standardized. The formula is as follows:

$$b_{ij} = \frac{a_{ij} - a_j^{\min}}{a_j^{\max} - a_j^{\min}} \text{ (Standardization of positive indicators)} \tag{9}$$

$$b_{ij} = \frac{a_j^{\max} - a_{ij}}{a_j^{\max} - a_j^{\min}} \text{ (Standardization of negative indicators)} \tag{10}$$

The data is normalized to form matrix B_{mn} after processing.

$$B_{mn} = \{b_{ij}\}_{m \times n} \tag{11}$$

Step 3: Calculate the information entropy E_j of the group j.

$$E_j = -(\ln m)^{-1} \sum_{j=1}^{m} P_{ij} \ln P_{ij} \tag{12}$$

$$P_{ij} = \frac{b_{ij}}{\sum_{i=1}^{m} b_{ij}} \tag{13}$$

Step 4: The weight is calculated according to the information entropy of each index.

$$W_j = \frac{1 - E_j}{n - \sum_{j=1}^{n} E_j} \tag{14}$$

TOPSIS Mothed. TOPSIS is "a method to identify the schemes closest to the ideal solution and furthest away from the negative ideal solution in a multi-dimensional computing space"(Qin et al., 2008) [11]. Its advantage lies in its simplicity and easy of programming. TOPSIS has been applied in many fields, such as supply chain management and logistics, design, engineering and manufacturing systems, business and marketing management (Velasquez, M., and Hester, P. T., 2013) [12]. The application of TOPSIS method in this paper is mainly based on two points: one is that the TOPSIS method has good application effect in transportation, logistics, commerce, marketing and other fields; the other is that the method can eliminate the interference of different dimensions in network topology index and economic index. The specific steps are as follows:

Step 1: Construct a weighted normalization matrix R_{ij}.

$$R_{mn} = \{r_{ij}\}_{m \times n} = W_j \times b_{ij} \tag{15}$$

Step 2: Calculate the optimal solution and the worst solution.

$$\text{The optimal solution } X^+ = \{r_1^+, r_2^+, \ldots r_n^+\}, \ r_j^+ = \max(r_{ij}) \tag{16}$$

$$\text{The worst solution } X^- = \{r_1^-, r_2^-, \ldots r_n^-\}, \ r_j^- = \min(r_{ij}), \tag{17}$$

Step 3: Calculate the distance from the weighted evaluation normalized vector to the optimal solution and the worst solution.

$$D_i^+ = \sqrt{\sum_{j=1}^{n} (r_{ij} - r_j^+)^2} \tag{18}$$

$$D_i^- = \sqrt{\sum_{j=1}^{n} (r_{ij} - r_j^-)^2} \tag{19}$$

Step 4: Calculate closeness.

$$G = \frac{D_i^-}{D_i^- + D_i^+} \tag{20}$$

Step 5: Use the value of G as the evaluation result. The larger the value, the better the evaluation result, and the smaller the evaluation value, the worse the result.

4   Data Acquisition and Result Analysis

4.1   Data Acquisition

Network Topology Index. During data processing, we merged the data of Beijing Capital Airport and Beijing Daxing airport as one node, and so did Shanghai. The data in this paper includes the data volume from March 1 to 7, 2021. For the strength index, the number of flights between airports in a week is selected as the weight for calculation. The calculation of the network topology index of each airline is shown in the following table.

Table 1. Main indicators of each airline

Index	Airlines					
	China Cargo Airlines	Longhao Airlines	China Post Airlines	SF Airlines	Jinpeng Airlines	Yuantong Airlines
Number of nodes	45	13	27	42	12	7

(*continued*)

Table 1. (*continued*)

Index	Airlines					
	China Cargo Airlines	Longhao Airlines	China Post Airlines	SF Airlines	Jinpeng Airlines	Yuantong Airlines
Average degree	4.09	2.15	3.19	3.29	1.33	2.29
Average strength	30.00	25.54	41.48	36.48	3.33	30.57
Clustering coefficient	0.81	0.30	0.74	0.64	0.78	0.30
Characteristic path length	2.03	2.03	2.04	2.20	1.2	1.86

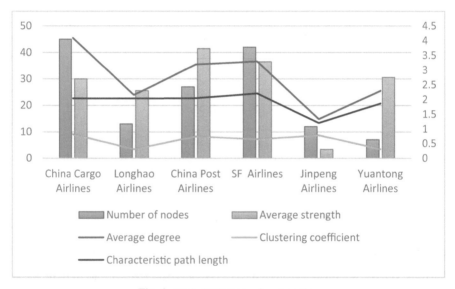

Fig. 4. Main indicators of each airline

From Table 1 and Fig. 4, it can be seen that China Cargo Airlines, China Post Airlines and SF Airlines have a large number of nodes, indicating that they have opened routes in more airports, and Yuantong Airlines has the least number of nodes, that is, fewer airports have opened on their routes. The two indicators of degree and strength generally have the same trend. The more edges a node has on the network, the more flights may be allocated to the node. Therefore, the greater the degree of the node, the greater its strength. China Cargo Airlines, China Post Airlines and SF Airlines are all relatively large in degree and strength, while Jinpeng Airlines has the smallest degree and strength. In terms of clustering coefficient, China Cargo Airlines is the largest, indicating that a node in China Cargo Airlines network has a higher degree of correlation with its neighboring nodes, while Longhao and Yuantong airlines have the smallest clustering coefficient. The characteristic path length is an indicator reflecting the convenience of transmission. The smaller it is, the more convenient the transmission. In terms of the characteristic path length, China Post Airlines is the largest and Jinpeng Airlines is the smallest.

Economic Index. The analysis of the aircraft growth of the 6 Chinese cargo airlines from 2017 to 2020, using the average value as the analysis data.

Table 2. Aircraft growth of different Cargo Airlines

Airlines	Years				
	2017	2018	2019	2020	Average
China Cargo Airlines	0	0	0	2	0.50
Longhao Airlines	3	2	1	2	2
China Post Airlines	2	0	3	2	1.75
SF Airlines	7	9	8	3	6.75
Jinpeng Airlines	1	0	2	0	0.75
Yuantong Airlines	3	3	1	1	2

Table 3. Cargo and mail turnover of different cargo airlines in 2018

Index	Airlines					
	China Cargo Airlines	Longhao Airlines	China Post Airlines	SF Airlines	Jinpeng Airlines	Yuantong Airlines
Cargo and mail turnover	280257.2	3136.7	15212.9	62924.2	97762.2	5246.6

Unit: 10000 tons-kilometers

4.2 Evaluation Results

Combine Table 1, Table 2 and Table 3 to form the original matrix data.

Table 4. Original matrix data

Airlines	Index					
	Network topology index				Economic index	
	Average degree	Average strength	Clustering coefficient	Characteristic path length	Cargo and mail turnover	Average aircraft increment
China Cargo Airlines	4.09	30.00	0.81	2.03	280257.2	0.50
Longhao Airlines	2.15	25.54	0.30	2.03	3136.7	2
China Post Airlines	3.19	41.48	0.74	2.04	15212.9	1.75
SF Airlines	3.29	36.48	0.64	2.20	62924.2	6.75
Jinpeng Airlines	1.33	3.33	0.78	1.20	97762.2	0.75
Yuantong Airlines	2.29	30.57	0.30	1.86	5246.6	2

The matrix is obtained according to the data in Table 4, and then the data is standardized to form a standard matrix to eliminate the impact of the difference between each index on the final result. The information entropy of each index is calculated by entropy weight method. As shown in Table 5.

Table 5. Information entropy of each index

Index	Network topology index				Economic index	
	Average degree	Average strength	Clustering coefficient	Characteristic path length	Cargo and mail turnover	Average aircraft increment
Information entropy	0.84641	0.88842	0.76758	0.89352	0.56762	0.67117

As shown in Table 6, the weight of each index can be calculated according to the formula.

Through the evaluation of the TOPSIS method, the optimal solution and the worst solution are calculated as follows:

$$X^+ = \{0.11250, 0.08173, 0.17024, 0.77991, 0.31670, 0.24085\}$$
$$X^- = \{0,0,0,0,0,0\}$$

Table 6. The weight of each indicator

Index	Network topology index				Economic index	
	Average degree	Average strength	Clustering coefficient	Characteristic path length	Cargo and mail turnover	Average aircraft increment
Weight	0.11250	0.08173	0.17024	0.07800	0.31670	0.24085

The final calculated ranking result are: China Cargo Airlines 0.339, SF Airlines 0.290, China Post Airlines 0.201, Jinpeng Airlines 0.185, Longhao Airlines 0.112 and Yuantong Airlines 0.111. Compared with the results only considering topology indicators, China Cargo Airlines, SF Airlines and China Postal Airlines are still ranked high, indicating that they are not only outstanding on existing networks, but also excellent in future network development.

4.3 Sensitivity Analysis

TOPSIS method does not consider the weight of each index when calculating, assuming that all indexes are equally important. Therefore, it cannot reflect the difference between the weight of existing network and future network characteristic indicators. The weight setting of network topology index and economic index is changed from 1:1 to 1:2 and then to 2:1, so as to further analyze the impact of weight change on each airline. These three weight changes represent that airlines pay more attention to the development of future network, pay equal attention to the current network structure and future network development, and pay more attention to the structure of existing network, which are expressed as the initial stage, growth stage and maturity stage of each airline.

Initial Stage. When the ratio is 1:2, it is the initial stage of the airline. And the weight is recalculated, as shown in Table 7 below.

Table 7. The weight of each indicator when the ratio is 1:2

Index	Network topology index				Economic index	
	Average degree	Average strength	Clustering coefficient	Characteristic path length	Cargo and mail turnover	Average aircraft increment
Weight	0.08475	0.06157	0.12825	0.05876	0.37868	0.28799

The calculation results are arranged as follows: China Cargo Airlines: 0.589, SF Airlines: 0.520, Jinpeng Airlines: 0.312, China Post Airlines: 0.270, Yuantong Airlines: 0.172, Longhao Airlines: 0.171.

Table 8. The weight of each indicator when the ratio is 1:1

Index	Network topology index				Economic index	
	Average degree	Average strength	Clustering coefficient	Characteristic path length	Cargo and mail turnover	Average aircraft increment
Weight	0.12713	0.09236	0.19238	0.08814	0.28401	0.21600

Growth Stage. When the ratio is 1:1, it is the growth stage of the airline. And the weight is recalculated, as shown in Table 8 below.

The calculation results are arranged as follows: China Cargo Airlines: 0.634, SF Airlines: 0.506, China Post Airlines: 0.410, Jinpeng Airlines: 0.382, Yuantong Airlines: 0.222, Longhao Airlines: 0.221.

Mature Stage. When the ratio is 2:1, it is the growth stage of the airline. And the weight is recalculated, as shown in Table 9 below.

Table 9. The weight of each indicator when the ratio is 2:1

Index	Network topology index				Economic index	
	Average degree	Average strength	Clustering coefficient	Characteristic path length	Cargo and mail turnover	Average aircraft increment
Weight	0.16951	0.12314	0.25650	0.11751	0.18934	0.14340

The calculation results are arranged as follows: China Cargo Airlines: 0.719, SF Airlines: 0.628, China Post Airlines: 0.568, Jinpeng Airlines: 0.451, Yuantong Airlines: 0.275, Longhao Airlines: 0.273.

According to the above three tables, the results of each index under different weights are different. No matter at any stage, China Cargo Airlines and SF Airlines have outstanding performance, while China Post Airlines has caught up from behind. The results of growth and maturity stages are consistent with those of TOPSIS method.

5   Conclusions and Recommendations

5.1   Conclusions

This paper uses network topology index and economic index to evaluate the current situation and development potential of the network, so as to effectively evaluate different freight airlines in China. From the analysis of network topology index, it is concluded that each airline has its own different characteristics, merit and demerit. China Cargo

Airlines has the most connected cities and has the greatest advantages. It also performs well in terms of flight density and network accessibility. Sf Airlines has a large number of navigable cities, and the flight density of its routes is good, but poor network accessibility and inconvenient transfer. Although China Post Airlines does not connect so many cities and has poor transit performance, the density of routes between the cities and airports that have already been connected is high, and the network connectivity is good. Jinpeng Airlines, Longhao Airlines and Yuantong Airlines are all connected to a relatively small number of airports. The network density of Longhao Airlines and Yuantong Airlines is general, but the network connectivity is not good. On the contrary, Jinpeng Airlines has the worst network density, but the connectivity is good, and the traffic between the two nodes is convenient. From the perspective of economic indicators, each airline has its own advantages and disadvantages, and only Longhao Airlines and Yuantong Airlines are relatively average.

5.2   Recommendations

1. Cargo airlines should reasonably divide their development stages. The development focus of different development stages is different. In the initial stage, attention is paid to market development based on freight turnover and increasing the number of aircraft to improve the ability of market supply capabilities. In the mature stage, attention is paid to connotative development, that is, the optimization of existing route network. In the growth stage, it is necessary to redevelop route network optimization, expand the market and increase market supply. Only in this way can we be in a relatively leading position in the market.
2. Cargo Airlines reasonably determine benchmarking enterprises in different stages. In the initial stage, China Cargo Airlines, SF Airlines and Jinpeng Airlines should be the benchmark enterprises, and in the growth and maturity stages, China Cargo Airlines, SF Airlines and China Postal Airlines should be the benchmark enterprises.
3. When introducing air cargo enterprises to establish bases, local governments should comprehensively consider the current network of air cargo enterprises and the economic indicators affecting the future network development. Under controllable conditions, the economic indicators affecting the future development of air cargo network should be the key factors to be considered.

Acknowledgement. This work is supported by National Natural Science Foundation of China (71403225).

References

1. Boccaletti, S., Latora, V., Moreno, Y., Chavez, M., Hwang, D.: Complex networks: structure and dynamics. Phys. Rep. **424**(4–5), 175–308 (2006)
2. da Costa, L.F., et al.: Analyzing and modeling real-world phenomena with complex networks: a survey of applications. Adv. Phys. **60**(3), 329–412 (2011)

3. Pan, K.-Y., Cao, Y.-H., Wei, H.-Y.: The study on distributing pattern and network structure of air freight airports in china(in Chinese). J. Econ. Geog. **27**(04), 653–657 (2007)

4. Xie, F.-J., Cui, W.-T.: Complex structural properties and evolution mechanism of air express network. J. Syst. Eng. (9), 114–119 (2014) (in Chinese)

5. Dang, Y.-R., Peng, L.-N.: Hierarchy of air freight transportation network based on centrality measure of complex networks. J. Transport. Syst. Eng. Inform. Technol. **12**(03), 109–114 (2012). (in Chinese)

6. Dang, Y.-R., Meng, C.-H.: Analysis on structure of air cargo network of China based on economy. J. Civil Aviation Univ. China **30**(01), 50–55 (2012). (in Chinese)

7. Li, H.-Q., Yuan, J.-L., Zhao, W.-C., Zhang, L.: Statistical characteristics of air cargo-transport network of China. J. Beijing Jiaotong Univ. (Soc. Sci. Edn.) **16**(02), 112–119 (2017). (in Chinese)

8. Mo, H.-H., Hu, H.-Q., Wang, J.: Air cargo carriers development and network evolution: a case study of China. J. Geographic. Res. **36**(08), 1503–1514 (2017). (in Chinese)

9. Hossain, M., Alam, S., Rees, T., Abbass, H.: Australian airport network robustness analysis: a complex network approach. In: Proc. 36th Australasian Transp. Res. Forum, pp. 1–21 (2013)

10. Wang, T.-C., Lee, H.-D.: Developing a fuzzy TOPSIS approach based on subjective weights and objective weights. Expert Syst. Appl. **36**(5), 8980–8985 (2009). (in Chinese)

11. Qin, X.S., Huang, G.H., Chakma, A., Nie, X.H., Lin, Q.G.: A MCDM-based expert system for climate-change impact assessment and adaptation planning – a case study for the Georgia Basin, Canada. Expert Syst. Appl. **34**(3), 2164–2179 (2008). (In Chinese)

12. Velasquez, M., Hester, P.T.: An analysis of multi-criteria decision making methods. Int. J. Operations Res. **10**(2), 56–66 (2013)

13. Yao, H.-G.: Empirical study on statistical characteristics of topological structure of aviation network of China. J. Logistics Technol. (13), 134–137 (2015) (in Chinese)

14. Chen, H.-Y., Li, H.-J.: Analysis of characteristics and applications of Chinese aviation complex network structure. J. Comput. Sci. **46**(6A), 300–304 (2019). (in Chinese)

Exploration of Non-legacy Creative Product Development Based on Information Technology

Kun Gao[1(✉)], Lijie Xun[2], and Zhenlu Wu[3]

[1] Chinese Song Art College, Guangdong Ocean University, Guangdong, China
63063456@qq.com
[2] Changzhou Institute of Technology, Jiangsu, China
[3] Software Engineering, Guangdong Ocean University, Guangdong, China
luke.woo@foxmail.com

Abstract. With the development of The Times, the development of information technology is accelerating, rapidly into the life, learning in all fields. Under the background of information technology, the dissemination and development of intangible cultural heritage and the development of non-heritage products have been updated. Therefore, a new way of developing intangible cultural heritage should be set up to make it highly compatible with the development of cultural and creative products, so as to build a new development pattern of mutual promotion, integration and reciprocity between intangible cultural heritage culture and cultural and creative products. This paper analyzes the concept of non-heritage products and the value of the combination of intangible cultural heritage and cultural creation, and discusses the development strategy of non-heritage products based on information technology for reference.

Keywords: Information technology · Non-legacy products · Product experience · E-commerce platform · Intangible town

1 Introduction

In the Internet era, the concept of "Internet Plus" has received unprecedented attention, especially when the concept of mass entrepreneurship and innovation is put forward. In this process, with the help of the "Internet plus" concept of compliance, different industries have achieved varying degrees of improvement. It can be said that the " +" in "Internet +" represents the infinite possibility of organic integration of information technology represented by Internet technology with different industries. In other words, relying on Internet thinking, in-depth innovation of industry development can be realized, and consumer experience and added value of products and services can be improved. By introducing the concept of "Internet +" into the field of intangible cultural heritage, the business structure of non-heritage creative products will be greatly changed, and the design mode, production mode and marketing mode of the industry will be reshaped, thus providing a better opportunity and path for the benign development of intangible cultural heritage culture.

© The Author(s) 2022
Z. Qian et al. (Eds.): WCNA 2021, LNEE 942, pp. 139–146, 2022.
https://doi.org/10.1007/978-981-19-2456-9_15

2   Concept of Non-legacy Creation Products

According to the Law of the People's Republic of China on Intangible Cultural Heritage, Intangible cultural heritage refers to "all kinds of traditional cultural expression forms handed down from generation to generation and regarded as part of their cultural heritage, as well as objects and places related to traditional cultural expression forms" [Zhu Bing. Main Content and System interpretation of Intangible Cultural Heritage Law of the People's Republic of China. China's Intangible Cultural Heritage, 2021(01): 6–14.] In the era of global integration, various cultures show a significant homogenization trend in the integration and collision, thus highlighting the uniqueness of intangible cultural heritage culture. Under the impact of the commodity economy, how to realize the better protection and national intangible cultural heritage is a realistic problem worthy of attention and thinking, how to grasp the social public cultural appeal, and in the process of implementation of heritage and its surrounding products and packaging, also is a key focal point question.

In recent years, with the continuous improvement of the national economic level, the social public is no longer satisfied with the rich material life, but attaches more importance to the rich spiritual world. Therefore, non-legacy products with unique forms of cultural expression and bearing unique cultural connotations have increasingly attracted the attention of the public. This consumption tendency also reflects the public's love and pursuit of a better spiritual life to a considerable extent. Different from ordinary commodities, non-heritage products are the design and creation inspiration that designers get from intangible cultural heritage. With unique visual symbols of regional culture as the design carrier, such cultural and creative products are endowed with profound cultural value connotation. Through the design, production and sales of non-heritage products, tourists can have a more profound sensory impression on intangible cultural heritage, and at the same time, it will help the inheritance and dissemination of intangible cultural heritage.

3   The Value of Combining Intangible Cultural Heritage with Cultural Creation

Intangible cultural heritage is the outstanding cultural achievements created by the Chinese people of all ethnic groups in the long period of social practice, which can be regarded as an important representative of the manifestation of national culture. When we look at intangible cultural heritage, we can see that it not only shows the extraordinary memory, but also shows the unique national thinking and cultural thinking mode. It can be said that these characteristics are extremely scarce in the era of global integration and the serious homogenization tendency of culture. Each intangible cultural heritage project contains unique value, but due to the lack of effective communication channels, some outstanding intangible cultural heritage skills and traditional crafts are declining, and related non-inheritors and traditional craftsmen are facing the dilemma of no successor. How to combine the consumption habit and aesthetic orientation of contemporary people to make the ancient intangible cultural heritage enter the public life with a new attitude is the question of the era of intangible cultural heritage protection.

It should be noted that intangible cultural heritage originated from the agricultural era, so although it has attracted the attention of the public, it is incompatible with the inherent requirements of the commodity society. If this phenomenon cannot be dealt with and solved, it will inevitably lead to various difficulties in the process of non-inheritance.

Since the rise of cultural and creative industry, along with the trend of global integration, the industry has spread rapidly in different countries and regions, and in this process, it has connected with other industries based on its unique cultural form and operation mode. Figure 1 shows the operating income of China's cultural and creative industry from 2012 to 2019. It can be seen that the data is increasing year by year.

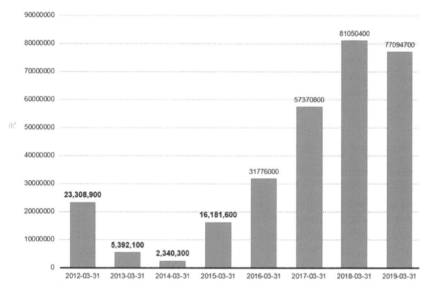

Fig. 1. Operating revenue of China's National Cultural Industry Enterprises (Chinese cultural and creative products) in 2021–2019. (The data comes from Ai Media website)

Culture is the foundation and carrier of cultural and creative industry, which can be called the source of cultural and creative industry. Non-legacy works are important forms of cultural expression. Through in-depth research and exploration of design materials by designers, intangible heritage will be ensured to become the design source and creative inspiration of non-legacy products. At the same time, non-legacy creation products and intangible cultural heritage are mutually reinforcing. The latter provides design and creation materials for the former, while the former provides communication carrier and opportunity for the latter.

4   The Development of Non-legacy Creative Products Based on Information Technology

It has many advantages to develop non-legacy creative products based on information technology. We can take advantage of the information technology, and through careful investigation and research activities, realize the accurate grasp of user demand, to carry out the "Internet +" the papers and the product experience, relying on the digital technology to the depth development of papers and works, electric business platform was used to optimize the papers and product development, build a legacy town, set up corresponding technical team. Below, the author will combine their own understanding and understanding, respectively on the following aspects to talk about the development of non-legacy creative products based on information technology.

4.1   Accurate Grasp of Users' Demands Through Careful Research Activities

Excellent wen gen products must be realized with precision to meet user demand, therefore, the formal product design development of the papers and, before the survey should be based on activities, realize to the demands of user research, the research content includes the user's age structure, professional distribution, gender, willingness to spend, consumption habits, consumer preferences, income level, etc., Only in this way can accurate analysis of user groups be achieved from the perspective of consumer psychology, thus providing scientific basis for the subsequent design and research and development of non-legacy creative products.

4.2   Developing the Experience of "Internet Plus" Non-legacy Creative Products

In general, museums and art galleries are places for the exhibition of non-legacy products. Although these venues provide opportunities for the public to contact and understand non-legacy products, it should also be noted that some museums and art galleries are too serious about the display and exhibition of non-legacy products. This may cause the audience with the papers and psychological distance between products being expanded, leading to the audience hard to display and exhibition of papers and products to generate understanding and explore enthusiasm, for this reason, on the papers and the product display and exhibition, should be adhering to the "Internet +" thinking, with the aid of modern information technology to achieve visual display of the papers and the products, In this way, the charm of non-legacy creation products will be highlighted to the greatest extent, and the audience will have enthusiasm for appreciation and interest in exploration, so as to deepen their interest in intangible cultural heritage culture.

At present, with the development of information technology, modern information technology means such as VR and AR are improving day by day. The advantage of these modern information technology means is that they can break the limitation of time and space and create a scene organically combining virtual and reality, so that the audience can get a more intuitive appreciation experience. VR technology relies on computer technology, information technology and simulation technology to achieve, with the help of this technology, the audience can get a sense of immersion. AR technology can realize the deep integration of virtual information and the real world, and rely on the way of simulation processing, so that the audience can get an immersive sensory experience.

For example, in the fourth Non-heritage Expo, designers showed the intangible heritage lifelike to every audience through the application of VR technology and AR technology, so that every audience got an audio-visual feast. The exhibition also relies on information technology to build a database covering a large number of traditional literature and art resources, and provides free query and download services for the public.

For example, when weifang kite is displayed with VR technology and AR technology, the intuitive display of kite making process can be realized, and the legend of kite origin can be displayed for the public with the aforementioned technology, and the audience can also experience the process of simulated kite flying. It can be said that such a comprehensive experience will leave a deep impression on the audience and generate a strong interest in non-heritage products and intangible cultural heritage culture in the process.

4.3 Relying on Digital Technology to Realize the In-Depth Development of Non-legacy Works

During the Shanghai World Expo, the China Pavilion used modern information technology to display the Riverside Scene at Qingming Festival. In this way, "Along the River During the Qingming Festival" is vividly and dynamically presented to the audience, thus making it the jewel of the China Pavilion during the World Expo. In recent years, the Palace Museum, Tencent and local museums have successively devoted themselves to the design and research and development of digital cultural and creative products. For example, relying on digital technology, the Palace Museum has produced cultural and creative products represented by Auspicious Signs in the Forbidden City, thus helping the public to have a more detailed understanding of the Palace Museum culture. This work is in the form of an APP. After the public installs this APP on their smartphones or tablets, they can appreciate various cultural relics of the Palace Museum with the help of information interaction technology. The mobile game APP "Search for Fairy" produced and launched by Tencent fully integrates traditional cultural elements in intangible cultural heritage, thus realizing the dissemination of traditional excellent culture in the form of game, and also giving young consumers, the target audience of mobile games, an opportunity to have an in-depth understanding of intangible cultural heritage and traditional culture.

4.4 Optimize the Development of Non-legacy Creative Products by Using E-commerce Platforms

Under the background of information technology, e-commerce platform plays an important role in the development of cultural and creative products. E-commerce platforms gather a large number of customer groups, which can further expand the customer group of non-legacy creative products, so that more young people can understand non-legacy creative products more conveniently and conveniently, and promote the publicity of non-legacy creative products.

In 2020, the number of videos related to national intangible heritage on Douyin increased by 188% year on year, and the cumulative broadcast volume increased by 107% year on year [Zhu Yinxia. Research on the communication effect of short videos of intangible heritage [D]. Nanchang University,2020.] E-commerce platforms have also brought huge sales for INTANGIBLE cultural heritage products. For example, Li Tinghuai, the representative inheritor of the national-level Ru porcelain firing technique, sold ru porcelain over 3 million yuan through Douyin e-commerce; Sun Yaqing, the representative inheritor of the state-level intangible cultural heritage fan-making technique, participated in more than 20 intangible cultural heritage e-commerce activities, with a total sales volume of more than 700,000 yuan. Visible, we can make full use of the advantage of electric business platform to optimize the papers and the product development. On May 3, for example, in 2021, in "trill 55 tide purchase season" "originality tide have fei" zone, trill electricity sale "shadow play printing T-shirt" and "yun" kite "condensed intangible craftsmanship in the two products, this is the trill genetic bearing electrical business hand in hand to the people, products, manufacturers such as power, together with the papers and the product. "Shadow play" printing T-shirt the papers and the products on sale in Japan, live trill platform, with the help of powerful propaganda trill platform, the once pushed on the papers and the product was a great success For young people who are keen on Douyin platform, they learn about Traditional Chinese shadow puppetry through watching live broadcast, enrich their knowledge, and deepen their understanding and appreciation of the history of national literature and art.

4.5 Build an Intangible Heritage Town and Set up a Corresponding Technical Team

Intangible cultural heritage town is a town form formed in a certain space with the help of intangible cultural heritage resources, which has functions such as industry, town, human resources and culture. In such small towns, a large number of non-genetic inheritors are gathered. Relying on the guiding effect of policies, art practitioners are attracted to such small towns, thus achieving the benign interaction and in-depth communication between non-genetic inheritors and literary and art creators, and thus achieving the goal of attracting talents. A relatively successful example in this regard is Wutong Mountain Art Town, which greatly improves the creative vitality of non-legacy products by attracting art designers to enter and opening art studios.

Peroration

Intangible cultural heritage is not only an important cultural heritage of the Chinese nation, but also a spiritual treasure belonging to the whole mankind and the whole world. It is not inherited, but the inheritance of traditional culture with a long history. In view of this, the protection of intangible cultural heritage is an important work. How to realize the effective inheritance and dissemination of intangible cultural heritage is related to the continuation of cultural blood. To do this, we need to do two things. First, regarding the protection and inheritance of intangible cultural heritage, relevant institutions should be aware of the significance and value of the "Internet+" concept for the protection and inheritance of intangible cultural heritage, and provide and create brand-new carriers for the protection and inheritance of intangible cultural heritage with the help of various modern information technology means. At the same time, the designers also shall be with the aid of modern information technology, as an effective way to design and research and development of papers and the product, in order to improve the papers and the products in the heart of the social public appeal, as a result, not only can achieve the purpose of the prosperity of socialist culture, will also realize the effective promotion of intangible culture, More importantly, the public will have a strong interest in intangible cultural heritage through the purchase and consumption of non-heritage products, thus contributing to better inheritance of intangible cultural heritage. Secondly, through the development of intangible cultural heritage + cultural creative products, it is an inevitable choice for non-inheritance and development to use non-heritage creative products to make intangible cultural heritage out of the minority and into life. Cultural and creative products are not only the embodiment of culture itself, but also a way of cultural inheritance. The integration of INTANGIBLE cultural heritage and cultural creation interprets the intangible cultural heritage culture and further promotes the development of art and culture. The intangible cultural heritage culture and cultural and creative products should be well integrated, with the help of products to spread traditional culture, constantly strengthen cultural confidence, promote non-inherited inheritance and development, make Chinese traditional culture long lasting, and help intangible cultural heritage realize the dream of cultural inheritance.

Acknowledgements. This paper is the research result of zhanjiang Tourism "Non-legacy Creative Products" Development Strategy Research, project number: ZJ21YB18, which is the 2021 planning project of Philosophy and Social Sciences of Zhanjiang city, Guangdong Province.

References

1. Xiao, Y., Yao, Y., Lin, J., Chen, W., Wang, L.: Design and development path analysis of Huaihua non-legacy creative brands. Art Apprec. (32), 41–42 (2021)
2. Zhao, J., Liu, M., Zheng, Z.: Research on non-heritage design in Dongguan based on experience vision. West Leather **201,43**(19), 71–72
3. Zhang, X.: Research on the design methods of cultural creative products under the background of intangible cultural heritage + cultural creation -- taking the forbidden city theme cultural creation products as an example. Art Apprec. (30), 99–100 (2021)
4. Gao, J., Xiang, Y.: Development of Non-heritage Products based on Guangzhou jade carving -- taking Yuefan series products as an example. Tiantian (04), 5–9 (2021)
5. Sun, N., Li, B.: Cultural Inheritance in the New era – The development and application of non-legacy creative products. Grand View (07), 65-66 (2021)
6. Yang, F.: Digital communication of cultural space in jiangsu characteristic towns – a case study of Qixia mountain non-legacy creative town. Beauty Times (I) (08), 4–6 (2021)
7. Wang, Y., Lu, Y.: Research on the design of cultural tourism products based on the activation of intangible cultural heritage – taking the design of cultural tourism products in northeast China as an example. Strait Sci. Industry **201,34**(04), 72–74
8. Wen, X., Liu, Z., Li, L.: Brand construction and exploration based on non-legacy creation: a case study of Tujia brocade in western hunan. Furniture Interior Decor. (09), 55–59 (2021)
9. Chen, Q.: Application of User experience audit design method in non-legacy creative product design – Taking Wenzhou Lanjierian as an example. Western Leather **201,43**(13), 69–71

Gait Planning of a Quadruped Walking Mechanism Based on Adams

Gangyi Gao[1]([⊠]), Hao Ling[2], and Cuixia Ou[3]

[1] College of Mecanical Engineering, Jingchu University of Technology, Jingmen, Hubei, China
87437396@qq.com
[2] Suzhou Newcity Investment and Development Co., Ltd., Suzhou, Jiangsu, China
[3] College of Mecanical Engineering, Hangzhou Dianzi University, Hangzhou, Zhejiang, China

Abstract. In this paper, the kinematics analysis and gait planning of quadruped walking mechanism are carried out. Firstly, a simplified four-legged mechanism model is established; then the kinematics of the walking mechanism is analyzed; On this basis, the gait planning of walking mechanism is studied, the forward motion (four step movement) gait is analyzed, the corresponding leg swing order of each gait is calculated; Finally, ADAMS software is used to simulate and analyze the gait planning.

Keywords: Quadruped mechanism · Motion analysis · Gait planning · Motion simulation

1 Introduction

The research of quadruped robot began in 1960s. With the development of computer technology, it has developed rapidly since 1980s. After entering the 21st century, the application research of quadruped robot continues to extend from structured environment to unstructured environment, from known environment to unknown environment. At present, the research direction of quadruped robot has been transferred to the gait planning which has certain autonomous ability and can adapt to complex terrain. Based on the kinematics research of the quadruped mechanism, the gait of the walking mechanism is planned, and the corresponding leg swing sequence of the forward motion (four steps) gait is obtained. Finally, the gait planning is simulated and verified by ADAMS software, and good results are achieved.

2 Kinematic Analysis

2.1 Simplified Model

In the initial kinematic analysis modeling of simplified model, it is not necessary to excessively pursue whether the details of the component geometry are consistent with the reality, because it often takes a lot of modeling time and increases the difficulty of kinematic analysis. The key at this time is to pass the kinematic analysis smoothly

© The Author(s) 2022
Z. Qian et al. (Eds.): WCNA 2021, LNEE 942, pp. 147–159, 2022.
https://doi.org/10.1007/978-981-19-2456-9_16

and obtain the preliminary results. In principle, as long as the mass, center of mass and moment of inertia of the simplified model are the same as those of the actual components. In this way, the simplified model is equivalent to the physical prototype. The simplified model is shown in Fig. 1.

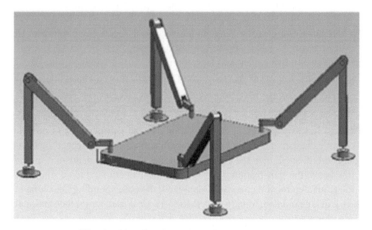

Fig. 1. Simplized quadruped-leg robot model

2.2 Establish Coordinate System

In order to clearly show the relative position relationship between the foot and the body of the walking mechanism and the three-dimensional space, three sets of coordinate systems are established, namely the leg coordinate system, the body coordinate system and the motion direction coordinate system.

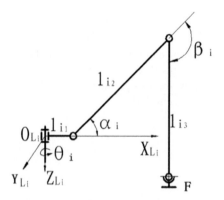

Fig. 2. Body diagram and the coordinate of walking system

Leg Coordinate System. $O_{Li}X_{Li}Y_{Li}Z_{Li}$ coordinate system is shown in Fig. 2, Coordinate origin O_{Li} is the axis of rotation of the hip joint and the bar L_{i1} intersection; axis

Z_{Li} is downward along the rotation axis of the hip joint; axis X_{Li} is in the leg plane and perpendicular to the axis Z_{Li}; axis Y_{Li} is determined by the right-hand rule. Select Z_{Li} down, the selection of downward is mainly to intuitively display the change of the height of the center of gravity. The walking mechanism has four legs. Therefore, there are four leg coordinate systems as shown in Fig. 3, $i = 1,2,3,4$.

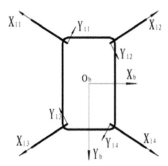

Fig. 3. Body coordinate and leg coordinate system of walking mechanism

Volume Coordinate System. The body coordinate system is a coordinate system fixed on the body and moving with the movement of the walking mechanism. As shown in Fig. 3, a three-dimensional coordinate system $O_b X_b Y_b Z_b$ is established. The coordinate origin O_b is located at the geometric center of the traveling mechanism; axis X_b starts from the coordinate origin, Along the horizontal direction of the body width of the traveling mechanism; Axis Z_b vertical down; axis Y_b is determined by the right-hand rule.

Motion Direction Coordinate System. Motion direction coordinate system when human beings walk, they always consider the difference between themselves and the target and how to move to reach the target. According to the thinking method of human walking, the motion direction coordinate system is established.$O_n X_n Y_n Z_n$. The establishment of the coordinate system of the motion direction system is as follows: The origin coincides with the origin of the volume coordinate system, axis Z_n coincides with the axis Z_b, axis X_n points in the direction of this movement,Y_n is determined by the right-hand rule. This coordinate places the planner on the walking mechanism itself and thinks that the movement of the walking mechanism is equivalent to the movement of his own legs, which brings a lot of convenience to the gait planner. It not only reduces many transformations in walking and greatly reduces the amount of calculation, but also for the operator, the walking mechanism is equivalent to himself. How much is the difference between himself and the target, How to move to reach the target is clear in the eyes of the operator. The motion direction coordinate is set to solve the motion relationship between the quadruped walking mechanism and the environment. It has a certain relationship with the earth directly. It can also be said to be the geodetic coordinate system of a certain motion. It only works when the walking mechanism moves along a certain motion direction.

2.3 Kinematic Calculation of Leg

As shown in Fig. 2. The robot has three driving joints, that is, three degrees of freedom. The three joint angles are $\theta_i/\alpha_i/\beta_i$. The length of each rod is shown in Fig. 2. The position of the foot end in the leg coordinate system is $F(x_{Fi}, y_{Fi}, z_{Fi})$, axis X_{Li} is always in the leg plane, $y_{Fi} = 0$.

Forward Kinematics Calculation of Leg. The forward kinematics of the leg calculates the forward kinematics of the leg, which refers to determining the position of the foot in the corresponding coordinate system according to the motion of the driving joint of the leg. The structural parameters and three joint angles of quadruped walking mechanism are shown in Fig. 3, foot end position:

$$x_{Fi} = l_{i1} + l_{i2}cos\alpha_i + l_{i3}cos(\beta_i - \alpha_i) \tag{1}$$

$$y_{Fi} = 0 \tag{2}$$

$$z_{Fi} = l_{i3}sin(\beta_i - \alpha_i) - l_{i2}sin\,\alpha_i \tag{3}$$

$$\theta_i = \theta_i \tag{4}$$

Inverse Kinematics Calculation of Leg. The inverse kinematics of the leg calculates the inverse kinematics of the leg, which refers to calculating the motion parameters of each driving joint of the leg according to the position of the foot in the coordinate system.
From Eq. (1):

$$x_{Fi} = l_{i1} + x_{fi} \tag{5}$$

$$x_{fi} = l_{i2}\cos\alpha_i + l_{i3}\cos(\beta_i - \alpha_i) \tag{6}$$

$$X_{fi}^2 + Z_{Fi}^2 = I_{i2}^2 + I_{i3}^2 + 2I_{i2}I_{i3}\cos\beta_i \tag{7}$$

$$\cos\beta_i = \frac{(x_{Fi} - I_{i1})^2 + Z_{Fi}^2 - I_{i2}^2 - I_{i3}^2}{2I_{i2}I_{i3}} \tag{8}$$

$$k_i = \cos\beta_i$$

$$\beta_i = \arccos k_i \tag{9}$$

from Eq. (3):

$$Z_{Fi} = -\sin\alpha_i(l_{i2} + l_{i3}\cos\beta_i) + \cos\alpha_i(l_{i3}\sin\beta_i) \tag{10}$$

$$\frac{Z_{Fi}}{\sqrt{(I_{i2}+l_{i3}\cos\beta_i)^2+(l_{i3}\sin\beta_i)^2}} = -\frac{I_{i2}+l_{i3}\cos\beta_i}{\sqrt{(I_{i2}+l_{i3}\cos\beta_i)^2+(l_{i3}\sin\beta_i)^2}}sin\,\alpha_i \\ + \frac{l_{i3}\sin\beta_i}{\sqrt{(I_{i2}+l_{i3}\cos\beta_i)^2+(l_{i3}\sin\beta_i)^2}}cos\,\alpha_i \tag{11}$$

$$\sin \gamma_i = \frac{I_{i3}\sin \beta_i}{\sqrt{(I_{12}+I_{i3}\cos \beta_i)^2+(I_{i3}\sin \beta_i)^2}}$$

$$\cos \gamma_i = \frac{I_{i2}+I_{i3}\cos \beta_i}{\sqrt{(I_{i2}+I_{i3}\cos \beta_i)^2+(I_{i3}\sin \beta_i)^2}}$$

$$\frac{Z_{Fi}}{\sqrt{(I_{12}+I_{i3}\cos \beta_i)^2+(I_{i3}\sin \beta_i)^2}} = -\sin \alpha_i \cos \gamma_i + \cos \alpha_i \sin \gamma_i$$

$$\sin(\gamma_i - \alpha_i) = \frac{Z_{FI}}{\sqrt{(I_{12}+I_{i3}\cos \beta_i)^2+(l_{i3}\sin \beta_i)^2}} \tag{12}$$

$$\gamma_i - \alpha_i = \arcsin\left(\frac{Z_{Fi}}{\sqrt{(I_{12}+I_{i3}\cos \beta_i)^2+(I_{i3}\sin \beta_i)^2}}\right) \tag{13}$$

$$\alpha_i = \gamma_i - \arcsin\left(\frac{Z_{FI}}{\sqrt{(I_{12}+I_{i3}\cos \beta_i)^2+(I_{i3}\sin \beta_i)^2}}\right) \tag{14}$$

$$\gamma_i = \arcsin\left(\frac{I_{i3}\sin \beta_i}{\sqrt{(I_{12}+I_{i3}\cos \beta_i)^2+(I_B\sin \beta_i)^2}}\right)$$

3   Analysis of Translational Gait of Walking Mechanism

Gait refers to the movement process of each leg of the walking mechanism according to a certain order and trajectory. It is precisely because of this movement process that the walking movement of the walking mechanism is realized. The walking mechanism discussed in this paper is in a static and stable walking state, that is, at any time, the walking mechanism has at least three legs supported on the ground. This state belongs to the slow crawling of the robot.

3.1   Static Stability Principle

The static stability of multi legged robot refers to the stability that the robot does not flip and fall when walking and maintains the balance of the body. If the vertical projection of the center of gravity of the robot is always surrounded by polygons formed by alternating footholds, the robot is statically stable. If the center of gravity of the robot exceeds the stability range, the robot will lose stability. As shown in Fig. 5, legs 1, 3 and 4 are set as support legs, and O is the center of gravity of the robot. Triangular area $\triangle ABC$ represents the stable area surrounded by three footholds of the robot. When the center of gravity o of the robot is located in this area, the robot is statically stable. If the center of gravity of the robot will exceed the stable area, it will lead to the instability of the robot. During static and stable walking, the vertical center of gravity of each part of the robot is required to always fall in the stable area, which makes the walking speed of the robot very slow, so it is called crawling or walking.

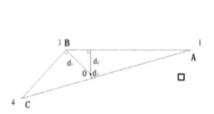

Fig. 4. Principle of the static stability

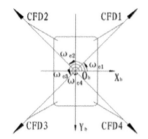

Fig. 5. All the CFD of walking mechanism

3.2 Critical Direction Angle

The critical direction angle ω_c refers to the angle between the critical forward direction (CFD) of the walking mechanism and the axis X_b of the body coordinate system. The critical forward direction indicates the straight line direction formed by the vertical projection of the quadruped walking mechanism doing translational crawling along the diagonal at the current and next foothold, which is through the vertical projection of the center of gravity of the walking mechanism in a gait cycle. Therefore, as shown in Fig. 5, four critical directions can be obtained. These direction angles and axis X_b and axis y_b of the body coordinate system divide the direction angle ω into eight regions to determine and select the leg swing sequence: $0 \leq \omega \leq \omega_{c1}, \omega_{c1} \leq \omega \leq \pi/2, \pi/2 \leq \omega \leq \omega_{c2}, \omega_{c2} \leq \omega \leq \pi, \pi \leq \omega \leq \omega_{c3}, \omega_{c3} \leq \omega \leq 3\pi/2, 3\pi/2 \leq \omega \leq \omega_{c4}, \omega_{c4} \leq \omega \leq 2\pi$.

3.3 The Swing Sequence of the Legs in the Translational Gait

Taking one of the eight areas as an example, this paper expounds the selection process of leg swing sequence. According to the walking direction, the principle of total leg swing is to meet the principle of static stability of the walking mechanism. In addition, since the walking machine is symmetrically distributed and has a simple structure, when setting the swing sequence of legs, the stability of the walking mechanism is judged according to the position of the center of gravity of the walking mechanism. It is assumed that the traveling mechanism is at an angle with the X direction ω As shown in Fig. 6, the initial attitude of the walking mechanism is represented by a dotted line, the solid line represents the attitude of the walking mechanism after movement, and the dotted line represents the stable triangle formed by the support points of each foot of the walking mechanism. A gait cycle of the walking mechanism is divided into four stages. In each stage, one leg is lifted and dropped, and then the body moves. It is represented by four diagrams in Fig. 6 (a), (b), (c) and (d). If the step size of a gait cycle is s, the moving distance of the body in each stage is s/4.

As shown in Fig. 6 (a), the traveling mechanism moves s/4 along the direction angle ω, in which it moves along the X direction and along the Y direction. It can be seen that after the movement, the center of gravity of the body is in $\triangle P_2P_3P_4$ and $\triangle P_1P_2P_4$. It can be seen that both leg 1 and leg 3 can be lifted. However, considering that the stability

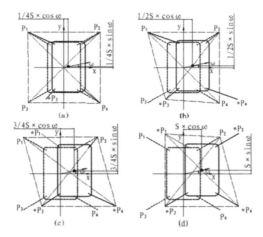

Fig. 6. Swinging leg selection of walking mechanism walk to the front

margin of leg 3 is greater than that of leg 1, leg 3 should be selected as the first swing leg.

As shown in Fig. 6 (b), after the leg 3 swings, the traveling mechanism moves s/4 again along the direction angle. If it moves $S \times \cos\omega/4$ and $S \times \sin\omega/4$ respectively along the X and Y directions, it moves cumulatively along the X direction $S \times \cos\omega/2$ and the Y direction $S \times \sin\omega/2$. According to the stability principle, only leg 4 can be selected as the second swing leg this time.

Similarly, the swing sequence of each leg of the walking mechanism in a gait cycle can be obtained when the walking mechanism moves at the direction angle in each area, as shown in Table 1.

Table 1. The legs' swing sequence in different walking direction

ω	Legs' swing sequence
$0 \leq \omega \leq \omega 1$	$3 \rightarrow 4 \rightarrow 1 \rightarrow 2$
$\omega 1 \leq \omega \leq \pi/2$	$3 \rightarrow 1 \rightarrow 4 \rightarrow 2$
$\pi/2 \leq \omega \leq \omega 2$	$4 \rightarrow 2 \rightarrow 3 \rightarrow 1$
$\omega 2 \leq \omega \leq \pi$	$4 \rightarrow 3 \rightarrow 2 \rightarrow 1$
$\pi \leq \omega \leq \omega 3$	$2 \rightarrow 1 \rightarrow 4 \rightarrow 3$
$\omega 3 \leq \omega \leq 3\pi/2$	$2 \rightarrow 4 \rightarrow 1 \rightarrow 3$
$3\pi/2 \leq \omega \leq \omega 4$	$1 \rightarrow 3 \rightarrow 2 \rightarrow 4$
$\omega 4 \leq \omega \leq 2\pi$	$1 \rightarrow 2 \rightarrow 3 \rightarrow 4$

3.4   The Swinging Sequence of Gait Legs with Fixed-Point Rotation

In order to make the walking mechanism have greater mobility, it is necessary to further design the fixed-point rotation gait of the walking mechanism. The rotation angle conforms to the right-hand rule. When the traveling mechanism turns left $\gamma > 0$, when it turns right $\gamma \leq 0$.

The swing sequence of the walking mechanism legs rotating around the geometric center of the walking mechanism is analyzed as follows:

As shown in Fig. 7, the selection of leg swing sequence is illustrated by taking the left turn of the body as an example. Because it rotates around the fixed point of the geometric center of the walking mechanism, the center of gravity of the body remains unchanged and is always at the geometric center of the body during the rotation. Assuming that the angle γ of a gait cycle is, a gait cycle of the walking mechanism is divided into four stages as shown in Fig. 7 (a), (b), (c), (d) and (e). The dotted line in the figure represents the stable triangle formed by the support points of each foot of the walking mechanism. The angle of each body rotation is. The solid line in figure (a) represents the initial attitude of the walking mechanism, the dotted line represents the attitude of the walking mechanism after a gait cycle, and the dotted line in Fig. 7 (b), (c), (d) and (e) represents the current attitude of the walking mechanism, The solid line represents the posture of the walking mechanism after a phase of gait rotation.

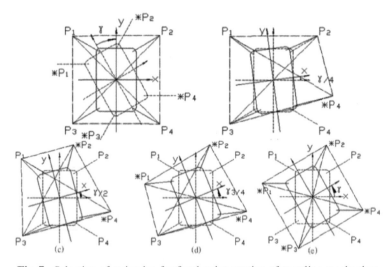

Fig. 7. Selection of swing leg for fixed-point rotation of traveling mechanism

The initial and final positions of the traveling mechanism are shown in Fig. 7 (a). Firstly, the initial posture of the walking mechanism is shown by the solid line in Fig. 7 (a). When the walking mechanism rotates to the left $\gamma/4$, it is feasible to lift any leg according to the stability principle. We select leg 4 as the first swing leg, as shown in Fig. 7 (b). After leg 4 swings, the posture of the walking mechanism is shown as the solid line in Fig. 7 (b). The traveling mechanism rotates to the left. According to the stability principle, only leg 2 can be selected as the second swing leg this time, as shown in Fig. 7 (c).

After leg 2 swings, the posture of the walking mechanism is shown by the solid line in Fig. 7 (c). The traveling mechanism rotates to the left $\gamma/4$ again. According to the stability principle, only leg 1 can be selected as the third swing leg this time, as shown in Fig. 7 (d).

After leg 1 swings, the posture of the walking mechanism is shown as the solid line in Fig. 7 (d). The traveling mechanism rotates to the left $\gamma/4$ again. According to the stability principle, only leg 3 can be selected as the fourth swing leg, as shown in Fig. 7 (e).

The final pose is shown by the solid line in Fig. 7 (e).

Similarly, the swing sequence of legs under fixed-point rotation gait is summarized in Table 2.

Table 2. The swinging sequence of gait legs rotating around a fixed point of the geometric center of the walking mechanism

Turn left	Turn right
1→3→2→4	1→2→4→3
2→1→3→4	2→4→3→1
3→4→2→1	3→1→2→4
4→2→1→3	4→3→1→2

4   Simulation (Take the Four Step Walking in Front as an Example)

In order to verify the rationality of the mechanism design and gait planning of the walking mechanism, the simulation analysis is carried out by using UG and ADAMS software. After the simplified model of the walking mechanism is created in UG software, the model is imported into ADAMS software by using ADAMS/exchange module, and other environments (such as ground, etc.) are built in ADAMS software to form the large framework of the virtual prototype, and then the constraints and forces are applied to these components to establish the virtual prototype of the walking mechanism (Table 3).

Table 3. Motion planning of walking mechanism walking straight ahead (four steps)

Movement steps	Action
Step1(0→1 s)	Leg 3 forward 1 m
Step2(1.5→2.5 s)	Leg 1 forward 1 m
Step3(3→4 s)	Body forward by 1 m
Step4(4.5→5.5 s)	Leg 4 forward 1 m
Step5(6→7 s)	Leg 2 forward 1 m

4.1 Determine Simulation Parameters

According to the kinematics research of the walking mechanism, the size of the walking mechanism leg mechanism is substituted into the inverse kinematics calculation formula of the leg, and the rotation angle of each driving joint is calculated, as shown in Table 4.

Table 4. Rotation angle of each driving joint when walking straight ahead (four steps)

Joint	Step1	Step2	Step3	Step4	Step5
l_{11}–l_{12}		+21.11	−21.11		
l_{12}–l_{13}		−15/+18.41	−3.14		
l_{13}–foot$_1$		+19.93	−19.93		
l_{21}–l_{22}			+34.84		−34.84
l_{22}–l_{23}			+0.47		−15.47/+15
l_{23}–foot$_2$			+6.19		−6.19
l_{31}–l_{32}	+21.11		−21.11		
l_{32}–l_{33}	−18.41/+15		+3.14		
L_{33}–foot$_3$	−19.93		+19.93		
l_{41}–l_{42}			+34.84	−34.84	
l_{42}–l_{43}			+0.47	+15/−15.47	
L_{43}–foot$_4$			+6.19	−6.19	

4.2 Simulation Result

Input the parameters in Table 4 into the functions of each corresponding driver. The simulation process of the walking mechanism walking straight ahead is shown in Fig. 8.

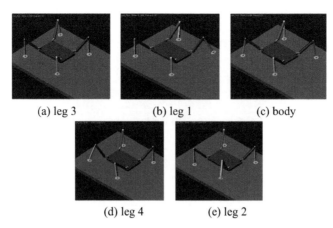

(a) leg 3   (b) leg 1   (c) body

(d) leg 4   (e) leg 2

Fig. 8. Screenshot of simulation process of walking mechanism walking straight ahead (four steps)

4.3 Analysis of Simulation Results

After simulation, the displacement of each point on the body of the walking mechanism along the axis Z_n direction is small, that is, the movement of the platform is relatively smooth and stable on the whole. However, there are still some problems, such as slight deviation of the motion trajectory and instability of individual steps, which are summarized as follows (Fig. 9):

1) When walking straight ahead (four steps), the mobile platform moves forward once after four steps, resulting in uncoordinated action when the platform moves forward. This is because the active drive is much more than the spatial degrees of freedom of the walking mechanism, resulting in redundant constraints. For this problem, you can try to take a two-step approach.
2) When walking, the platform tilts slightly in individual steps. The reason is that the center of gravity of the whole mobile platform is too close to the edge of its stable triangle, resulting in the reduction of stability margin. To solve this problem, by modifying the motion parameters of the corresponding joints, the distance between the center of gravity of the walking mechanism and the edge of the stable area surrounded by the three supporting feet is increased (as shown in Fig. 4, the value of the shortest distance d1 among the three distances d1d2d3 is increased), the stability margin of the walking mechanism is increased, so as to greatly improve the walking stability of the platform.
3) Similar methods can be used to study the gait of walking mechanism in front of walking (two-step movement), right front $45°$ walking (one-step movement) and right front $45°$ walking (two-step movement).

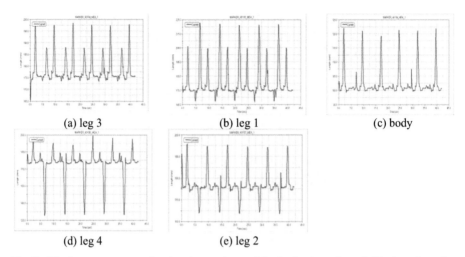

(a) leg 3   (b) leg 1   (c) body

(d) leg 4   (e) leg 2

Fig. 9. Displacement curve of each point on the machine body along the axie Z_n direction when the walking mechanism moves straight ahead (four steps)

Acknowledgements. Fund project: Hubei Provincial Department of Education Science Research Program Project (B2020196); Jingmen City Science and Technology Program Project (2020YFYB051).

References

1. Zhanghao: Space analysis and trajectory planning of Quadruped Robot. Equipment Manuf. Technol. (09) (2020)
2. Yang, J., Sun, H., Wang, C.H., Chen, X.D.: Review of quadruped robot research. Navigation, Positioning and Timing (05) (2019)
3. Zhou, L., Cai, Y.: Simulation of bionic quadruped robot. Mech. Drive (09) (2013)
4. Yuan, G., Li, L.: Optimal design of walking trajectory of Quadruped Robot. Comput. Simul. (10) (2018)
5. Luo, Q.S.: Bionic Quadruped Robot Technology. Beijing University of Technology Press (2016)
6. Xu, Z.D.: Structural Dynamics. Science Press (2007)
7. Luo, H.Y., Wei, L., Li, Z., Zeng, S.: Motion planning and gait transformation of bionic quadruped robot. Digital Manuf. Sci. (01) (2018)
8. Zhou, K., Li, C., Li, C., Zhu, Q.: Motion planning method of Quadruped Robot for unknown complex terrain. J. Mech. Eng. **56**(02), 210 (2020)
9. Li, Y., Li, B., Rong, X., Meng, J.: Structure design and gait planning of hydraulically driven quadruped bionic robot. J. Shandong Univ. (Eng. Edn.) (05) (2011)

10. Li, H.K., Li, Z., Guo, C., Dai, Z., Li, W.: Diagonal gait planning based on quadruped robot stability. Machine Design (01) (2016)
11. Mai, Y.J., Yuan. H.B., Guo, J., Pang, X.: Design and gait analysis of bionic quadruped robot. Machinery Manuf. (12) (2019)

Design and Implementation of Full Adder Circuit Based on Memristor

Ning Tang[1], Lei Wang[1,2(✉)], Tian Xia[1,2], and Weidong Wu[3]

[1] NARI Group Corporation / State Grid Electric Power Research Institute, Nanjing 211106, China
453927489@qq.com
[2] Nanjing University of Aeronautics and Astronautics, Nanjing 210000, China
[3] North Information Control Institute Group Co., Ltd., Nanjing 211153, China

Abstract. In order to break through the traditional von Neumann architecture of computing and memory cell separation and speed up the computing speed, it is necessary to realize in memory computing, and memristor is an excellent carrier to realize in memory computing. Then, the development, principle, characteristics and application prospect of memristor are briefly introduced, and the characteristic curve of memristor is obtained by simulating the model of memristor. The principle and characteristics of memristor are explained more intuitively. Then, based on the memory resistor, the simple logic circuit design principle is described. The logic structure can be realized by using the memory resistor as the calculation element and adding a CMOS inverter, so as to realize the simple logic circuit. The paper designs the simple logic circuit including gate, gate, or gate by spice software, and simulates the circuit of gate, gate, gate, or gate. Then, based on the above logic gate, the circuit design of adder is carried out, the circuit diagram and design scheme are given, and the simple description and SPICE simulation are given. The design scheme is reviewed and summarized, its advantages and disadvantages are analyzed, and the optimization and improvement scheme is proposed.

Keywords: Full adder · Memristor · Logic computing

1 Introduction

One-bit full adder is considered as an important case study of MRL (Memristor Ratio Logic) family [1]. The full adder consists of two half adder, while the half adder can be composed of an exclusive-OR gate and an AND gate. Based on the basic AND gate, OR gate and exclusive-OR gate, we can implement the circuit design of the adder [2].

In order to provide a standard cell design method, the standard cell is a NAND (NOR) logic gate. In a stable state, no current flows out from the output node because the output node of the AND (OR) logic gate is connected to the metal oxide semiconductor gate [3]. In this method, each standard cell needs to have two connections between the complementary metal oxide semiconductor layer and the memristor layer, one for intermediate level conversion and one for output. This method is robust, although it is inefficient in terms of power consumption and area compared with the optimized circuit.

© The Author(s) 2022
Z. Qian et al. (Eds.): WCNA 2021, LNEE 942, pp. 160–165, 2022.
https://doi.org/10.1007/978-981-19-2456-9_17

In the optimized circuit, CMOS phase inverter is applied only when signal recovery is needed or logic function needs signal inversion.

The research shows that for MRL logic family, linear memristor devices without current threshold is preferred, unlike other digital applications, which need threshold and nonlinearity [4–6]. Compared with nonlinear memristor devices, MRL gate based on linear memristor devices has faster speed, smaller size and lower power consumption. Memristor ratio logic series opens opportunities for additional memristor and complementary metal oxide semiconductor integrated circuits and improves logic density [7–11]. This enhancement can provide more computing power for processors and other computing circuits.

2 Design and Implementation of Adder Circuit Based on Memristor and Its SPICE Simulation

The schematic diagram of one-bit full adder used in this case study is shown in Fig. 1 below. One-bit full adder consists of six OR logic gates based on memristor, three AND logic gates based on memristor and four complementary metal oxide semiconductor phase inverters.

According to the schematic diagram of adder circuit in Fig. 1, the circuit can be built by Hspice software for simulation. The adder calculation formula used in this paper is as follows:

$$S = A \oplus B \oplus C_{IN} \tag{1}$$

$$C_{OUT} = A \cdot B + A \oplus B \cdot C_{IN} \tag{2}$$

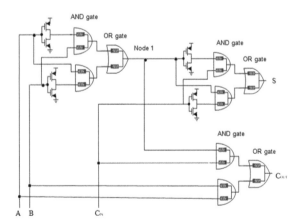

Fig. 1. Schematic diagram of adder circuit

The practical meanings represented by each item in the above formula are: A stands for summand, B stands for addend, C_{IN} stands for low carry, S stands for carry, C_{OUT} stands for sum.

2.1 Analysis of Simulation Results

According to the circuit schematic diagram of adder shown in Fig. 1, simulation analysis is carried out by using Hspice. In this scheme, a voltage of 4 V (high level, i.e., 1) is applied to port A, a voltage of 0 V (low level, i.e., 0) is applied to port B, and a voltage of 3 V (high level, i.e., 1) is applied to C_{IN} as an example to show the simulation results and analyze them.

The truth table of adder is shown in Table 1 below.

Table 1. Truth table of full adder

A	B	C_{IN}	C_{OUT}	S
0	0	0	0	0
0	0	1	0	1
0	1	0	0	1
0	1	1	1	0
1	0	0	0	1
1	0	1	1	0
1	1	0	1	0
1	1	1	1	1

A voltage of 4 V (high level, i.e., 1) is applied to port A , and a voltage of 0 V (low level, i.e., 0) is applied to port B. The curve of voltage and time of node 1 after the first exclusive-OR gate is shown in the following Fig. 2. It can be seen that when a voltage of 4 V is applied to port A and a low level is applied to port B, the curve of voltage and time of node 1 after the first exclusive-OR gate is basically consistent with the curve of output voltage of exclusive-OR gate when a high level and a low level are input above.

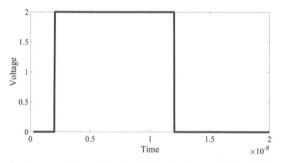

Fig. 2. The curve of voltage and time of node 1 after the first exclusive-OR gate when a voltage of 4 V (high level, i.e., 1) is applied to port A , and a voltage of 0 V (low level, i.e., 0) is applied to port B.

When a voltage of 3 V (high level, i.e. 1) is applied to port C_{IN}, the curve of output voltage and time of port S is shown in the following Fig. 3. It can be seen that the output voltage of port S decreases continuously from 0.2ns to 1.2ns, and the speed of taking effect is the fastest at 0.7s. In this period, it can be approximately considered that a high-level pulse voltage of 2 V is input from node 1 and a voltage of 3 V is applied to port C_{IN}, and the change characteristic curve of the output voltage of port S is basically consistent with the output voltage curve of exclusive-OR gate when two high levels are input above. When 1.72 V is taken as the threshold voltage, the output voltage is equal to 1.72 V, which is regarded as the output low level (0).

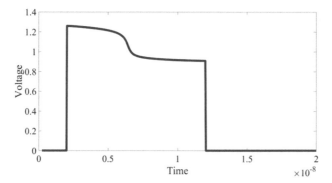

Fig. 3. Curve of output voltage and time of port S

When a 3 V voltage (high level, i.e., 1) is applied to the C_{IN}, the curve of the output voltage and time of port C_{OUT} is shown in the following Fig. 4. It can be seen that the output voltage of port C_{OUT} with 2.11 V remains stable at about 2.11 V during 0.2ns to 1.2ns, which can be regarded as an AND gate inputting a 2 V high level and a 3 V high level. Another AND gate inputs a 4 V high level and a low level, and the output voltages of the two AND gates can be regarded as high level (1) and low level (0) respectively, and then pass through an OR gate to obtain a curve. When 2.11 V is taken as the threshold voltage, the output voltage is equal to 2.11 V, which is regarded as the output high level (1).

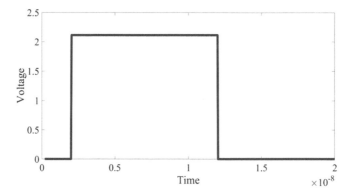

Fig. 4. Curve of output voltage and time of port C_{OUT}

In other cases, the output level basically meets the requirements of the truth table of the adder, which will not be discussed in this paper.

3   Analysis and Improvement of This Scheme

For the optimization method, when cascaded MRL gates based on memristor are connected, the current can flow from the output node to the input of the next logic gate. In this case, the currents flowing through two memristor devices of one gate are not equal, and the smaller current may drop below the current threshold of memristor devices, resulting in partial switching of logic gates. This phenomenon will reduce the output voltage and may cause the logic to fail after a single logic level.

One method to eliminate possible logic faults is to increase the voltage of high logic state to ensure that all currents in the circuit are greater than the current threshold of the device. The increase of voltage is limited by complementary metal oxide semiconductor process, because high voltage may lead to breakdown of complementary metal oxide semiconductor transistor (for example, drain and leakage of grid induction [12]), and also consume more power.

Another method to eliminate logic faults is to amplify signals with CMOS logic gate to prevent steady-state current leakage and perform signal recovery. In this case study, both methods are used. The voltage increases and the signal recovery is implemented by a complementary metal oxide semiconductor inverter. Note that these signal degradation problems are circuit-related, that is, the degree of signal degradation depends on the logic circuit structure and the parameters of memristor devices.

Memristor ratio logic is a hybrid complementary metal oxide semiconductor memory logic family. Compared with CMOS logic, this logic series uses less chip area. By using the standard cell library composed of NOR and NAND logic gates, the design workload of MRL circuit can be reduced. However, the standard cell limits the flexibility of the design process and the opportunity of saving area. Other optimization criteria, such as increasing the operating voltage and minimizing the number of connections between CMOS and memristor layer, are also possible.

4   Conclusion

In this paper, a one-bit adder is designed with 18 memristors and 4 CMOS phase inverters. The circuit design diagram of the scheme is given, and the principle, design ideas and possible problems of the scheme are introduced. The designed full adder is simulated by Hspice software, and the output voltage values under various conditions are obtained and compared with the truth table. Then, according to the content of the design scheme, the advantages and disadvantages of the scheme are found out, and the shortcomings are optimized and improved.

Acknowledgements. This work is supported by the State Grid Corporation Science and Technology Project Funded "Key technology and product design research and development of power grid data pocket book" (1400-202040410A-0–0-00).

References

1. Kvatinsky, S., Wald, N., Satat, G., Kolodny, A., Weiser, U.C., Friedman, E.G.: MRL — Memristor Ratioed Logic. In: 2012 13th International Workshop on Cellular Nanoscale Networks and their Applications, pp. 1–6 (2012)
2. Yadav, A.K., Shrivatava, B.P., Dadoriya, A.K.: Low power high speed 1-bit full adder circuit design at 45nm CMOS technology. Int. Conf. Recent Innov. Signal Proc. Emb. Sys. (RISE) **2017**, 427–432 (2017)
3. Xu, X., Cui, X., Luo, M., Lin, Q., Luo, Y., Zhou, Y.: Design of hybrid memristor-MOS XOR and XNOR logic gates. Inter. Conf. Elec. Devi. Sol.-Sta. Circ. (EDSSC) **2017**, 1–2 (2017)
4. Liu, B., Wang, Y., You, Z., Han, Y., Li, X.: A signal degradation reduction method for memristor ratioed logic (MRL) gates. IEICE Electron. Express, p. 12 (2015)
5. Cho, K., Lee, S.-J., Eshraghian, K.: Memristor-CMOS logic and digital computational components. Microelec. J. 214–220 (2015)
6. Cho, K., Lee, S.J., Eshraghian, K.: Memristor-CMOS logic and digital computational components. Microelec. J. 214–220 (2015)
7. Teimoory, M., Amirsoleimani, A., Ahmadi, A., Ahmadi, M.: A hybrid memristor-CMOS multiplier design based on memristive universal logic gates. In: 2017 IEEE 60th International Midwest Symposium on Circuits and Systems (MWSCAS), pp. 1422–1425 (2017)
8. Mirzaie, N., Lin, C.-C., Alzahmi, A., Byun, G.-S.: Reliability-aware 3-D clock distribution network using memristor ratioed logic. IEEE Trans. Compo. Pack. Manuf. Technol. **9**(9), 1847–1854 (2019). Sept.
9. Escudero, M., Vourkas, I., Rubio, A., Moll, F.: Memristive logic in crossbar memory arrays: variability-aware design for higher reliability. IEEE Trans. Nanotechnol. **18**, 635–646 (2019)
10. Liu, G., Zheng, L., Wang, G., Shen, Y., Liang, Y.: A carry lookahead adder based on hybrid CMOS-memristor logic circuit. IEEE Access **7**, 43691–43696 (2019)
11. Hoffer, B., Rana, V., Menzel, S., Waser, R., Kvatinsky, S.: Experimental demonstration of memristor-aided logic (MAGIC) using valence change memory (VCM). IEEE Trans. Electron Devices **67**(8), 3115–3122 (2020). Aug.
12. Kvatinsky, S., Friedman, E.G., Kolodny, A., Weiser, U.C.: TEAM: ThrEshold adaptive memristor model. IEEE Trans. Circuits Syst. I Regul. Pap. **60**(1), 211–221 (2013). Jan.

Multi-level Network Software Defined Gateway Forwarding System Based on Multus

Zhengqi Wang[1,2(✉)], Yuan Ji[1,2], Weibo Zheng[1,2], and Mingyan Li[3]

[1] NARI Group Corporation (State Grid Electric Power Research Institute), Nanjing, Jiangsu, China
wzqwzq@mail.ustc.edu.cn
[2] Nanjing NARI Information and Communication Technology Co., Ltd., Nanjing, Jiangsu, China
[3] State Grid Henan Electric Power Research Institute, Zhengzhou, Henan, China

Abstract. In order to solve the problem that the data forwarding performance requirements of the security gateway are becoming higher and higher, the difficulty of operation and maintenance is increasing day by day, and the physical resource configuration strategy is constantly changing, a multi-level network software defined gateway forwarding system based on Multus is proposed and implemented. On the basis of kubernetes' centralized management and control of the service cluster, different types of CNI plugins are dynamically called for interface configuration, At the same time, it supports the multi-level network of kernel mode and user mode, separates the control plane and data plane of the forwarding system, and enhances the controllability of the system service. At the same time, the load balancing module based on user mode protocol stack is introduced to realize the functions of dynamic scaling, smooth upgrade, cluster monitoring, fault migration and so on without affecting the forwarding performance of the system.

Keywords: Software-defined · Kubernetes · Forward system · Multus

1 Introduction

With the advancement of the construction of the Internet of things, the terminal equipment presents the development trend of large scale, complex structure and diverse types. The security services are facing many new problems [1]. First, the number of IOT network terminal equipment is increasing day by day, and the number of terminals is increasing exponentially. The requirements for the data forwarding performance of the border security gateway are becoming higher and higher. It is necessary to continuously expand and upgrade the equipment cluster, and the difficulty of operation and maintenance is increasing day by day. Second, with the continuous increase of security services, different types of services have different requirements for resources, resulting in the continuous dynamic change of the resource allocation strategy. The original gateway equipment of different types can not adapt to the dynamic changes of services, resulting in the shortage

Z. Qian et al. (Eds.): WCNA 2021, LNEE 942, pp. 166–176, 2022.
https://doi.org/10.1007/978-981-19-2456-9_18

of resources for some services and a large number of idle resources for other services. Limited physical resources need to be allocated more effectively and reasonably.

The development of docker technology [2] has set off a new change in the field of cloud platform technology, which enables various applications to be quickly packaged and seamlessly migrated on different physical devices [3]. The release of applications has changed from a lot of environmental restrictions and use dependencies to a simple image, which can be used indiscriminately on different types of physical devices. However, container is only a virtualization technology, and simple installation and deployment is far from being able to be used directly. We also need tools to arrange the applications and containers on so many nodes.

Kubernetes [4] container cluster management platform based on docker has developed rapidly in recent years. It is an open source system for automatic deployment, expansion and management of container applications, which greatly simplifies the process of container cluster creation, integration, deployment and operation and maintenance [5]. In the process of building container cluster network, kubernetes realizes the interworking between container networks through container network interface (CNI) [6]. Different container platforms can call different network components through the same interface. This protocol connects two components: container management system (i.e. kubernetes) and network plugins (common such as flannel [7], calico [8]). The specific network functions are realized by plugins. A CNI plugin usually includes functions such as creating a container network namespace, putting a network interface into the corresponding network space, and assigning IP to the network interface [9].

For the gateway forwarding system, because it involves a large number of packet forwarding services, the underlying logic is mostly implemented based on the Intel DPDK (data plane development kit) [10] forwarding driver. DPDK's application program runs in the userspace, uses its own data plane library to send and receive data packets, bypasses the data packet processing of Linux kernel protocol stack, and obtains high packet data processing and forwarding ability at the expense of generality and universality. Therefore, for the virtualization deployment of gateway forwarding system applications, the selection of CNI plugins has strong particularity. The current mainstream CNI plugins are uniformly deployed by kubernetes management plane, and their management of network interface is based on Linux kernel protocol stack, which is not suitable for DPDK forwarding driven gateway business applications. In addition, the software defines that the gateway forwarding system is composed of data plane and control plane. The data plane is responsible for the analysis and forwarding of data packets based on DPDK forwarding driver, which belongs to performance sensitive applications. The control plane is responsible for receiving control messages and configuring the network system and various protocols. For control plane message, due to the small amount of data, the Linux kernel protocol stack can be used for communication during cluster deployment to obtain more universality. To sum up, when the software defined gateway forwarding system for cluster is deployed, it calls different CNI container network plugins to configure the network interfaces according to different use scenarios, and develops CNI network plugins based on DPDK forwarding driver for the corresponding DPDK forwarding interface, which are the two major problems to be solved urgently for such systems to support virtualization deployment.

In this paper, a multi-level network software defined gateway forwarding system based on Multus is proposed and implemented, and the CNI plugin and load balancing module based on DPDK network interface are implemented to ensure that the application performance based on DPDK is not affected. At the same time, for the control plane interface, because the kernel protocol stack is used to communicate with kubernetes, this paper constructs a multi-level network based on Multus, dynamically calls different types of CNI plugins for interface configuration, realizes the cluster deployment scheme compatible with kubernetes kernel protocol stack, and enhances the controllability of system services, It realizes the functions of dynamic scaling, smooth upgrade, cluster monitoring, fault migration and so on.

2   Design of Multi-level Network Gateway Forwarding System Based on Multus

With the development of nfv technology, virtual network devices based on X86 and other general hardware are widely deployed in the data center network. These virtual network devices carry the software processing of many high-speed network functions (including tunnel gateway, switch, firewall, load balancer, etc.), and can deploy multiple different network services concurrently to meet the diversified, complex and customized business needs of users. OVS (open vswitch) [11] and VPP (vector packet processor) [12] are two virtual network devices widely used in industry.

OVS is an open multi-layer virtual switch, which can realize the automatic deployment of large-scale networks through open API interfaces. However, the definition of flow table rules is complex, which can be realized only by modifying its core software code, and its packet processing performance is not as good as that of traditional switches. VPP is an efficient packet processing architecture. The packet processing logic developed based on this architecture can run on a general CPU. In terms of packet processing performance, VPP is based on DPDK userspace forwarding driver and adopts vector packet processing technology, which can greatly reduce the overhead of data plane processing packets, and the comprehensive performance is better than OVS. Therefore, in the multi-level network software defined gateway forwarding system proposed in this paper, we choose VPP as its receiving and contracting management framework.

The overall architecture of the system is shown in Fig. 1, in terms of configuration management, it is mainly divided into the management of various gateway services and the management of container resources. The management of gateway service mainly includes business configuration management, policy management, remote debugging management, log audit management, etc. the business developer is responsible for packaging the management process into the container image of the business. When the service is pulled up, it can communicate with the master node to complete the business-related configuration. Container resource management is related to cluster deployment, mainly including deployment cluster management, resource scheduling management, service scheduling management, operation monitoring management, etc. this part of management is related to the operation status of service cluster. It is the basis for providing functions such as dynamic scaling, smooth upgrade, cluster monitoring and fault migration. Kubernetes cluster management framework is responsible for it. The secure service

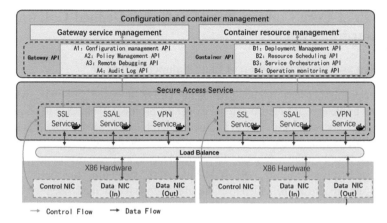

Fig. 1. The overall architecture of the software-defined gateway forwarding system

process will be uniformly packaged as a business image and loaded into the host machine that can be deployed by the kubernetes management framework for scheduling by kubernetes. When you need to create or expand a certain type of service, you can create several service containers corresponding to the service in the host of the existing cluster. Similarly, when a certain kind of service resources are surplus and need to shrink, only a few service containers need to be destroyed. Compared with the traditional scheme of purchasing customized physical equipment at a high price and manually joining the network cluster, its cost and operation portability have been greatly improved. In the traditional kubernetes solution, Kube proxy component provides load balancing services for all business pods to realize the dynamic selection of traffic. Besides, we need a load balancing component based on DPDK user mode protocol stack, which will be introduced in Sect. 3.1.

The last module is the hardware network card driver responsible for sending and receiving data packets. The DPDK based userspace forwarding driver at the bottom of the VPP forwarding framework avoids two data copies from the user space of the traditional protocol stack to the kernel state by creating a memif interface, as shown in Fig. 2. Therefore, the network card responsible for forwarding traffic on the service data plane needs to load the DPDK forwarding driver, while the network card responsible for forwarding messages on the control plane can communicate through the kernel protocol stack. In the overall architecture shown in Fig. 1, the data plane network card and the control plane network card should adopt a multi-level network management scheme based on the Multus CNI plugin to meet the communication requirements of kubernetes cluster management and the high-speed forwarding requirements of various gateway service data packets.

3 Design of Core Components of Software Defined Gateway Forwarding System

3.1 Design and Implementation of Load Balancing Module

This paper proposes user mode load balancing DPDK-lb based on DPDK, which uses DPDK user mode forwarding driver to take over the protocol stack, so as to obtain higher data message processing efficiency. The overall architecture of DPDK-lb is shown in Fig. 3. DPDK-lb hijacks the network card, bypasses the kernel protocol stack, parses the message based on the user mode IP protocol stack, and supports common network protocols such as IPv4, routing, ARP, ICMP, etc. At the same time, the control plane programs dpip and ipadm are provided to configure the load balancing strategy of DPDK-lb. In order to optimize the performance, DPDK-lb also supports CPU binding processing, realizes the lock free processing of key data, avoids the additional overhead required by context switching, and supports the batch processing of data messages in TX/RX queue.

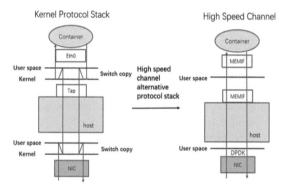

Fig. 2. Forwarding performance optimization of VPP memif interface

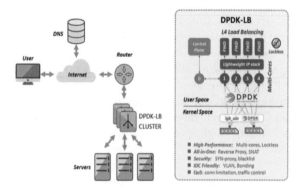

Fig. 3. Overall architecture of DPDK-lb load balancing

3.2   Design and Implementation of Multi-level CNI Plugin

Due to the gateway forwarding service based on VPP includes control message and data message, the data message runs in the user mode protocol stack, and all the above CNI plugins need to use the kernel protocol stack to analyze the data packet, so it can not meet the networking requirements of the data plane of the system. The control message is mainly used to update the service flow table and the distributed configuration management of kubernetes cluster. It is necessary to realize the cross host communication of pod in different network segments. Therefore, for the control plane, you can choose the mainstream CNI plugins that support overlay mode. As a result, in the software defined gateway forwarding system with the separation of control plane and data plane, the responsibilities of control plane and data plane are different, and the selection criteria of network plugins are also different. It is difficult to support the network communication of the system through a single CNI plugin. In order to meet the requirement of creating multiple network interfaces using multiple CNI plugins, Intel implemented a CNI plugin named Multus [13]. It provides the function of adding multiple interfaces to the pod. This will allow the pods connecting to multiple networks by creating multiple different interfaces, and different CNI plugins can be specified for different interfaces, so as to realize the separation control of network functions, as shown in Fig. 4.

Before using the Multus plugin, kubernetes container cluster deployment can only create a single network card eth0, and call the specified CNI plugin to complete interface creation, network setting, etc. When using Multus, we can create eth0 for the control plane of pod to communicate with the master node of kubernetes. At the same time, we can create net0 and net1 data plane network interfaces, and configure the data plane by using userspace CNI plugins to achieve cascade use of multi-level CNI plugins. Kubernetes calls Multus for interface management, and Multus calls the self-developed userspace CNI plugin to realize data plane message forwarding. In this way, it not only meets the separation of control plane and data plane required in the software defined gateway system, but also ensures that in the process of data plane message forwarding, the DPDK forwarding driver based on VPP completes the forwarding operation of data message without copying from operating system kernel state to userspace. To sum up,

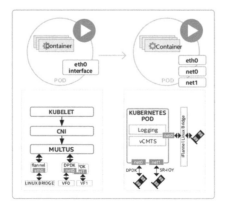

Fig. 4. Comparison before and after using Multus

Multus' multi-level CNI plugin scheme is very applicable in the software defined gateway forwarding system.

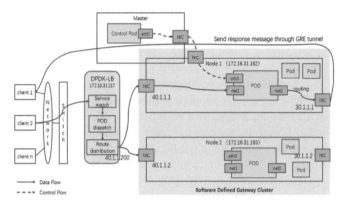

Fig. 5. Software defined gateway experimental networking

3.3 Design and Implementation of Userspce CNI Plugin

In order to register the new service pod on load balancing, it is not enough to only use flannel to complete the control plane network configuration, but also rely on the userspace CNI plugin mentioned above. The plugin needs to complete two types of work: first, create several service interfaces on the local pod, assign the created interfaces to the corresponding IP, and then access the specific network on the host to ensure that the data plane traffic can reach. Second, after the interface is created in the pod, because the kernel protocol stack is not used, it is necessary to configure the interface in the VPP forwarding framework in the pod (such as completing memif port pairing, assigning memif port address, etc.), and connect the newly created interface to the current data plane container network. Memif interfaces created in VPP appear in pairs and communicate by sharing large page memory. Therefore, the memif interfaces in the pods will find two corresponding virtual memif interfaces on the VPP of the host. By using these two pairs of memif interfaces, we can realize the data plane communication from the host message to the service pod.

The traffic of the system cluster is shown in Fig. 5. Taking the working node as an example, the service pod creates three network interfaces, eth0 is used for control plane message communication with the master node, the network card is created and configured by flannel, and net1 and net2 are the two data plane network interfaces required by the service, which are created and configured by the userspace network plugin. All data packages (red in the figure) are taken over by the userspace protocol stack, which improves the overall data message processing capacity of the system. Flannel provides network services for control messages related to configuration and cluster management (blue in the figure), which realizes the functions of dynamic expansion, smooth upgrade, cluster monitoring, fault migration and so on.

4   Experimental Scheme and Results

In this paper, we limit the resources of a single service pod to 1GB of large page memory. We will conduct three groups of comparative experiments. Firstly, we will compare and test whether there is a gap between the service capability provided by a single pod in the software defined gateway system and that provided by the traditional gateway device when it is limited to 1GB of available memory. Then, we will compare the maximum number of pods (16) run by a single physical device in the way of software defined gateway with the traditional way of running the service by a single device, so as to judge whether the performance of the original system is affected under the same hardware conditions after the introduction of kubernetes cluster management scheme. Finally, we will completely release the cluster system, no longer limit physical resources, and verify the overall performance and feasibility of the system. In the experiment, the connection request of real customers is simulated, and the number of access users is increasing. The overall resource consumption of the system is observed through the Prometheus component provided by kubernetes. The scheme comparison of the three experiments is shown in Table 1 and the results is shown in Fig. 6.

Table 1. Comparison of three experimental schemes

	Group 1	Group 2	Group 3
Software defined gateway cluster	Single pod (1GB memory limit)	Single node (POD dynamic scaling)	Two nodes cluster
Traditional physical gateway device	Single device (available physical memory limit 1 GB)	Single physical device	Single physical device

The experimental results are shown in Fig. 6. In the first group of experiments, 1GB memory can server about 7500 client terminals. When the number of clients reaches 7000, the connection failure begins to occur. The scheme provided in this paper is almost the same as that of traditional equipment. Therefore, the way of providing services through virtualization has no impact on the performance of the original service.In the second group of experiments, the scheme in this paper and the traditional single device begin to fail when the number of users is close to 110000. When the number of users is close to 120000, they can no longer accept more user access due to memory constraints. The overall performance of this scheme is not inferior to or even slightly better than that of the original single equipment. In the third group of experiments, when the number of users is close to 120000, the memory occupancy rate of each device in the cluster is about 50%. Eight pods are scheduled on each of the two nodes, and each pod provides services for nearly 7500 users. At this time, nearly 50% of the resources of the physical machine node can be used for the deployment of other services. When the number of clients continues to increase, kubernetes will continue to evenly allocate new resources on the two nodes and create new pods to provide services for more users. Until the number of users is close to 240000, the physical node tends to be saturated. However, the traditional single physical device can no longer provide services for so many users.

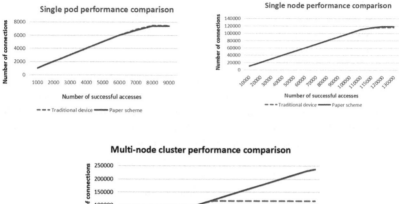

Fig. 6. Comparison of three groups of experiments

It can be seen that when the number of physical machines in the cluster continues to increase, the ability of the whole system to provide services will increase linearly. When the number of users decreases in a certain period of time, the physical machine resources are released and can dynamically provide services for other services. Therefore, the service provider only needs to ensure that the total amount of equipment for multiple services is sufficient. Since the peak usage of each service is different, the proportion of physical resources occupied by different services will be dynamically adjusted by kubernetes.

5 Conclusion

In this paper, a multi-level network software defined gateway forwarding system based on Multus is proposed and implemented, and the CNI plugin and load balancing module based on DPDK network interface are implemented. The created gateway service container is based on VPP packet processing framework, and the corresponding DPDK interface can be created to associate with the host interface, It ensures that the packet processing efficiency of the data forwarding application based on DPDK is not affected. At the same time, for the control plane interface of the gateway forwarding system, because the kernel protocol stack is used to communicate with kubernetes, this paper constructs a multi-level network based on Multus, dynamically calls different types of CNI plugins for interface configuration according to the use scenario and attribute configuration of relevant interfaces, and realizes the cluster deployment scheme compatible with kubernetes kernel protocol stack, The controllability of system services is enhanced, and the functions of dynamic expansion, smooth upgrade, cluster monitoring, fault migration and so on are realized.

This paper was partly supported by the science and technology project of State Grid Corporation of China: "Research on The Security Protection Technology for Internal and External Boundary of State Grid information network Based on Software Defined Security" (No. 5700-202058191A-0–0-00).

References

1. Huang, Y., Dong, Z., Meng, F.: Research on security risks and countermeasures in the development of internet of things. Inf. Secur. Commu. Priva. **000**(005), 78–84 (2020)
2. Nderson, C.: Docker. IEEE Softw. **32**(3), 102–103 (2015)
3. Yu, Y., Li, B., Liu, S.: Research on the portability of docker. Comp. Eng. Softw. (07), 57–60 (2015)
4. https://kubernetes.io/docs/home/
5. Li, Z., et al.: Performance overhead comparison between hypervisor and container based virtualization. In: 2017 IEEE 31st International Conference on Advanced Information Networking and Applications (AINA). IEEE (2017). https://doi.org/10.1109/AINA.2017.79
6. Networking Analysis and Performance Comparison of Kubernetes CNI Plugins: Advances in Computer, Communication and Computational Sciences. In: Proceedings of IC4S 2019 (2020). https://doi.org/10.1007/978-981-15-4409-5_9
7. https://docs.openshift.com/container-platform/3.4/architecture/additional_concepts/flannel.html
8. Sriplakich, P., Waignier, G., Meur, A.: CALICO documentation, pp. 1116–1121 (2008)
9. Kapocius, N.: Performance studies of kubernetes network solutions. In: 2020 IEEE Open Conference of Electrical, Electronic and Information Sciences (eStream). IEEE (2020). https://doi.org/10.1109/eStream50540.2020.9108894
10. https://www.DPDK.org/
11. Pfaff, B., et al.: The design and implementation of open vswitch. In: 12th USENIX Symposium on Networked Systems Design and Implementation. USENIX Association, Berkeley, pp. 117-130 (2015)
12. Barach, D., et al.: High-speed software data plane via vectorized packet processing. IEEE Commun. Mag. **56**(12), 97–103 (2018). https://doi.org/10.1109/MCOM.2018.1800069
13. https://github.com/k8snetworkplumbingwg/multus-cni

An Improved Chicken Swarm Optimization Algorithm for Feature Selection

Haoran Wang, Zhiyu Chen, and Gang Liu[✉]

School of Computer Science and Engineering, Changchun University of Technology, Changchun 130012, Jilin, China
lg@ccut.edu.cn

Abstract. In recent years, feature selection is becoming more and more important in data mining. Its target is that reduce the dimensionality of the datasets while at least maintaining the classification accuracy. There are some researches about chicken swarm optimization algorithm (CSO) applied to feature selection, the effect is extraordinary compared with traditional swarm intelligence algorithms. However, there is a complex search space in the challenging task feature selection, the CSO algorithm still has a default that quickly gets stuck in the local minimum problem. An improved chicken swarm optimization algorithm (ICSO) is proposed in this paper, which introduces the Levy flight strategy in the hen location update strategy and the nonlinear strategy of decreasing inertial weight in the chick location update strategy to increase the global search ability and avoid getting stuck in the local minimum problem. Compared with the other three algorithms on eighteen UCI datasets shows that the ICSO algorithm can greatly reduce the redundant features while ensuring classification accuracy.

Keywords: Chicken swarm optimization algorithm · Feature selection · Swarm intelligence algorithm

1 Introduction

Feature selection problem, also named as feature subset selection problem, refers to the selection of N features in the range of the existing M features to optimize the system's specific objectives, thereby reducing the data dimension and improving the performance of learning algorithms. In recent years, with the development of big data, industrial internet, and financial data analysis, more and more high-dimensional datasets are used in various fields of information systems, such as financial analysis, business management, and medical research. The dimensional disaster brought about by high-dimensional datasets makes feature selection an urgent and important task.

Feature selection methods can be divided into filter, wrapper, embedded, and ensemble [1]. The filter feature selection algorithm and learning algorithm are not related to each other. All features are sorted by specific statistical or mathematical attributes, such as Laplacian scores, Constraint scores, Fisher scores, Pearson correlation coefficients, and finally, a subset of features is selected by sorting. The wrapper feature selection

© The Author(s) 2022
Z. Qian et al. (Eds.): WCNA 2021, LNEE 942, pp. 177–186, 2022.
https://doi.org/10.1007/978-981-19-2456-9_19

algorithm encapsulates the selected learner looks like a black box, evaluates the performance of the selected feature according to its predictive accuracy on the feature subset, and gets the better subset with search strategy to obtain an approximate optimal subset. The embedded feature selection algorithm is embedded in the learning algorithm, with the training process of the classification algorithm is over, a subset of features can be obtained, such as ID3, C4.5, CART, etc. The features used in training are the result of feature selection. The ensemble feature selection algorithm draws on the idea of ensemble learning, which trains multiple feature selection methods and ensembles the results of all feature selection methods to achieve better performance than a single feature selection method. By introducing Bagging, many feature selection algorithms can be improved to be the ensemble.

Swarm intelligence optimization algorithms are often used to solve the feature selection problem and achieved good results. For example, genetic algorithm (GA) [2], ant colony algorithm (ACO) [3], and particle swarm optimization algorithm (PSO) [4], and so on. The Chicken swarm optimization algorithm (CSO) [5] proposed in 2014 is a kind of swarm intelligence optimization algorithm, which is inspired by the foraging behavior of the flock, is obtained a good optimization effect by grouping and updating the population, and has been applied in some fields. Hafez et al. [6] proposed a new feature selection method by using the CSO algorithm as part of the evaluation function. Ahmed et al. [7] applied logistic and tend chaotic mapping to help CSO explore the search space better. Liang, et al. [8] proposed a hybrid heuristic group intelligence optimization algorithm for cuckoo search-chicken swarm optimization (CSCSO) to optimize the excitation amplitude and spacing between the excitation amplitude of the linear antenna array (LAA) and the array of arrays of the circular antenna array (CAA). CSCSO has better solution accuracy and convergence speed in the optimization of LAA and CAA radiation patterns.

In this paper, an improved chicken swarm optimization algorithm (ICSO) is raised, which brings in the Levy flight strategy in the hen location update strategy and the nonlinear strategy of decreasing inertial weight in the chick location update strategy to enhance the ability of global search and decrease the probability of the algorithm falling into a local minimum. There are 18 UCI datasets are applied to compare the of effectiveness the algorithm in this paper with the other 3 algorithms. It's apparent that the algorithm in this paper has huge advantages.

2   Chicken Swarm Optimization Algorithm (CSO)

The chicken swarm optimization algorithm simulates the hierarchy of the chicken swarm and the competitive behavior in foraging. Within the algorithm, the chicken swarm is split into many subgroups, every as well as a rooster, many hens, and chicks. Completely different subgroups of the chicken swarm are subject to specific hierarchical system constraints, and there's competition within the foraging method. Positions of chickens are updated according to their respective motion rules. The behavior of chickens in the chicken swarm optimization algorithm is idealized with four rules, they are as follows:

i.   The chicken swarm is divided into many subgroups, there are three types of chick in every subgroup: a rooster, several hens, and chicks.

ii.   There are three types of chickens: rooster with the best fitness value, chick with the worst fitness value, and the others. The three types of chickens correspond to the roosters, the chicks, and the hens. It's worth noting that all the hens can freely choose the subgroup to which they belong. At the same time, the mother-child relationship between hens and chicks is also randomly established.

iii. The hierarchal order, dominance relationship, and mother-child relationship in a subgroup will change every period, but in the period all the relationships will keep unchanged.

iv.   All the chickens in the flock follow the rooster in their subgroup to find food and prevent other chickens from competing for food. The chicks follow the hens for food while assuming the chicks can eat food whichever the chickens find. Among them, chickens with better fitness have more advantages in finding food.

Assuming that the search space is D-dimensional, the total number of chickens in the entire chicken swarm is N, the number of roosters is N_R, the number of hens is N_H, the number of chicks is N_C, and mother hens is N_M. Let $x_{i,j}^t$ represents the position of the ith chicken, the t is the tth iteration, the j is the jth dimension searching space, where $i \in (1, 2, \ldots, N), j \in (1, 2, \ldots, D), t \in (1, 2, \ldots, T)$, the maximal iterative number is T.

(a) Rooster location update strategy. The roosters are the chickens with the best fitness value in the chicken swarm. The roosters with better fitness have the advantage over the roosters with poor fitness, so they can find food quickly than the roosters with poor fitness. At the same time can search for food on a larger scale in its position, realize the global search. Meanwhile, the roosters' location update is influenced by the location of other roosters randomly selected. The position update formulas of the rooster are as follows:

$$x_{i,j}^{t+1} = x_{i,j}^t * \left(1 + Randn\left(0, \sigma^2\right)\right) \tag{1}$$

$$\sigma^2 = \begin{cases} 1, & \text{if } f_i \le f_k, \\ \exp\left(\frac{f_k - f_i}{|f_i| + \varepsilon}\right), & \text{otherwise,} \end{cases} \quad k \in [1, N], k \ne i \tag{2}$$

where $Randn(0, \sigma^2)$ obey a normal distribution with standard deviation σ. k is the index of a rooster randomly selected from the rooster group. f_i is the fitness value of the corresponding rooster x_i. ε is the smallest constant to avoid the divide 0.

(b) Hen location update strategy. The search ability of hens is slightly worse than that of the roosters. Hens search food following their group-mate roosters, so the location update of the hens is affected by the position of their group-mate roosters. At the same time, due to their food stealing and competition between them, other roosters and hens also affect the location update. The location update formulas of the hen are as follows:

$$x_{i,j}^{t+1} = x_{i,j}^t + S1 * Rand * \left(x_{r1,j}^t - x_{i,j}^t\right) + S2 * Rand * \left(x_{r2,j}^t - x_{i,j}^t\right) \tag{3}$$

$$S1 = \exp\left(\frac{f_i - f_{r1}}{abs(f_i) + \varepsilon}\right) \tag{4}$$

$$S2 = \exp(f_{r2} - f_i) \tag{5}$$

where Rand is a uniform random number between 0 and 1. $abs(\cdot)$ is an absolute value operation. r_1 is the index of the rooster, and the ith hen search food following it. r_2 is an index of the roosters or hens randomly chosen from the whole chicken swarm, and $r_1 \neq r_2$.

(c) Chick location update strategy. The chicks have the worst search ability. They follow their mother hen, and the search range is the smallest. The chicks realize the mining of the local optimal solution. The search range of the chicks is affected by the position of their mother hen, and their position update formula is as follows:

$$x_{i,j}^{t+1} = x_{i,j}^t + FL * \left(x_{m,j}^t - x_{i,j}^t \right) \tag{6}$$

where m is an index of the mother hen, and the ith chick follows it to search for food. FL is a random value selected in the range [0, 2], and its main role is to keep the chick searching for food rounding its mother.

3 Improved Chicken Swarm Optimization Algorithm (ICSO)

Although the CSO algorithm can improve the population utilization rate through a hierarchical mechanism, the effectiveness of its location update method is low, which leads to a decrease in the overall search ability of the algorithm. Given this, this paper proposes an improved chicken swarm optimization algorithm (ICSO), which is based on the grouping idea of the CSO algorithm. The ICSO algorithm improves the position update method of the hens and the chicks respectively to enhance the algorithm's global search ability and decrease the probability of the algorithm falling into the local minimum.

3.1 Hen Location Update Strategy of ICSO

Levy flight is a strategy in the random walk model. In Levy flight, short-distance exploratory local search is alternated with occasional long-distance walking. Therefore, some solutions are searched near the current optimal value, which speeds up the local search; the other part of the solution can be searched in a space far enough from the current optimal value to ensure that the system will not fall into a local optimal [9, 10]. In the CSO algorithm, the number of hens is the largest in three types, so the hens play an important role in the entire population [11]. Inspired by this, the Levy flight search strategy is introduced to the hen location update formula, which can hold back falling into the local minimum while increasing the global search ability of the algorithm in a way. The improved location update formula of the hen is as follows:

$$x_{i,j}^{t+1} = x_{i,j}^t + S1 * Rand * \left(x_{r1,j}^t - x_{i,j}^t \right) + S2 * Rand * Levy(\lambda) \otimes \left(x_{r2,j}^t - x_{i,j}^t \right) \tag{7}$$

where $\otimes$ is point-to-point multiplication. $Levy(\lambda)$ is a random search path.

3.2 Chick Location Update Strategy of ICSO

In the CSO algorithm, the chicks only are affected by their mother hen, not by the rooster in the subgroup. Therefore, the location update information of the chicks only comes from their mother hen, and the location information of the rooster is not used. In this case, once the mother hen of a chick falls into the local optimal solution, the following chicks are easy to fall into the local optimal solution. Using a nonlinear strategy of decreasing inertial weight to update the position of the chick allows the chick to learn from itself while allowing the chick to be affected by the rooster in the subgroup, which can prevent the algorithm from falling into a locally optimal solution as soon as possible. The improved location update formulas of the chick are as follows:

$$x_{i,j}^{t+1} = w * x_{i,j}^t + FL * \left(x_{m,j}^t - x_{i,j}^t \right) + C * \left(x_{r,j}^t - x_{i,j}^t \right) \tag{8}$$

$$w = wmin * \left(\frac{wmax}{wmin} \right)^{\left(\frac{1}{1+10*\frac{t}{T}} \right)} \tag{9}$$

where w is the self-learning coefficient of the chick, which is very similar to the inertial weight in particle swarm optimization algorithm. $wmin$ is the minimum inertial weight, $wmax$ is the maximum inertial weight, t is the current number of iterations, and T is the maximum iteration. Let C denote the learning factor, which means that the chick is affected by the rooster in the subgroup. r is the index of the rooster which is the chick's father.

3.3 Experimental Results and Analysis

To verify the effectiveness of the ICSO algorithm, a comparison experiment is set up. The algorithms in comparison are chicken swarm optimization algorithm (CSO), genetic algorithm (GA), and particle swarm optimization algorithm (PSO).

3.4 Fitness Function

Each particle in the chicken swarm corresponds to a solution of feature selection. The particles are coded by real numbers, as shown in Eq. (10). Each solution X contains n real numbers, and n represents the total number of features of the corresponding dataset, where each dimension x_i represents whether to select this feature. To form a feature subset, it is necessary to perform a decoding process before decoding. The position of the particle can be converted into a subset of the following features:

$$X = [x_1, x_2, \ldots, x_n] \tag{10}$$

$$A_d = \begin{cases} 1, x_d > 0.5 \\ 0, else \end{cases} \tag{11}$$

where A_d represents the feature subset decoded from the d-dimension of each solution. A_d can be selected as 0 or 1, according to the value x_d of the d-dimensional feature of the particle: if $A_d = 1$, it means that the d-dimensional feature is selected; otherwise, the dimensional feature is not selected.

The purpose of feature selection is to find a combination that has the highest classification accuracy and the smallest number of selected features. Although it is a combination, the classification accuracy is the first consideration. The fitness function is to maximize classification accuracy over the test sets given the train data, as shown in Eq. (12) at the same time keeping a minimum number of selected features.

$$Fitness(i) = \alpha * ACC(i) + (1 - \alpha) * \left(\frac{FeatureSum(i)}{FeatureAll} \right) \tag{12}$$

where α is a constant less than 1 and bigger than 0, which controlling the importance of classification accuracy to the number of selected features. The bigger the α, the more important the classification accuracy. $ACC(i)$ is the classifier accuracy of the particle i. $FeatureSum(i)$ is the number of features corresponding to the particle i. $FeatureAll$ is the total amount number of features in the dataset.

3.5 Parameters Setting

In this paper, all comparative experiments work on a PC that has 8GB of memory, and the programming environment is Python 3.8.5. Let set 50 is the population size, the α in the fitness function is set to 0.9999, 20 independent running experiments are performed on the datasets, and setting 500 is the maximum number of iterations. The KNN (K = 5) classifier is used to test the classification accuracy of the selection scheme corresponding to each particle. The hyperparameter settings of each algorithm are shown in Table 1. The information of the eighteen UCI datasets is described in Table 2. Most datasets are two-class, as well as there are multi-class datasets. It can be seen intuitively that the largest number of features is 9 and the lowest is 309 in datasets.

Table 1. Hyperparameter settings

Algorithm	Hyperparameters
ICSO	$N_R = 0.2N$, $N_H = 0.6N$, $N_C = N\text{-}N_R\text{-}N_H$, $N_M = 0.1N$, G = 10, $wmax = 0.9$, $wmin = 0.4$, C = 0.4
CSO	$N_R = 0.2N$, $N_H = 0.6N$, $N_C = N\text{-}N_R\text{-}N_H$, $N_M = 0.1N$, G = 10
PSO	w = 0.729, c1 = c2 = 1.49445
GA	Crossover_prob = 0.7, Mutation_prob = 0.25

Table 2. Datasets description

Dataset	Number of features	Number of instances	Number of classes
Wine	13	178	3
Lymphography	18	148	4
LSVT	309	126	2
Breast Cancer	9	699	2
WDBC	30	569	2
Zoo	16	101	7
House-votes	16	435	2
Heart	13	270	2
Ionospher	34	351	2
Chess	36	3196	2
Sonar	60	208	2
Spect	22	267	2
German	24	1000	2
Arrhythmia	279	456	16
Glass	9	214	6
Australia	14	690	2
Biodeg	40	1055	2
Spambase	56	4601	2

3.6 Results and Analysis

Table 3 shows the experimental results of the ICSO algorithm and the other three comparison algorithms on eighteen datasets. Where bold fonts represent the largest mean classification accuracy among all algorithms. It can be seen intuitively from Fig. 1 that the ICSO algorithm has obtained the best results on eighteen test datasets. And the mean accuracy of the ICSO algorithm is more excellent than the CSO algorithm, the mean accuracy of the CSO algorithm is more excellent than the PSO algorithm, the mean accuracy of the PSO algorithm is more excellent than the GA algorithm, the mean accuracy of the GA algorithm in feature selection is the worst. Through observation and calculation, the datasets with poor mean accuracy on full features, such as Wine, LSVT, Arrhythmia, etc., after the ICSO algorithm feature selection, the mean accuracy increases by 20% ~ 50%. Datasets with better mean accuracy on full features, such as Breast Cancer, WDBC, Zoo, etc., after the ICSO algorithm feature selection, the mean accuracy was improved by less than 10%. The experimental results fully verify the superiority of the ICSO algorithm.

Table 3. Mean accuracy for the different algorithms

Dataset	Full Feature	GA	PSO	CSO	ICSO
Wine	0.7407	0.7565	0.9759	0.9796	**0.9815**
Lymphography	0.8000	0.7344	0.8967	0.9067	**0.9100**
LSVT	0.5526	0.5645	0.8184	0.8579	**0.8658**
Breast Cancer	0.9561	0.9383	0.9708	**0.9756**	**0.9756**
WDBC	0.9591	0.9383	**0.9708**	**0.9708**	**0.9708**
Zoo	0.8710	0.8129	0.9435	0.9452	**0.9532**
House-votes	0.9714	0.9057	0.9950	0.9957	**1.0000**
Heart	0.6420	0.6951	0.8827	**0.8870**	**0.8870**
Ionospher	0.8585	0.8533	0.9637	0.9755	**0.9759**
Chess	0.9416	0.7941	0.9777	0.9760	**0.9766**
Sonar	0.9413	0.8310	0.9802	**0.9849**	**0.9849**
Spect	0.7219	0.7430	0.9201	0.9198	**0.9201**
German	0.6633	0.6810	0.7793	0.7791	**0.7795**
Arrhythmia	0.5238	0.4048	0.6881	0.7310	**0.7476**
Glass	0.5846	0.5938	**0.6923**	**0.6923**	**0.6923**
Australia	0.6908	0.7169	0.8780	0.8780	**0.8787**
Biodeg	0.8328	0.8232	0.9120	**0.9135**	**0.9159**

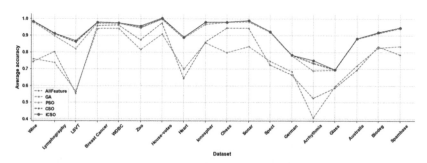

Fig. 1. Mean accuracy line chart

Table 4 lists the mean features and dimension standard deviation of the four algorithms after feature selection for each dataset. It can be seen intuitively that, compared with the GA algorithm and the PSO algorithm, the CSO algorithm and the ICSO algorithm have obvious dimensionality reduction effects, and the dimensional standard deviation is low, indicating that the algorithm stability is relatively high. The experimental results directly verify that the ICSO algorithm has a strong superiority in eliminating redundant features, and can achieve better classification accuracy on datasets, while greatly reducing the number of redundant features.

Table 4. Mean and Std dimension after different algorithm feature selection

Dataset	GA		PSO		CSO		ICSO	
	Mean	Std	Mean	Std	Mean	Std	Mean	Std
Wine	6.65	1.31	5.00	0.00	5.00	0.00	5.00	0.00
Lymphography	8.75	1.87	7.35	2.26	4.45	1.53	4.25	0.89
LSVT	155.75	10.67	29.55	8.75	13.05	6.00	14.85	6.06
Breast Cancer	4.25	1.41	5.05	0.22	5.00	0.00	5.00	0.00
WDBC	15.50	3.32	3.85	0.36	3.95	0.22	3.95	0.22
Zoo	8.90	2.00	5.70	0.90	5.65	0.91	6.15	0.96
House-votes	7.75	1.92	4.85	1.28	4.75	0.43	5.25	0.77
Heart	6.20	1.78	6.00	1.34	5.85	0.65	5.85	0.65
Ionospher	15.35	3.32	5.85	1.82	5.00	1.00	5.00	0.95
Chess	18.20	2.54	20.95	2.42	16.65	3.05	17.40	3.09
Sonar	29.15	4.64	16.10	2.62	14.35	2.43	14.45	3.06
Spect	11.00	2.28	1.20	0.87	1.00	0.00	1.30	1.31
German	11.70	2.55	11.00	3.39	8.50	2.52	7.75	2.05
Arrhythmia	136.75	6.84	62.60	8.11	27.85	16.92	26.70	9.02
Glass	4.85	1.24	4.05	0.22	4.00	0.00	4.00	0.00
Australia	6.00	1.48	5.35	1.19	5.35	1.19	5.70	0.90
Biodeg	20.80	3.17	14.35	2.13	12.80	1.94	12.25	1.70
Dataset	28.45	4.93	29.85	3.73	25.75	3.18	23.95	3.84

4   Conclusions

Swarm intelligence optimization achieved good results in the feature selection problem. In the chicken swarm optimization algorithm, there is a weakness in that it is still easy to fall into the local minimum. To overcome this, this paper proposes an improved chicken swarm optimization algorithm. On the basis of the population grouping update mechanism of the CSO algorithm, the ICSO algorithm introduces the Levy flight strategy in the hen location update strategy and the nonlinear strategy of decreasing inertial weight in the chick location update strategy to enhance the algorithm's global search ability and decrease the probability of the algorithm falling into the local minimum. It can be seen from the experimental results that compared with the other three related algorithms, the ICSO algorithm can tremendously decrease the redundant features while ensuring classification accuracy in the feature selection.

References

1. Li, Z., Du, J., Nie, B., Xiong, W., Huang, C., Li, H.: Feature selection methods. Comp. Eng. Appl. **55**, 10–9 (2019)
2. Yang, J., Honavar, V.: Feature subset selection using a genetic algorithm feature extraction. Construction and Selection (Springer), pp. 117–36 (1998)
3. Sreeja, N., Sankar, A.: Pattern matching based classification using ant colony optimization based feature selection. Appl. Soft Comput. **31**, 91–102 (2015)
4. Kennedy, J., Eberhart, R.: Particle swarm optimization. In: Proceedings of ICNN'95-international conference on neural networks, vol 4, pp. 1942–8. IEEE (1995)
5. Meng, X., Liu, Y., Gao, X., Zhang, H.: A New Bio-inspired Algorithm: Chicken Swarm Optimization. Adv. Swarm Intell. Lec. Notes Comp. Sci. **8794**, 86–94 (2014)
6. Hafez, A.I., Zawbaa, H.M., Emary, E., Mahmoud, H.A., Hassanien, A.E.: An innovative approach for feature selection based on chicken swarm optimization. In: 2015 7th International Conference of Soft Computing and Pattern Recognition (SoCPaR), pp. 19–24 (2015)
7. Ahmed, K., Hassanien, A.E., Bhattacharyya, S.: A novel chaotic chicken swarm optimization algorithm for feature selection. 2017 Third International Conference on Research in Computational Intelligence and Communication Networks (ICRCICN), pp. 259–64 (2017)
8. Liang, S., Feng, T., Sun, G.: Sidelobe-level suppression for linear and circular antenna arrays via the cuckoo search–chicken swarm optimization algorithm. IET Microw. Anten. Prop. **11**, 209–218 (2017)
9. Yahya, M., Saka, M.: Construction site layout planning using multi-objective artificial bee colony algorithm with Levy flights. Autom. Constr. **38**, 14–29 (2014)
10. Reynolds, A.: Cooperative random lévy flight searches and the flight patterns of honeybees. Physics letters A **354**, 384–388 (2006)
11. Liang, X., Kou, D., Wen, L.: An improved chicken swarm optimization algorithm and its application in robot path planning. IEEE Access **8**, 49543–49550 (2020)

A Method of UAV Formation Transformation Based on Reinforcement Learning Multi-agent

Kunfu Wang, Ruolin Xing, Wei Feng, and Baiqiao Huang[✉]

System Engineering Research Institute of China State Shipbuilding Corporation, BeiJing, China
bq_huang@126.com

Abstract. In the face of increasingly complex combat tasks and unpredictable combat environment, a single UAV can not meet the operational requirements, and UAVs perform tasks in a cooperative way. In this paper, an improved heuristic reinforcement learning algorithm is proposed to solve the formation transformation problem of multiple UAVs by using multi-agent reinforcement learning algorithm and heuristic function. With the help of heuristic back-propagation algorithm for formation transformation, the convergence efficiency of reinforcement learning is improved. Through the above reinforcement learning algorithm, the problem of low efficiency of formation transformation of multiple UAVs in confrontation environment is solved.

Keywords: Multi UAV formation · Formation transformation · Agent · Reinforcement learning

1 Introduction

With the development of computer, artificial intelligence, big data, blockchain and other technologies, people have higher and higher requirements for UAV, and the application environment of UAV is more and more complex. The shortcomings and limitations of single UAV are more and more prominent. From the functional point of view, a single UAV has only part of the combat capability and can not undertake comprehensive tasks; From the safety point of view, a single UAV has weak anti-jamming ability, limited flight range and scene, and failure or damage means mission failure. Therefore, more and more research has turned to the field of UAV cluster operation. UAV cluster operation is also called multi UAV cooperative operation, which means that multiple UAVs form a cluster to complete some complex tasks together [1]. In such a multi UAV cluster, different UAVs often have different functions and play different roles. Through the cooperation among multiple UAVs, some effects that can not be achieved by a single UAV can be achieved. Based on the reinforcement learning algorithm of multi-agent agent, this paper introduces the heuristic function, and uses the heuristic reinforcement learning of multi-agent agent to solve the formation transformation problem of multi UAV formation in unknown or partially unknown complex environment, so as to improve the solution speed of reinforcement learning.

© The Author(s) 2022
Z. Qian et al. (Eds.): WCNA 2021, LNEE 942, pp. 187–195, 2022.
https://doi.org/10.1007/978-981-19-2456-9_20

2   Research Status of UAV Formation

With the limited function of UAV, facing the increasingly complex combat tasks and unpredictable combat environment, the performance of a single UAV can not meet the operational requirements gradually. UAV more in the way of multi aircraft cooperative operation to perform comprehensive tasks. Multi UAV formation is an important part of multi UAV system, and it is the premise of task assignment and path planning. But it has also been challenged in the dynamic environment of high confrontation, including: (1) the multi UAV formation constructed by the existing formation method can not be satisfied both in formation stability and formation transformation autonomy (2) When formation is affected, it is necessary to adjust, the formation transformation speed is not fast enough, the flight path overlaps and the flight distance is too long.

The process of multi UAV system to perform combat tasks includes: analysis and modeling, formation formation, task allocation, path allocation, and task execution. When encountering emergency threat or task change, there are formation transformation steps. Among them, the formation method of UAV is always used as the foundation to support the whole task. The formation control strategy of UAV is divided into centralized control strategy and distributed control strategy [2]. The centralized control strategy requires at least one UAV in the UAV formation to know the flight status information of all UAVs. According to these information, the flight strategies of all UAVs are planned to complete the combat task. Distributed control strategy does not require UAVs in formation to know all flight status information, and formation control can be completed only by knowing the status information of adjacent UAVs (Table 1).

Table 1. Parison of advantages and disadvantages between centralized control and distributed control

Name	Advantage	Disadvantage
Centralized Control Strategy	Simple and complete theory	Lack of flexibility, fault tolerance, communication pressure
Distributed Control Strategy	High flexibility and low communication requirements	It is difficult to realize and is likely to be disturbed

The advantages of centralized control strategy are simple implementation and complete theory; The disadvantages are lack of flexibility and fault tolerance, and the communication pressure in formation is high [3]. The advantage of distributed control strategy is that it reduces the requirement of UAV Communication capability and improves the flexibility of formation; The disadvantage is that it is difficult to realize and the formation may be greatly disturbed [4].

Ru Changjian et al. designed a distributed predictive control algorithm based on Nash negotiation for UAVs carrying different loads in the mission environment, combined with the multi-objective and multi person game theory and the Nash negotiation theory of China. Zhou shaolei et al. established the UAV virtual pilot formation model and introduced the neighbor set, adopted distributed model predictive control to construct the reconfiguration cost function of multi UAV formation at the same time, and proposed an improved quantum particle swarm optimization algorithm to complete the autonomous reconfiguration of multi UAV formation. Hua siliang et al. studied the communication topology, task topology and control architecture of UAV formation, analyzed the characteristics of task coupling, collision avoidance and dynamic topology of UAV formation reconfiguration, and proposed a model predictive control method to solve the UAV formation reconfiguration problem. Wang Jianhong transformed the nonlinear multi-objective optimization model based on autonomous reconfiguration of multi UAV formation into a standard nonlinear single objective optimization model, and solved the optimal solution through the interior point algorithm in operational research. Mao Qiong et al. proposed a rule-based formation control method aiming at the shortcomings of existing methods in UAV formation control and the characteristics of limited range perception of UAV system [5–8].

3 Agent and Reinforcement Learning

3.1 Agent

The concept of agent has different meanings in different disciplines, and so far there has been no unified definition. In the field of computer, agent refers to the computer entity that can play an independent role in the distributed system. It has the following characteristics:

1) Autonomy: it determines its own processing behavior according to its own state and perceived external environment;
2) Sociality: it can interact with other agents and work with other agents;
3) Reactivity: agent can perceive the external environment and make corresponding response;
4) Initiative: be able to take the initiative and show goal oriented behavior;
5) Time continuity: the process of agent is continuous and circular;

A single agent can perceive the external environment, interact with the environment and other agents, and modify its own behavior rules according to experience, so as to control its own behavior and internal state. In the multi-agent system, there are agents who play different roles. Through the dynamic interaction, they make use of their own resources to cooperate and make decisions, so as to achieve the characteristics that a single agent does not have, namely, emergence behavior. Each agent can coordinate, cooperate and negotiate with each other. In the multi-agent system, each agent can arrange their own goals, resources and commands reasonably, so as to coordinate their own behaviors and achieve their own goals to the greatest extent. Then, through coordination and cooperation, multiple agents can achieve common goals and realize multi-agent cooperation. In the agent model, the agent has belief, desire and intention. According to the target information and belief, the agent can generate the corresponding desire and make the corresponding behavior to complete the final task (Fig. 1).

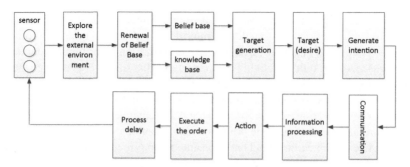

Fig. 1. Agent behavior model

When there are multiple agents in a system that can perform tasks independently, the system is called multi-agent system. In the scenario of applying multi-agent system to deal with problems, the focus of problem solving is to give full play to the initiative and autonomy of the whole system, not to emphasize the intelligence of a single agent. In some scenarios, it is often impossible to simply use the reinforcement learning algorithm of single agent to solve the problem of multi-agent (Fig. 2).

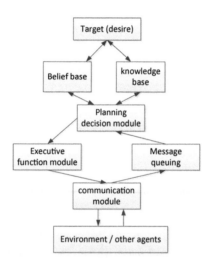

Fig. 2. The structure of agent in combat simulation architecture

According to the classification of Multi-Agent Reinforcement learning algorithm, it can be divided into the following categories according to the types of processing tasks

(1) Multi agent reinforcement learning algorithm in the case of complete cooperation. All the participants in the system have the same optimization goal. Each agent makes its own action by assuming that the other agents choose the optimal action in the current state, or makes some combination action through the cooperation mechanism to obtain the optimal goal.

(2) Multi agent reinforcement learning algorithm under complete competition. The goals of all participants in the system are contrary to each other. Each agent assumes that the other agents make the actions to minimize their own benefits in the current state, and make the actions to maximize their own benefits at this time.

(3) Reinforcement learning algorithm of multi-agent agent under mixed tasks. It is the most complex and practical part in the current research field.

3.2   Reinforcement Learning

The standard reinforcement learning algorithm mainly includes four elements: environment, state, action and value function. The problem can be solved by constructing mathematical model, such as Markov decision process (Fig. 3).

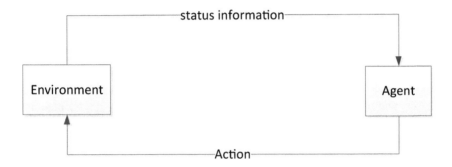

Fig. 3. Basic concept map of reinforcement learning

At present, the research on agent reinforcement learning algorithm has built a perfect system and achieved fruitful results. However, the processing ability and efficiency of a single agent are always limited. It is an effective way to solve the problems in complex environment by using the Multi-Agent Reinforcement learning algorithm. When there are multiple agents in a system that can perform tasks independently, the system is called multi-agent system. In the scenario of multi-agent system, the key point of problem solving is to give full play to the initiative and autonomy of the whole system, not the intelligence of single agent. In some scenarios, it is difficult to use the reinforcement learning algorithm of single agent to solve the problem of multi-agent. Therefore, the research and attention of experts and scholars on the reinforcement learning algorithm of multi-agent is improving.

4   A Method of UAV Formation Transformation Based on Reinforcement Learning Multi-agent

4.1   Description of UAV Formation Transformation Model

The core model of reinforcement learning: Markov decision-making process is usually composed of a quadruple: $M = (S, A, P_{sa}, R)$. S represents the states in finite space; A represents the actions in finite space; P_sa represents the probability set of state transfer, that is, in the current s ∈ S state, the probability that action a ∈ A will be transferred to other states after action a ∈ A is selected; R represents the return function, which is usually a function related to state and action, which can be expressed as r(s, a). The agent takes action a under state s, and performs the following actions. The expected return can be obtained as follows:

$$R_{sa} = E\left[\sum_{k=0}^{\infty} \gamma^k r_{k+1} | S = s, A = a\right] \tag{1}$$

γ is a discount factor with a value between 0 and 1, which makes the effect of the later return on the return function smaller. It simulates the uncertainty of the future return and makes the return function bounded.

In this paper, four tuples (S, A, P, R) are used to represent the Markov decision process model for formation transformation of multiple UAVs. Where S is the state space set of UAV, A is the action space set of UAV, P is the state transition probability of UAV, and R is the action return function of UAV.

Let the UAV move in the constructed two-dimensional grid, and use $Z(Z > 0)$ to represent a positive integer, then the two-dimensional grid space is Z^2, and the UAV coordinate in the two-dimensional grid space is (x_{ti}, y_{ti}), indicating the state s of UAV $s_{ti} \in Z^2$, and toward the corresponding target point $G_i(i = 1, 2, 3, .. N)$ motion, the target point of each UAV will be given in advance according to the conditions. During the flight of UAV I, action set $A_i(s) = \{up, down, left, right, stop\}$.

4.2 A Method of UAV Formation Transformation Based on Reinforcement Learning Multi Agent Agent

The fundamental goal of reinforcement learning is to find a strategy set ⟨S, A⟩ so that the expected return of agent in any state is the largest. The agent can only get the immediate return of the current step each time. We choose the classical Q-learning algorithm state action value function Q(s, a) instead of R_{sa}. According to a certain action selection strategy, the agent makes an action in a certain state and gets immediate feedback from the environment. The Q value increases when it receives positive feedback, and decreases when it receives negative feedback. Finally, the agent will select the action according to the Q value. The action selection function of traditional Q-learning algorithm is as follows:

$$\pi(s) = \begin{cases} \arg\max[Q(s, a)], & \text{if } q < 1 - \varepsilon \\ a_{randon} & \text{otherwise} \end{cases} \tag{2}$$

ε is a parameter of $\varepsilon - greedy$, When the random number q is less than $1 - \varepsilon$ Choose the behavior a that makes the Q value maximum, otherwise choose the random behavior a. In the practical algorithm design, the iterative approximation method is usually used to solve the problem:

$$Q^*(s, a) = Q(s, a) + \alpha[r(s, a) + \gamma maxQ(s', a) - Q(s, a)] \tag{3}$$

where a is the learning factor, the larger the value of a is, the less the results of previous training are retained; $maxQ(s', a)$ is the prediction of Q value, as shown in algorithm 1:

Algorithm 1 Q-learning algorithm

Input: iteration times T, state set S, learning rate a, exploration rate ϵ, Discount factor γ

Output: state action value function Q (S, A)

1. Initialize the Q values of all States and actions
2. For i = 1 to T do:
3.   Initialize state s as the first state
4.   While the final state is not reached:
5.   use ε − greedy selects action A according to the current state S
6.   Perform action A in current state S, get new status S′and reward r (S, A)
7.   Update Q value:Q (S, A) = Q (S, A) + α[r (S, A) + γmaxQ (S′, A) − Q (S, A)]
8.   S = S′
9.   End Wbile
10. End For
11. Return Q (S, A)

In this paper, the multi UAV formation problem based on reinforcement learning can be described as: UAV interacts with the environment, learning action strategy, so that the whole UAV group can reach their respective target points with the minimum consumption steps without collision. In the process of learning the optimal action strategy, when all UAVs arrive at the target point, the group will get a positive feedback r_+, otherwise it will get a negative feedback r_-.

The reinforcement learning algorithm of multi-agent needs to change the action of each agent in each state to a_{si} (i = 1, 2, . . . n) is regarded as a joint action $\rightarrow a_{si}$ can be considered. The learning process of the algorithm is complex, consumes more resources and is difficult to converge. Therefore, we introduce heuristic function H to influence the action selection of each agent. Formula 1.2 can be changed as follows:

$$\pi^H (s) = \begin{cases} argmax[Q(s, a) + \beta H(s, a)], & if \ q < 1 - \varepsilon \\ a_{randon'} & otherwise \end{cases} \quad (4)$$

where β is the real number that controls the effect of the heuristic function on the algorithm. The heuristic function H needs to be large enough to affect the agent's action selection, and it should not be too large to prevent the error that affects the result. when β is 1, the mathematical expression of heuristic function H can be defined as:

$$\pi^H (s) = \begin{cases} argmax[Q(s, a) + \beta H(s, a)], & if \ q < 1 - \varepsilon \\ a_{randon'} & otherwise \end{cases} \quad (5)$$

where δ is a relatively small real number, which makes the heuristic function H larger than the difference between Q values and does not affect the learning process of reinforcement learning. The whole process of improved heuristic reinforcement learning is as follows (Fig. 4):

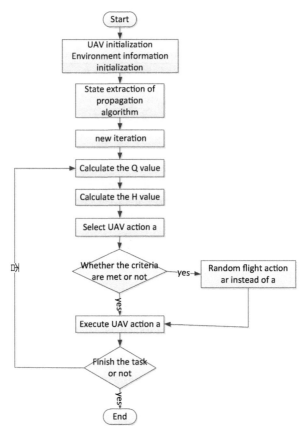

Fig. 4. The whole flow chart of improved heuristic reinforcement learning

5 Summary

In this paper, a reinforcement learning based multi-agent UAV formation transformation method is proposed. The heuristic algorithm is used to improve the traditional reinforcement learning algorithm, and the optimal path without collision is planned for the multi UAV system in the formation transformation stage, which solves the problem that the reinforcement learning algorithm consumes a lot of computing resources when facing the multi-agent problem.

References

1. Jia, Y., Tian, S., Li, Q.: Recent development of unmanned aerial vehicle swarms. Acta Aeronautica ET Astronautica Sinica 1–12 [2020–02–19]
2. Li, L., Xu, Y., Jiang, Q., Wang, T.: New development trends of military UAV equipment and technology in the world in 2018. Tactical Missile Technol. **02**, 1–11 (2019)
3. Wang, Q.-Z., Cheng, J.-Y., Li, X.-L.: Method research on cooperative task planning for multiple UCAVs. Fire Cont. Comm. Cont. **43**(03), 86–89+94 (2018)

4. Chen, X., Serrani, A., Ozbay, H.: Control of leader-follower formations of terrestrial UAVs. IEEE Conf. Deci. Cont. **1**(1), 498–503 (2004)
5. Jie, Y., et al.: UAV Form. Cont. Based Impr. APF. **3160**, 358–364 (2014)
6. Ili, P., Wang, H., Li, X.: Improved ant colony algorithm for global path planning. Advances in Materials, Machinery, Electronics I (2017)
7. Marsella, S., Gratch, J.: Evaluating a computational model of emotion. Autonomous Agents and Multi-Agent Systems (S1387–2532) **11**(1), 23–43 (2006)
8. Martins, M.F., Bianchi Reinaldo, A.C.: Heuristically-accelerated reinforcement learning: a comparative analysis of performance. In: 14th Annual Conference on Towards Autonomous Robotic Systems (TAROS) (2013)

A Formalization of Topological Spaces in Coq

Sheng Yan, Yaoshun Fu, Dakai Guo, and Wensheng Yu[✉]

School of Electronic Engineering, Beijing Key Laboratory of Space-Ground Interconnection and Convergence, Beijing University of Posts and Telecommunications, Beijing 100876, China
wsyu@bupt.edu.cn

Abstract. It is a wish for Wu Wen-tsun to implement the mechanical proving of theorems in topology. Topological spaces constitute a fundamental concept of general topology, which is significant in understanding the essential content of general topology. Based on the machine proof system of axiomatic set theory, we presented a computer formalization of topological spaces in Coq. Basic examples of topological spaces are formalized, including indiscrete topological spaces and discrete topological spaces. Furthermore, the formal description of some well-known equivalent definitions of topological spaces are provided, and the machine proof of equivalent definitions based on neighborhood system and closure is presented. All the proof code has been verified in Coq, and the process of proof is standardized, rigorous and reliable.

Keywords: Coq · Formalization · Axiomatic set theory · General topology · Topological spaces

1 Introduction

The formal verification of mathematical theorems profoundly incarnates the basic theories of artificial intelligence, have also attracted more attention from researchers [1].

Some famous mathematical theorems have been already formalized. In 2005, Gonthier and Werner have given the formal proof of the "Four-color Theorem" in Coq [2]. After six years, formal verification of "Odd Order Theorem" has been achieved by Gonthier in [3]. Hales provided formal proof of "Kepler Conjecture" in Isabelle/HOL [4]. There has a list about Formalizing 100 Theorems on the web [5], which will keep track of theorems from this list that have been formalized.

The theorem prover Coq is a tool used to verify whether the proofs of theorems are correct, and the theorem can be taken from general mathematics, protocol verification or safety programs. The Coq system is extremely powerful and expressions in reasoning and programming. Moreover, the process of proofs is built interactively in Coq with the aid of tactics [6]. There are various tactics of available in Coq, which make it become the mainstream tool in the field of interactive theorem proving in the world [7].

Topological spaces constitute a fundamental concept of general topology. There are many ways to create the definition of topological spaces [8]. During the early periods

© The Author(s) 2022
Z. Qian et al. (Eds.): WCNA 2021, LNEE 942, pp. 196–204, 2022.
https://doi.org/10.1007/978-981-19-2456-9_21

of general topology, some scholar defined the topological spaces by axioms of neighborhood systems or axioms of closure. With the development of general topology, it was revealed that the definition of topological spaces from various basic concepts is equivalent, and one of the convenient tools for exploring topological spaces is to use the axioms of open sets [9].

Being such an elementary concept in general topology, the definition of topological spaces appears in several formalization works with a variable degree of details and generality. A definition of topological spaces has been already formalized by Schepler in Coq contribution libratory based on type theory [10]. The topological spaces theory has been developed based on theorem prover Coq by Wang in [11]. Another work involved the formal description of topological spaces has been carried out by Hölzl in [12], which formalize the development process of space in the history of mathematics, including topological space, metric space and Euclidean space.

This paper presented a computer formalization of topological spaces in Coq. The formal proof of two basic examples in topological spaces is given, including indiscrete topological spaces and discrete topological spaces. The key points of our work are to realize the formal description of equivalent definitions of topological spaces, and to present the machine proof of equivalent definitions based on neighborhood system and closure.

In the paper structure, we briefly give the formalization of set theory in Sect. 2, which act as preliminaries for the formalization of topological space. Section 3 introduces the concepts of topological spaces in Coq based on the axioms of the open sets. We present the formal proof of equivalent definitions of topological spaces based on neighborhood system and closure in Sect. 4. The conclusions are given in Sect. 5.

2 Formalization of Set Theory

Set theory is the foundation of modern mathematics [13]. The author has done the work about the formalization of axiomatic set theory in [14]. A formalization of naive set theory is introduced based on the axiomatic set theory.

To make our source code more readable, some mathematical symbols are added by using the command Notation, including the quantification symbol '∀' and '∃', logical symbol '¬', '∨' and '∧', symbol '→'and'↔'.

Some basic logical properties are essential in our formal system. In fact, we only need the law of the excluded middle, and some other logical properties can be proved by using it [15]. We can formalize some of the frequently used logical properties as follows:

```
Axiom classic : ∀ M : Prop, M ∨ ¬M.
Proposition NNPP : ∀ M, (¬ (¬ M) ↔ M).
Proposition inp : ∀ M N : Prop, (M ↔ N) → (¬ M → ¬ N).
```

The most difference between our work and present formalization efforts in Coq with topological spaces is the type representations of sets and with members of sets. The type of sets and with members of sets is Class in our system, which can formalize as follows:

```
Parameter Class : Type.
```

The symbols '∈' and '{...:...}' are two primitive constants besides the symbol '=', which formalize as follows:

```
Parameter In : Class → Class → Prop.
Parameter Classifier : ∀ M : Class → Prop, Class.
```

We admit there is no set belonging to itself in our system [14]. The formal description of the Axiom of Extent and Classification axiom-scheme in our paper is given as follows:

```
Axiom ExtAx : ∀ X Y : Class, X = Y ↔ (∀ x, x ∈ X ↔ x ∈Y).
Axiom ClaAx : ∀ x (M : Class → Prop), x ∈ \{ M \} ↔ (M x).
```

Now, we can introduce the definition and properties of set theory. The properties are used repeatedly in the process of proving the rest theorems. Due to space reasons, the formal code of definition and properties is not presented here, and the entire source code file of our paper is available online: https://github.com/BalanceYan/TopologicalSpaces.

3 Topological Spaces in Coq

3.1 Formalization of Topological Spaces

We can realize the definition of topological spaces from open sets, neighborhood systems, closed sets, closure, interior, bases and subbases. In this paper, we presented the definition of topological spaces through the axioms of open sets.

In mainstream mathematics [9], a topological space is defined as a pair of (X, T) where X is a set and T is a subset family of X, and (1) $X, \emptyset \in T$; (2) If $A, B \in T$, then $A \cap B \in T$; (3) If $T1 \subset T$, then $\bigcup T1 \in T$. And T is a topology for X, the elements of the topology T are called open relative to T. The previous conditions are called the axioms of open sets. The formal code of topological space is as follows:

```
Definition Topology X cT := cT ⊂ cP(X) ∧ X ∈ cT ∧ ∅ ∈ cT ∧
  (∀ A B, A ∈cT → B ∈cT → A ∩ B ∈ cT) ∧
  (∀ cT1, cT1 ⊂ cT →∪cT1 ∈ cT).
```

Therefore, we can draw a conclusion: The set X is always open; $\emptyset$ is always open; the intersection of any two members of T is always open; the union of the elements of any subset family of T is always open.

3.2 Basic Examples of Topological Spaces

To better understand the definition of topological spaces, we present two basic examples of topological spaces, including indiscrete topological spaces and discrete topological spaces.

The family T has only two elements X and $\emptyset$, which is the indiscrete topology for the set X; we called topological space (X, T) an indiscrete topological space. A formal description of these properties is given as follows:

```
Definition Indiscrete X := [X] ∪ [∅].
Example IndiscreteP : ∀ X, Topology X (Indiscrete X).
```

The family T contains all subsets of X; it is called the discrete topology for the set X. A formal description of these properties is given as follows:

```
Definition Discrete X := cP(X).
Example DiscreteP : ∀ X, Topology X (Discrete X).
```

The reader can find the complete formal proof of the basic examples in the source code file. In addition, limitary complement topological space and countable complement topological space also is basic examples of topological spaces. The reader can further explore and formal proof more examples based on our formal system.

4   Equivalent Definition of Topological Space

4.1   Based on Neighborhood System

In this section, we give a brief account of the formal description of the neighborhood in topological spaces, and also an overview of the most basic properties of the neighborhood.

A set A in a topological space (X, T) is a neighborhood of a point x iff A contains an open set to which x belongs. The neighborhood system of a point is the family of all neighborhoods of the point. The formal description of these definitions is as follows:

Definition TNeigh x A X cT := Topology X cT ∧ x ∈ X ∧ A ⊂ X ∧
∃ V, V ∈ cT ∧ x ∈ V ∧ V ⊂ A.
Definition TNeighS x X cT := \{ A, TNeigh x A X cT \}.

1 Theorem *A set is open iff it is a neighborhood of each of its point.*

Theorem Theorem1 : ∀ A X cT, Topology X cT → A ⊂ X →
(A ∈ cT ↔ ∀ x, x ∈ A → A ∈ TNeighS x X cT).

2 Theorem *If X is a topological space, U_x is the neighborhood system of a point x, then: (1) if x ∈ X, then $U_x \neq \emptyset$; if A ∈ U_x, then x ∈ A; (2) if A, B ∈ U_x, then A ∩ B ∈ U_x; (3) if A ∈ U_x and A ⊂ B, then B ∈ U_x; (4) if A ∈ U_x, then exists B ∈ U_x satisfies the conditions (i) B ⊂ A and (ii) if y ∈ B, then B ∈ U_y.*

Theorem Theorem2a : ∀ x X cT, Topology X cT → x ∈ X →
TNeighS x X cT ≠ ∅ ∧ (∀ A, A ∈ TNeighS x X cT → x ∈ A).

Theorem Theorem2b : ∀ x X cT, Topology X cT → x ∈ X →
(∀ A B, A ∈ TNeighS x X cT → B ∈ TNeighS x X cT →
A ∩ B ∈ TNeighS x X cT).

Theorem Theorem2c : ∀ x X cT, Topology X cT → x ∈ X →
∀ A B, A ∈ TNeighS x X cT → B ⊂ X → A ⊂ B →
B ∈ TNeighS x X cT.

Theorem Theorem2d : ∀ x X cT, Topology X cT → x ∈ X →
∀ A, A ∈ TNeighS x X cT → ∃ B, B ∈ TNeighS x X cT ∧
B ⊂ A ∧ (∀ y, y ∈ B → B ∈ TNeighS y X cT).

3 Theorem *If x ∈ X, U_x is a subset family of a set X which x appoint, and U_x satisfies the conditions in Theorem 2. Then, there exists a unique topology T and U_x is the neighborhood system of a point x in a topological space (X, T).*

Theorem Theorem3 : ∀ f X, Mapping f X cP(cP(X)) →
(∀ x, x ∈ X → f[x] ⊂ cP(X) ∧
f[x] ≠ ∅ ∧ (∀ A, A ∈ f[x] → x ∈ A) ∧
(∀ A B, A ∈ f[x] → B ∈ f[x] → A ∩ B ∈ f[x]) ∧
(∀ A B, A ∈ f[x] → B ⊂ X → A ⊂ B → B ∈ f[x]) ∧
(∀ A, A ∈ f[x] → ∃ B, B ∈ f[x] ∧ B ⊂ A ∧
(∀ y, y ∈ B → B ∈ f[y]))) → exists! cT,
(Topology X cT ∧ ∀ x, x ∈ X → f[x] = TNeighS x X cT).

Theorem 2 shows that the properties of the neighborhood can prove by the axioms of open sets. Theorem 3 achieved the construction of topology from the neighborhood system. Thus, the formal proof of equivalent definition of topological space was completed.

4.2 Based on Closure

We first present the definition of accumulation points, derived sets, closed sets and closure, and formal verification of the basic properties of these definitions.

A point x is an accumulation point of a subset A of a topological space (X, T) iff every neighborhood of x contains point of A other than x.

Definition Condensa x A X cT := Topology X cT ∧ A ⊂ X ∧ x ∈ X ∧
    ∀ U, TNeigh x U X cT → U ∩ (A - [x]) ≠ ∅

The set of all accumulation points of a set A is called the derived set, is denoted by $d(A)$.

Definition Derivaed A X cT := \{ λ x, Condensa x A X cT \}.

A subset A of a topological space (X, T) is closed if the derived set of A contained in A.

Definition Closed A X cT :=
    Topology X cT ∧ A ⊂ X ∧ Derivaed A X cT ⊂ A.

The closure of a subset A of a topological space (X, T) is the union of the set A and derived set of A, *is denoted by A^-.*

Definition Closure A X cT := A ∪ Derivaed A X cT.

4 Theorem *If A is a subset of a topological space X, then: (1) $d(\emptyset) = \emptyset$; (2) if $A \subset B$, then $d(A) \subset d(B)$; (3) $d(A \cup B) = d(A) \cup d(B)$; (4) $d(d(A)) \subset A \cup d(A)$.*

Theorem Theorem4a : ∀ X cT, Topology X cT → Derivaed ∅ X cT = ∅.
Theorem Theorem4b : ∀ A B X cT, Topology X cT → A ⊂ X →
    B ⊂ X → A ⊂ B → Derivaed A X cT ⊂ Derivaed B X cT.
Theorem Theorem4c : ∀ A B X cT, Topology X cT → A ⊂ X →
    B⊂ X → Derivaed (A∪ B) X cT = Derivaed A X cT ∪ Derivaed B X cT.
Theorem Theorem4d : ∀ A X cT, Topology X cT → A ⊂ X →
    Derivaed (Derivaed A X cT) X cT ⊂ A ∪ Derivaed A X cT.

5 Theorem *If F is a family of all closed sets of a topological space X, then: (1) X, $\emptyset$ $\in$ F; (2) if A, B $\in$ F, then A $\cup$ B $\in$ F; (3) if $\emptyset \neq F_1 \subset F$, then $\cap F_1 \in F$.*

Theorem Theorem5a : ∀ X oT, Topology X oT →
   X ∈ cF X cT ∧ ∅ ∈ cF X cT.

Theorem Theorem5b : ∀ A B X cT, Topology X cT →
   A ∈ cF X cT → B ∈ cF X cT → A ∪ B ∈ cF X cT.

Theorem Theorem5c : ∀ cF1 X cT, Topology X cT → cF1 ≠ ∅ →
   cF1 ⊂ cF X cT → ∩cF1 ∈ cF X cT.

6 Theorem *If A and B is a subset of a topological space X, then (1) $\emptyset^- = \emptyset$; (2) A $\subset$ A$^-$; (3) (A $\cup$ B)$^- = $ A$^- \cup$ B$^-$; (4) A$^{--} = $ A$^-$.*

Theorem Theorem6a : ∀ X cT, Topology X cT → ∅ = Closure ∅ X cT.

Theorem Theorem6b : ∀ A X cT, Topology X cT → A ⊂ X →
   A ⊂ Closure A X cT.

Theorem Theorem6c : ∀ A B X cT, Topology X cT → A ⊂ X →
   B ⊂ X → Closure (A ∪ B) X cT = Closure A X cT ∪ Closure B X cT.

Theorem Theorem6d : ∀ A X cT, Topology X cT → A ⊂ X →
   Closure (Closure A X cT) X cT = Closure A X cT.

The mapping c^* from the power set of X to the power set of X is called the closure operator on X, and (1) $c * (\emptyset) = \emptyset$; (2) A $\subset c * (A)$; (3) $c*(A \cup B) = c * (A) \cup c * (B)$; (4) $c * (c * (A)) \subset c * (A)$. These four conditions are called Kuratowski closure axioms.

Definition Kuratowski X c := Mapping c cP(X) cP(X) ∧
   (c[∅] = ∅) ∧ (∀ A, A ∈ cP(X) → A ⊂ c[A]) ∧
   (∀ A B, A ∈ cP(X) → B ∈ cP(X) → c[A ∪ B] = c[A] ∪ c[B]) ∧
   (∀ A, A ∈cP(X) → c[c[A]] = c[A]).

7 Theorem *If c^* is a closure operator on the set X, then there exists a unique topological T in a topological space (X, T); and if A $\subset X$ then $c^*(A) = A^-$.*

Theorem Theorem7 : ∀ X c, Kuratowski X c →
   exists! cT, Topology X cT ∧ (∀A, A⊂ X → c[A] = Closure A X cT).

Theorem 7 presented the construction of topology from Kuratowski closure axioms. The machine proof of equivalent definition of topological space was completed once again.

4.3 Based on Other Concepts

We can also realize the construction of topological spaces by using closed sets, interior, bases, subbases, neighborhood bases and neighborhood subbases. Take the interior, for example, the definition of the interior and the formal verification of the basic properties of the interior are first presented. Then, we set up the topological space by the properties of the interior and realize the machine proof of equivalent definition of topological spaces. Interested readers can construct topological spaces by other concepts based on our work to enhance their understanding.

5   Conclusions

Topological spaces are one of the prominent concepts of general topology. We introduced a definition of topological spaces in Coq based on set theory, which allows us to state and prove basic examples and theorems in topology spaces. We implemented the formal description of equivalent definitions of topological spaces and presented machine proof of theorems about equivalent definitions of topological spaces from neighborhood system and closure. Our code was developed under Coq 8.9.1. The complete source file is accessible at: https://github.com/BalanceYan/TopologicalSpaces.

Furthermore, we will construct topological spaces by other concepts and formalize more theorems in general topology based on present works.

Acknowledgment. This work is supported by National Natural Science Foundation of China (No. 61936008).

References

1. Wiedijk, F.: Formal proof - getting started. Not. Am. Math. Soc. **55**, 1408–1414 (2008)
2. Gonthier, G.: Formal proof - the four color theorem. Not. Am. Math. Soc. **55**, 1382–1393 (2008)
3. Gonthier, G., Asperti, A., Avigad, J., et al.: Machine-checked proof of the Odd Order Theorem. In: Blazy, S., Paulin-Mohring, C., Pichardie, D. (eds.) ITP 2013. LNCS, vol. 7998, pp. 163–179. Springer, Heidelberg (2013). https://doi.org/10.1007/978-3-642-39634-2_14
4. Hales, T.C., Adams, M., Bauer, G., et al.: A formal proof of the Kepler conjecture. Forum Math. Pi **5**, e2 (2017)
5. Formalizing 100 Theorems. http://www.cs.ru.nl/~freek/100/
6. Bertot, Y., Castéran, P.: Interactive Theorem Proving and Program Development – Coq' Art: The Calculus of Inductive Constructions. Spring, Berlin (2004). https://doi.org/10.1007/978-3-662-07964-5
7. Harrison, J., Urban, J., Wiedijk, F.: History of interactive theorem proving. Handb. Hist. Log. **9**, 135–214 (2014)
8. You, S.J., Yuan, W.J.: The equivalent definition of topology. J. Guangzhou Univ. (Nat. Sci. Ed.) **3**, 492–495 (2004)
9. Kelley, J.L.: General Topology. Springer, New York (1955)
10. Schepler, D.: Topology: general topology in Coq (2011). https://github.com/coq-community/topology
11. Wang, S.Y.: FormalMath: a side project about formalization of mathematics (Topology) (2021). https://github.com/txyyss/FormalMath/tree/master/Topology
12. Hölzl, J., Immler, F., Huffman, B.: Type classes and filters for mathematical analysis in Isabelle/HOL. In: Blazy, S., Paulin-Mohring, C., Pichardie, D. (eds.) ITP 2013. LNCS, vol. 7998, pp. 279–294. Springer, Heidelberg (2013). https://doi.org/10.1007/978-3-642-39634-2_21
13. Enderton, H.B.: Elements of Set Theory. Spring, New York (1977)

14. Yu, W.S., Sun, T.Y., Fu, Y.S.: Machine Proof System of Axiomatic Set Theory. Science Press, Beijing (2020)
15. Yu, W.S., Fu, Y.S., Guo, L.Q.,: Machine Proof System of Foundations of Analysis. Science Press, Beijing (2021)

A Storage Scheme for Access Control Record Based on Consortium Blockchain

Yunmei Shi[1,2](✉), Ning Li[1,2], and Shoulu Hou[1,2]

[1] Beijing Key Laboratory of Internet Culture and Digital Dissemination Research, Beijing Information Science and Technology University, Beijing 100101, China
sym@bistu.edu.cn
[2] School of Computer, Beijing Information Science and Technology University, Beijing 100101, China

Abstract. The heterogeneous access control information scattered around different organizations or units is difficult to be gathered and audited, however, easy to be maliciously tampered. Aiming at the problems, this paper presents a blockchain-based storage scheme to store access control information, which can protect information privacy and facilitate the audit work. This is achieved by exploiting consortium blockchain, cryptography technology. Based on the scheme, we define the format of Access Control Record (ACR), design upload and download protocols, and realize the signature and encryption process for ACRs in a simulation environment. Theoretical analyses demonstrate that the proposed storage scheme needs lower storage cost and has higher efficiency compared with existing schemes, and can resist typical malicious attacks effectively.

Keywords: Access control record · Blockchain · Storage scheme · Privacy preservation

1 Introduction

Generally, the access control information produced by application systems is stored and managed by respective organization or unit separately, which bring great troubles for information collection and audit. Besides, the access control information from different applications often has different formats, which also bring burdens to audit works. In addition, from the security perspective, the scattered access control information has a greater security risk.

Blockchain has the characteristics of persistency, immutability and auditability. Owing to its advantages, blockchain technology is applied to access control fields in literatures [2–7]. These literatures treat the blockchain as a credible storage entity to store access control rights or access control polices, or make it provide trusted computing as well as information storage, in which smarts contracts are utilized to authenticate visitors, verify access rights or access behaviors. Whatever the case, these literatures mainly focus on the security related to access control policies or access control models. Obviously these researches have different motivations from ours, but they give us good ideas to solve our problems.

© The Author(s) 2022
Z. Qian et al. (Eds.): WCNA 2021, LNEE 942, pp. 205–218, 2022.
https://doi.org/10.1007/978-981-19-2456-9_22

Blockchain uses a universal ledger, and every node in the blockchain has the same one. That means the data stored in blockchain is maintained by all nodes. If the information in one node is tampered or destroyed, data authenticity cannot be affected, unless over 51% nodes are tampered. Since the distributed ledger in blockchain is tamper-resistant and strongly anti-attack, the blockchain network is very suitable for storing the access control information. Blockchain is divided into three types: public, private and consortium blockchain. Compared with the first two types, consortium blockchain can provide higher security for access control information, and is suitable for centralized and unified information supervision of administrative agency.

Unfortunately, the data stored in blockchain is often in plaintext. when an unauthorized intruder gets the access control information, he can easily analyze someone's behaviors and working habits. The intrusion may lead to disastrous consequences, especially when the stolen information is related to important persons.

Aiming at the problems, we propose an ACR storage scheme based on consortium blockchain to ensure information reality and validity by using the auditability and immutability of blockchain technology, and preserve information privacy by using identity authentication and confidentiality mechanisms.

2   Related Work

2.1   Blockchain and Access Control

Blockchain technology uses distributed and decentralized computing and storage architecture, which solves the security problems caused by trust-based centralized model, and avoids data to be traced or tampered. At present, the researches on blockchain technology mainly focus on computing and storage power, furthermore, they can be classified into three types: only considering the security storage, only using the trusted computing capability, and combination of both [1].

For the researches and applications involving with access control and blockchain technology, a common approach is that a blockchain is regarded as a trusted entity to save access control policies and provide trusted computing through smart contracts.

Zhang Y et al. proposed an access control scheme based on Ethereum smart contracts which are responsible for checking the behaviours of the subject, and determine whether to authorize the access request according to predefined access control policies and dynamic access right validation [2]. Damiano et al. introduced blockchain to save access control policies, instead of traditional relational database [3]. Alansari et al. used blockchain to store access control policies, and utilize blockchain and trusted hardware to protect the policies [4, 5]. Liu H et al. presented an access control mechanism based on the hyper ledger, in which the policy contract provides access control polices for admin users, the access contract implements an access control method for normal users [6]. Wang et al. proposed a model for data access control and an algorithm based on blockchain technology. The model was divided into five layers, in which the contract layer provides smart contract services with major function of offering access control polices [7]. Only the accounts that meet specific attributes or levels are permitted to access data. Zhang et al. proposed a EMR (Electronic Medical Record) access control scheme based on blockchain, which uses smart contracts to implement access control policies. Only the users with permissions can access data [8].

The above studies mainly focus on saving access control policies through the blockchain and using smart contracts to manage the access control policies or authorization of user access control. Unfortunately, these studies rarely consider how to use blockchain technology to store the comprehensive information caused by various access control policies, user authority and user access behaviour for future audit and supervision.

2.2 Blockchain and Privacy Preservation

To reach consensus on the transactions among the nodes of blockchain network, all transactions are open, and that means the participants in the blockchain can easily view all transactions in the blockchain. However, not all transaction information is expected to be obtained by all participants, thereby causing a huge hidden security danger for privacy preservation.

Zhu et al. divided the privacy in blockchain into two categories: identity privacy and transaction privacy [9]. Transaction privacy refers to the transaction records stored in the blockchain and the knowledge behind them. Many researchers have carried out relevant researches on transaction privacy preservation.

In the medical field, the researches mainly focus on the sharing of patient information. Peterson et al. applied the blockchain technology to the sharing and exchange of medical records, which not only realize data sharing, but also protect patients' privacy and security [10]. Shae and Tsai proposed a blockchain platform architecture to help medical clinical trials and precision medicine [11]. Wang et al. used a blockchain to store patient medical records and other files to realize cross-domain data sharing, and encrypt transaction data through asymmetric encryption technology to protect patient data privacy [12]. Zhai et al. applied blockchain technology to EMR sharing. In their proposed EMR sharing model, private and consortium blockchain are utilized simultaneously to store encrypted EMR by users and safety index records of EMR respectively [13]. Based on type and identity, they combine distributed key generation technology and proxy re-encryption scheme to realize data sharing among users, thus preventing data modification and resisting attacks. Xu et al. utilized the blockchain network to store electronic health records to realize safe sharing of medical data effectively [14]. In order to strengthen privacy protection for users' data, they used cryptography technology, and achieve good security and performance.

In the above literatures, cryptography technology is used to protect the data security in transactions, and achieve good privacy preservation effect. However, these researches on blockchain and access control mainly focus on access control policy storage and user authorization with blockchain technology, few literatures research on how to store access control information in blockchain and how to protect its privacy.

Aiming at these problems, we obtain the access control related information from the user login logs, access control policies, user authorization records and etc. to build ACR based on ABAC (Attribute-Based Access Control) model, then upload the encrypted ACR to blockchain to guarantee the security and auditability of the access control information.

3   ACR Storage and Privacy Preserving Scheme

3.1   ACR Definition

It can improve information security to store access control related information into blockchain, such as user login records, access control policies, authorization record. However, it exists the following problems. First, due to the different access control mechanisms adopted by the participants in the blockchain network, the format of access control information is prone to be inconsistent, reducing the audit efficiency. Second, the log information recorded by system access control module is limited, and it cannot describe the whole access control behaviours of users.

This paper designs the format of ACR based on ABAC model, which integrate contents of access control related information from different sources to achieve fine grained management of access control and user behaviour tracking. ACR is defined as follows:

ACR (LogID, LoginUser, Time, ACA, PI, APUser, UserRights, Remarks)

The definition of the fields in ACR is as follows:

LogID: is the log number.

LoginUser: is the login name.

Time: is the login date and time of users.

ACA (Access Control Activities): represents access control activities related to users.

PI (Policy Information): means the access control policies related with users.

APUser (Access-Permitted User): user name assigned permissions

UserRights: rights owned by users

Remarks: is comments.

ACR originates from access control related information generated by diverse applications in various organizations, and is the preprocessed and aggregated results of the information. It can comprehensively contain the user's operation behaviour based on access control policy, thus facilitating the future data audit.

3.2   ACR Storage Scheme Based on Blockchain

The storage scheme, illustrated in Fig. 1, is mainly divided into three parts: networks of organizations or units, consortium blockchain network to store ACRs, and the authority responsible for audit work.

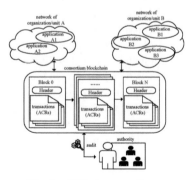

Fig. 1. ACR storage scheme.

As mentioned above, ACRs are gathered from various organizations or units, and then uploaded to the blockchain. When uploading ACRs, a smart contract is triggered, which executes a transaction according to its rules, and transfer the ACRs to the blockchain according to consensus mechanism. ACR stored in the blockchain acquires its immutability and traceability with the help of the tamper-resistant nature of blockchain.

In order to reduce the cost of uploading ACRs to blockchain, we set a threshold in the storage scheme. That means only when the number of ACR reaches a predetermined value, the ACRs can be uploaded by a smart contract, otherwise, they will wait until the number reaches the threshold.

Generally, blockchain can be categorized into three types: public blockchain, consortium blockchain and private blockchain. Each node in a public blockchain is anonymous and can join and leave freely. From respective of safety, this kind of open management mechanism is unsuitable for organizations. Besides, the public blockchain uses PoW (Proof of Work) consensus mechanism, which relies on computing power competition to guarantee the consistency and security of the data in blockchain. From this perspective, the public blockchain is also inappropriate for organizations or units. A consortium blockchain is initiated by organizations or units, and each node couldn't join or exit the network until authorized. This feature ensures the data not to be tampered or erased, which can satisfy the data storage requirements in some extent. A private blockchain is regarded as a centralized network since it is fully controlled by one organization [15], and strictly speaking, it is not decentralized.

Based on its distinctive characteristic, we choose the consortium blockchain in our scheme. The data saved in the consortium blockchain is not open, and only shared among the participants of the federation to ensure the data security.

Figure 2 shows the ACR upload and download process in more detail.

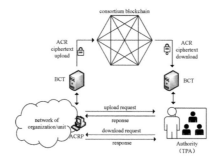

Fig. 2. Upload and download process of ACR.

The component in Fig. 2 is demonstrated as follows:

1) ACRP (Access Control Record Provider) is responsible for managing access control information from organizations or units. Firstly, ACRP preprocesses and integrates the access control information to produce ACRs, then uploads them to BCT.
2) BCT (BlockChain Terminal) is a node of consortium blockchain. The node is used to realize the decentralized application of Ethereum, and isolates users and application

systems in the internal network. Before uploading ACR, the BCT administrator need to create an account in the wallet and connect BCT to the blockchain network.

3) TPA (Third Party Auditor), located in the authority, is in charge of ACR audit.

Blockchain cannot guarantee the perfect privacy preservation due to the intrinsic constraint [15], including privacy leakage, and the data privacy needs extra protection mechanism.

In our scheme, ACRP need to encrypts ACRs, then upload to BCT. BCT executes a transaction through a smart contract, and adds the execution results to the consensus process. After consensus, the transaction information with ACR ciphertext will be recorded in a universal ledger to ensure the data consistency in the blockchain.

To improve efficiency and reduce cost, some ACRs, named ACR set, are packed in one transaction. In this way, when ACR uploaded, ACR set only need to be signed one time, avoiding each ACR is signed separately. Obviously, it can greatly reduce the total cost to pack ACR set in one transaction. Meanwhile, ACR set can reduce the transferring time and the traffic between nodes in the blockchain, mitigating the burden of network.

Once TPA needs to audit ACR, it first sends a download request to corresponding ACRP. After receiving the request, ACRP first verifies the identity of TPA, then sends a response message.

Finally, TPA sends a request for downloading BCT to acquire the ACR ciphertext from the blockchain, and get the plaintext by decrypting data with symmetric keys. Then, the audit process can be carried out.

3.3   Upload and Download Protocols

Based on the scheme discussed in the previous section, we design the upload and download ACR protocols.

ACR Upload Protocol. 1) ACRP sends an upload request to TPA, and provides the identity information in the following format.

$M_{1(ACRP \to TPA)}$: $\{ID_{Provider}, R_1 \| T_1, PriKey\_sign_{Provider}(R_1)\}$

$ID_{Provider}$ is an identification of ACPR, which can uniquely identify an ACRP.

R_1 is a random number, which is used to provide necessary information for authenticating ACRP.

$PriKey\_sign_{Provider}(R_1)$ is a signature value with ACPR's private key. The signature is sent with other fields of the message to TPA. Once the message is received, TPA validates the signature to verify ACRP's identity by using ACRP's public key.

T_1 is a timestamp, which indicates message generation time. The timestamp is used to confirm the refresh interval, and it can prevent replay attacks.

2) After the identity of ACRP is verified, TPA will send the response messages to ACPR. The response message carries the corresponding symmetric key, and can be described as follows.

$M_{2(TPA \to ACRP)}$: $\{PriKey\_sign_{Auditor} \ (R_1), \ T_2, \ PubKey\_Encrypt_{Provider}(key(a), Hash(R_1 \| T_2)\}$

PriKey_sign$_{Auditor}$ (R_1), the signature with TPA's private key, is used to verify TPA's identity.

T_2 is a timestamp, and has the same meaning as T_1 in message M_1.

Key(a) is the symmetric key provided by TPA, which is used to encrypt the data. *Hash(R_1||T_2)* is used to enhance the transmission security of the symmetric key. For security, these two parameters are encrypted with ACPR's public key.

3) ACRP signs the hash of ACR with its private key.

PriKey_sign$_{Provider}$ (Hash(ACR))

The hash value of ACP can help the TPA retrieve ACR when auditing, which is abbreviated as *HASH_ID$_{ACR}$*, and the signature value of *HASH_ID$_{ACR}$* is denoted by *Sign_Hash(ACR)*.

4) Primary encryption.

ACRP encrypts both ACR and the result of previous step with its symmetric key *key(p)*. The encrypted data is denoted by *Sym_Encrypt(ACR, Sign_Hash(ACR))*.

Sym_Encrypt$_{key(p)}$(ACR, Sign_Hash(ACR))

5) Secondary encryption.

ACRP uses symmetric encryption algorithm to encrypt the result of last round, and the symmetric key used is *key (a)* provided by TPA.

Sym_Encrypt$_{key(a)}$(Sym_Encrypt(ACR, Sign_Hash(ACR))

The encrypted result is denoted by *Sym_Encrypt (Sym_Encrypt(ACR, Sign_Hash(ACR)))*.

6) ACRP transfers encrypted message containing encrypted ACR and hash value to BCT.

M$_{3(ACRP \to BCT)}$:{Sym_Encrypt (Sym_Encrypt(ACR, Sign_Hash(ACR)))}

7) BCT publishes encrypted ACR to the blockchain network.

After receiving the ACR ciphertext, BCT publishes the encrypted data to each node in the blockchain network through smart contract and consensus mechanism.

ACR Download Protocol. 1) TPA sends a download request to ACRP and provides its identity information for authentication.

M$_{4(TPA \to ACRP)}$: {ID$_{Auditor}$, R_2||T_3, PriKey_sign$_{Auditor}$(R_2)}

The message is designed the same as the request message of the upload protocol. The parameters of the message are defined as follows:

ID$_{Auditor}$ is the identification of TPA to which can uniquely identify a TPA.

R_2 is a random number. Both *ID$_{Auditor}$* and *PriKey_sign$_{Auditor}$(R_2)* are used to realize the authentication of ACPR. When receiving the message, ACRP parses it and get the signature *PriKey_sign$_{Auditor}$(R_2)*. If the verification result is the same as R_2, it shows that the request message is truly sent by TPA.

T_3 is a timestamp to ensure the refresh interval.

2) When receiving the request, ACRP verifies TPA's identity, and then responds to the sender.

$M_{5(ACRP \rightarrow TPA)}$: $\{PriKey\_sign_{Provider}(R_2),$ $T_4,$ $PubKey\_Encrypt_{Auditor}(key_{(p)},$ $Hash(R_2 \| T_4)\}$

$PriKey\_sign_{Provider}(R_2)$ is the signature value of ACRP for verifying the identity of ACRP.

T_4 is also a timestamp, which effect is similar to T_3.

$key_{(p)}$ is the symmetric key produced by ACRP which will be used to encrypt the ACR. Both $Hash(R_2 \| T_4)$ and $key_{(p)}$ are encrypted simultaneously to ensure the key is uneasy to be cracked.

3) TPA sends a request of downloading ACR from BCT.

$M_{6(TPA \rightarrow BCT)}$: $\{ID_{Auditor}, R_3 \| T_5, PriKey\_sign_{Auditor}(R_3)\}$

The message is similar to the request of TPA sending to BCT, and the main differences between them are the destination address and some values of the fields in the messages. The first field of the message is $ID_{Auditor}$, which is the identification of TPA. R_3 is a random number, and T_5 is a timestamp. R_3 is signed with the private key of TPA to confirm the message is sent by TPA.

4) BCT transfers ACR ciphertext to TPA

$M_{7(BCT \rightarrow TPA)}$: $\{Sym\_Encrypt \, (Sym\_Encrypt(ACR, \, Sign\_Hash(ACR)))\}$

The message M_7 contains the ciphertext of twice symmetric encryptions to ACR.

5) TPA parses the message and decrypts the ciphertext.

$Decrypt_{key(p), \, key(a)} \, \{Sym\_Encrypt \, (Sym\_Encrypt(ACR, \, Sign\_Hash(ACR)))\}$

TPA decrypts the ACR ciphertext with $key \, (a)$ and $key \, (p)$ to obtain the plaintext of ACR. Then, TPA can audit ACR data. Since the data is preprocessed and integrated before transferred to blockchain, and saved in the universe formats, it is much easier to audit ACR rather than the original data scattered over different applications and organizations.

4 Experiment and Analysis

4.1 Experiment Environment

In test experiment, we adopt a simulation environment. For a simulation environment, it needs to provide developing and running environment for smart contracts, including program language, operation carrier such as virtual machine and etc.

Common simulation test environment adopts EVM (Ethereum Virtual Machine) as the execution environment of smart contracts and Ropsten as the blockchain network. Ropsten is a blockchain test network officially provided by Ethereum, which provides EVM for executing smart contracts.

We build test environment through Ropsten and Lite-server, and EVM is supported by Ropsten, as shown in Fig. 3. ACR information is submitted to Ropsten test blockchain network through user interface. The Lite-server, located between Ropsten and UI, is responsible for the interaction with Ropsten and UI. Lite-server acts as the role of BCT.

Lite-server supports web3.js, which is a JavaScript library that encapsulates the RPC communication interface of Ethereum, and provides a series of rules, definitions and functions required for interacting with Ethereum. Ethereum wallet provides users querying services for digital currency balance and transaction information, and helps users save the Ethereum private key.

The administrator signs and encrypts ACR from UI (User Interface), then submits to Lite-server. Lite-server utilizes the smart storage contract and functions provided by web3.js to store ACR ciphertext into Ropsten.

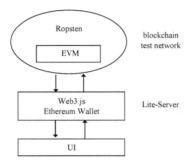

Fig. 3. Diagram of simulation test environment

We design a smart contract for storing ACR. The smart contract is developed in Truffle and programed with solidity programming language. Truffle, based on JavaScript, is a development and test framework of Ethereum, and supports smart contracts written with solidity language.

The smart contract realizes the function of storing ACR, which is called storage contract. By using the interface provided by web3.js, the storage contract is passed to the compiler, compiled into binary code, and deployed to the blockchain.

4.2 Experiment

The information administrator of organizations or units unifies and aggregates the information from access control logs, access control polices and authorization records. The finally integrated access control information is ACR, which will be encrypted and uploaded to blockchain. In the experiment, we get hundreds of ACRs from access control information. Table 1 shows a piece of ACR.

Table 1. A piece of sample of ACR.

Fields	Contents
LogID	61d75430-b444-460c-bfa4-5ec62c188c9e
LoginUser	Admin
Time	2015-2-25 131455
ACA	{"Message ":"Create Policy Policy-Test", "Subsystem":"Policies > Access Control > Access Control > Firewall Policy Editor", "Time":"2015-2-11 144834", "LoginUser":"admin"}
PI	Subject:Administrator; Resource:video; Action:query; Effect:Allow; Environment Time: [15, 17]
APUser	Admin
UserRights	Update
Remarks	ip:192.168.0.109

4.3 Analysis

Efficiency Analysis

Time Cost and Ciphertext Size. Literature [16] proposes a data encryption scheme for multi-channel access control of ad hoc network, and literature [17] presents a scheme for data access control, named DAC-MACS. Based on the two schemes, we conduct the comparison on the efficiency, and the results are shown in Table 2.

D is the size of a unit ciphertext. n is the number of ciphertext attribute. $Cert_{PID}$ represents pseudonym certificate. $T_{Encrypt}$ and $T_{Decrypt}$ are the time consumed by encryption and decryption for a unit of ciphertext, respectively.

The time cost of scheme 1 for encryption and decryption is the same as that of our scheme, however, the amount of ciphertext in scheme 1 is larger than that of our scheme.

The proposed scheme has shorter encryption and decryption time, and smaller ciphertext size, as compared to scheme 2. The reason is that scheme 2 employs the CP-ABE algorithm, and the number of ciphertext attributes affects the encryption and decryption cost, and the size of ciphertext. Whereas, the proposed scheme is independent of the number of ciphertext attributes.

Table 2. Comparison of time cost and ciphertext size.

Scheme	Encryption cost	Decryption cost	Ciphertext size
Scheme 1 [16]	$T_{Encrypt}$	$T_{Decrypt}$	$Cert_{PID} + D$
Scheme 2 [17]	$nT_{Encrypt}$	$nT_{Decrypt}$	$(3n + 1)D$
Proposed scheme	$T_{Encrypt}$	$T_{Decrypt}$	D

Storage Cost. In Ethereum, every participant should pay cost for each storage transaction, and the cost is measured with gas. Supposing the storage smart contract is triggered to commit a transaction whenever BCT receives an ACR ciphertext, it will definitely leads to great gas cost.

Table 3. Gas consumed during uploading ACRs.

Number of ACR	Gas used	
	With threshold	Without threshold
3	1,264,329	2,914,227
8	3,177,383	7,771,272
17	5,676,277	16,513,953
21	6,617,922	20,399,589

In order to reduce the cost of uploading ACRs to blockchain network, we set a threshold. If the ACR number from organizations or units is less than the threshold, the storage contract is not executed, until the number reaches the threshold. Table 3 shows the storage gas cost measured in the uploading ACRs experiments. In the experiment, we set the threshold with 7. The second column in Table 3 lists the gas cost with threshold constraints, and the third one is that without threshold. Obviously, the storage cost with threshold is much lower than the other one.

The comparison experiment shows that our ACR storage scheme can effectively reduce the storage cost by setting threshold.

Security Analysis. Security means ACR security, including storage security and transmission security.

Blockchain technology has the nature of immutability. The blockchain consists of a series of blocks, and each block holds the hash value of its previous block. If an attacker attempts to change the hash value of a block, he must have at least 50% computing power of the blockchain network. It's almost impossible, therefore, the ACR stored in the blockchain is immutable.

According to the features of blockchain technology, the encrypted ACR is visible to all participants, however, it is almost impossible to get the plaintext of double encryption ACR for malicious attackers without decryption keys.

The above analyses show that the ACR stored in blockchain has high storage security. For transmission security, detailed analyses will be introduced next.

For the sake of secutiy analysis, we collect the messages mentioned in Sect. 3.3 in Table 4.

Table 4. Messages of upload and download protocols.

Message	Contents	Sender	Receiver
M_1	$\{ID_{Provider}, R_1 \| T_1, PriKey\_sign_{Provider}(R_1)\}$	ACRP	TPA
M_2	$\{PriKey\_sign_{Auditor}(R_1), T_2, PubKey\_Encrypt_{Provider}(key(a), Hash(R_1 \| T_2)\}$	TPA	ACRP
M_3	$\{Sym\_Encrypt(Sym\_Encrypt(ACR, Sign\_Hash(ACR)))\}$	ACRP	BCT
M_4	$\{ID_{Auditor}, R_2 \| T_3, PriKey\_sign_{Auditor}(R_2)\}$	TPA	ACRP
M_5	$\{PriKey\_sign_{Provider}(R_2), T_4, PubKey\_Encrypt_{Auditor}(key(p), Hash(R_2 \| T_4)\}$	ACRP	TPA
M_6	$\{ID_{Auditor}, R_3 \| T_5, PriKey\_sign_{Auditor}(R_3)\}$	TPA	BCT
M_7	$\{Sym\_Encrypt(Sym\_Encrypt(ACR, Sign\_Hash(ACR)))\}$	BCT	TPA

Resist Replay Attack. The header of each block in blockchain contains a timestamp, and it is invalid for an attacker to replay a block during the creation of the block. Since the virtual currency used in blockchain in privacy preservation scheme has no physical value, replay attack against blockchain fork is meaningless for our scheme.

During the procedure of ACR upload and download, attackers may try to replay M_2 or M_5 to steal the symmetric keys for encryption. However, both M_2 and M_5 contain random number and timestamp. The random number makes M_2 and M_5 different in each round of communication, while the timestamp guarantees the message freshness.

Resist Man-in-the-Middle Attack. Man-in-the-middle attack is that attackers intercept the message sent by each side of the communication and try to tamper with and resend the message. There are three messages, M_1, M_2 and M_3, involved in uploading ACR. M_1 and M_2 are mainly composed of the random number newly generated, timestamp and signature, and M_3 contains the ciphertext of ACR, so it doesn't work to resend the messages. Without the private key for authentication, even if M_1 or M_2 is tampered and resent, the message cannot pass validation. The ciphertext in M_3 has the hash value of ACR and the signature of the sender, these protective measures can effectively ensure data integrity.

The messages for downloading ACR, including M_4, M_5, M_6 and M_7, adopt the same design ideas as those in the upload protocol, therefore, they can also effectively resist man-in-the-middle attack.

Resist Fake Attack. The attacker impersonates one participant of the blockchain and tries to obtain the plaintext of ACR. During the procedure of upload or download ACR, ACRP or TPA needs to use its own private key to sign random numbers in the messages to ensure data integrity and sender identity The attacker cannot complete the identity authentication without the private key, let alone obtain the plaintext data. Even if the attacker retransmits the intercepted message, it is impossible for the attacker to get any helpful information to crack the ACR ciphertext.

5 Conclusion and Future Work

In this paper, we propose a scheme for ACR storage and privacy preservation based on the consortium blockchain, and design the protocols of uploading and downloading ACR. The scheme has several main advantages. First, ACR provides a unified format which can integrates heterogeneous access control information. Then, the proposed scheme guarantees the secure storage of ACR based on the immutability of blockchain. Finally, the scheme protects ACR privacy by using the cryptography technology. The experimental results and theoretical analyses show that the scheme can guarantee the security and confidentiality of ACR, and bring great convenience for audit work.

Although the prososed scheme is effective for ACR storgae and privacy protection, it still exist some issues which need further research and discussion, for example, how to efficiently search ciphertext in blockchain, how to protect the privacy of transaction addresses. In future work, we will carry out in-depth studies on these issues.

Acknowledgement. The work described in this paper was supported by the National Key Research and Development Program of China (2018YFB1004100).

References

1. Shi, J.S., Li, R.: Survey of blockchain access control in internet of things. J. Softw. **30**(6), 1632–1648 (2019)
2. Zhang, Y., Kasahara, S., Shen, Y., et al.: Smart contract-based access control for the internet of things. IEEE Internet Things J. **6**(2), 1594–1605 (2018)
3. DiFrancescoMaesa, D., Mori, P., Ricci, L.: Blockchain based access control. In: Chen, L.Y., Reiser, H.P. (eds.) DAIS 2017. LNCS, vol. 10320, pp. 206–220. Springer, Cham (2017). https://doi.org/10.1007/978-3-319-59665-5_15
4. Alansari, S., Paci, F., Sassone, V., et al.: A distributed access control system for cloud federations. In: ICDCS 2017: International Conference on Distributed Computing Systems, pp. 2131–2136. IEEE (2017)
5. Alansari, S., Paci, F., Margheri, A., et al.: Privacy-preserving access control in cloud federations. In: 2017 10th International Conference on Cloud Computing, pp. 757–760. IEEE (2017)
6. Liu, H., Han, D., Li, D.: Fabric-IoT: a blockchain-based access control system in IoT. IEEE Access. **8**, 18207–18218 (2020)
7. Wang, X.L., Jiang, X.Z., Li, Y.: Model for data access control and sharing based on blockchain. J. Softw. **30**(6), 1661–1669 (2019)
8. Zhang, Y.B., Cui, M., Zheng, L.J., et al.: Research on electronic medical record access control based on blockchain. Int. J. Distrib. Sens. Netw. **15**(11), 1–13 (2019)
9. Zhu, L.H., Gao, F., et al.: Survey on privacy preserving techniques for blockchain technology. J. Comput. Res. Dev. **54**(10), 2170–2186 (2017)
10. Peterson, K., Deeduvanu, R., Kanjamala, P., et al.: A blockchain-based approach to health information exchange networks (2016). https://www.healthit.gov/sites/default/files/12-55-blockchain-based-approach-final.pdf. Accessed 1 Oct 2020

11. Shae, Z., Tsai, J.J.P.: On the design of a blockchain platform for clinical trial and precision medicine. In: Proceedings of the 2017 IEEE 37th International Conference on Distributed Computing Systems (ICDCS), pp. 1972–1980. IEEE (2017)

12. Wang, H., Song, Y.: Secure cloud-based EHR system using attribute-based cryptosystem and blockchain. J. Med. Syst. **42**(8), 1–9 (2018). https://doi.org/10.1007/s10916-018-0994-6

13. Zhai, S.P., Wang, Y.J., Cen, S.J.: Research on the application of blockchain technology in the sharing of electronic medical records. J. Xidian Univ. **47**(5), 103–112 (2020)

14. Xu, W.Y., Wu, L., Yan, Y.X.: Privacy-preserving scheme of electronic health records based on blockchain and homomorphic encryption. J. Comput. Res. Dev. **55**(10), 2233–2243 (2018)

15. Zheng, Z.B., Xie, S.A., et al.: An overview of blockchain technology: architecture, consensus, and future trends. In: 2017 IEEE 6th International Congress on Big Data, pp 557–564. IEEE (2017)

16. Li, M.F.: Multi-channel access control simulation of self-organizing network based on blockchain. Computer Simulation. **36**(5), 480–483 (2019)

17. Yang, K., Jia, X., Ren, K., et al.: DAC-MACS: effective data access control for multi-authority cloud storage systems. In: International Conference on Computer Communications 2013, pp. 2895–2903. IEEE (2013)

Design of Intelligent Recognition System Architecture Based on Edge Computing Technology

Lejiang Guo[✉], Lei Xiao, Fangxin Chen, and Wenjie Tu

The Department of Early Warning Surveillance Intelligent, Air Force Early Warning Academy, Hubei 430019, China
radar_boss@163.com

Abstract. With the increasing acceleration of 5G network construction, artificial intelligence, Quantum technology, edge computing provides content distribution and storage computing services near the network edge which greatly reducing the delay of data processing and service delivery. Starting with the process of information support and decision planning, it analyzes the relationship between edge computing, Quantum and the massive military data. It puts forward an intelligence system architecture design based on edge computing and Quantum. Combined with the openness and flexibility of the system architecture, this paper realizes the mix between data platform and data. It realizes the connection with the existing intelligence system which improves the efficiency of existing data and expands the scenario of edge computing.

Keywords: Edge calculation · Information support · Integration analysis · Content distribution

1 Introduction

With the rapid development of information technology and the in-depth improvement of new military reform, the era of information war has entered. Information war has the following main characteristics: the dominant element of combat power changes from material energy to information energy; the winning idea of war has changed from entity destruction to system attack; the release mode of combat effectiveness has changed from quantity accumulation to system integration; the range of battlefield space becomes full dimensional. Compared with the traditional technology, the new generation information and communication technology has lower delay, Edge computing solves the problem of data volume and time delay. It is the platform integrating the key capabilities of application, storage and network.

Edge computing and Quantum technology can greatly improve the intelligence capacity. First, it greatly improves the efficiency of intelligence information process. In modern war, the amount of battlefield datalake is largely huge unstructured data. If we use conventional methods to process these massive information. Using big data

© The Author(s) 2022
Z. Qian et al. (Eds.): WCNA 2021, LNEE 942, pp. 219–224, 2022.
https://doi.org/10.1007/978-981-19-2456-9_23

to process intelligence information, the theoretical time-consuming can reach the second level and the processing speed jumps exponentially, which can greatly improve the intelligence information acquisition and processing ability. Second, more valuable information can be founded. Under the constraints of investigation means, battlefield environment and other factors, the technology can quickly and automatically classify, sort, analyze and feed back the information from multiple channels. It separates the high-value military intelligence of the target object from a large number of relevant or seemingly unrelated, secret or public information to effectively solve the problem of intelligence Insufficient surveillance and reconnaissance system. Third, it can improve command and decision-making ability. The use of big data analysis technology can provide intelligent and automatic auxiliary means for the decision analysis, it improve the intelligent degree of the system and effectiveness of decision-making, so as to greatly improve the command efficiency and overall combat ability.

2   Characteristics of Edge Calculation

Edge Calculation defines three domains including device domain, data domain and application domain. The layers are the calculation objects of edge calculation. Device domain establishes TPM (trusted platform modules), which integrates the encryption key in the chip into the chip that can be used for device authentication in the software layer. If encode/decode of non shared key path occurs in TPM, the problems can be easily solved. Data domain e communicates with more edge gatewayswhich provide access to the authentic network. Application domain realizes interworking through Data domain or centralized layer. Edge computing is nearby the data source, it can firstly analyze and intelligently process the data in real time, which is efficient and secure. Both edge computing and cloud computing are actually a processing method for computing and running big data. Connectivity and location in Edge computing is based on connectivity. Because of the various connected data and application scenarios, edge computing is required to have rich connection functions.

When the network edge is a part of the network, little information can be used to determine the location of each connected device. It realizes a complete set of business use cases. In the interconnection scenario, edge gateways provide security which constraints and support the digital diversity scene of the industry.

High bandwidth and low delay of edge computing is nearby the datalake, simple data processing can be carried out locally. Since the edge service runs close to the terminal device, the delay is greatly reduced. Edge computing is often associated with the Internet of things which participate in a large amount of data generated network.

Distribution and proximity in Edge computing. Because edge computing is close to the data receiving source, it can obtain data in real time, analyze and process, In addition, edge computing can directly access devices, so it is easy to directly derive specific commercial applications. Integration and efficiency in edge computing distance is close, and the data filtering and analysis can be realized. With the real-time data, edge computing can process value data. On the other hand, edge computing having challenges including real-time data and collaboration data.

3 Information System Architecture Design Based on Edge Computing

According to the operational needs, the system dynamically connects various warning radar, reconnaissance satellite, aerial reconnaissance and message, image, video, electromagnetic. Depending on the supportive requirements, the information products are sent to the authorized users at different levels such as the command post according to the subscription and elationship formulated by the useras shown in Fig. 1.

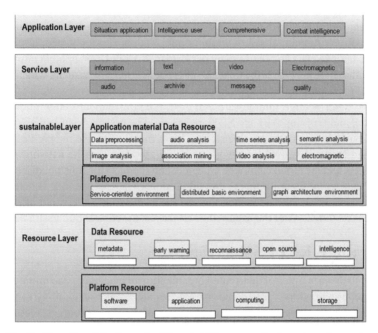

Fig. 1. Overall architecture of military intelligence analysis and service system

The support layer is the basic layer of the overall architecture providing a platform and business support environment for intelligence big data analysis and processing and service-oriented applications. It includes platform support and application support. The platform support part provides a platform environment for system construction and operation, including service-oriented support environment, data storage, distributed infrastructure, cluster computing environment and storm stream processing environment. The service-oriented support environment supports system development with a service-oriented architecture. The data storage module is used to support the storage and management of massive intelligence data resources. Storm big data processing frameworks provide a distributed parallel processing environment for massive big data. The application support part provides basic business support for the construction and operation of the system, and it provides common function module support for the service layer and application layer, including basic services such as data preprocess, image analysis, message analysis, audio analysis, video analysis, electromagnetic analysis, association mining, timing analysis, semantic analysis, knowledge reasoning and so on.

Application Layer is a cost effective edge computing gateway launched by inhand for the field of industrial device. With a variety of broadband services deployed worldwide, the product provides uninterrupted interconnection and connection available everywhere. It supports many mainstream industrial protocols. At the same time, it can connect with many mainstream cloud platforms so that field devices can be easily put into the cloud; It has an open edge computing platform, supports user secondary development, and realizes data optimization, real-time response and intelligent analysis at the edge of the Internet of things. The excellent product features, easy deployment and perfect remote management function help enterprises with digital transformation. It is used to transmit equipment or environmental safety warning information. If not avoided, it may lead to equipment damage, data loss, equipment performance degradation or other unpredictable results. As shown in the Fig. 1, the upper layer is application deployment, which is mainly responsible for deploying edge applications and creating an edge ecosystem of APP/vnf. The middle layer is edge middleware and API, creating standard edge platforms and middleware, and unifying API and SDK interfaces. The bottom layer is the layer which interfaces with the open source edge stack. This is mainly to solve the problem of weak network and restart. Even with network tunneling, the fact that the network instability of edge nodes and the cloud cannot be changed, and there is still constant disconnection. The edge autonomy function meets two edge scenarios. The network is disconnected between the center and the edge, and the service of the edge node is not affected. The edge node is restarted. After the restart, the services on the edge node can still be restored.

4   Characteristics Analysis Performance

According to the principles of distributed organization management and unified resource sharing, the system adopts distributed operation management technology to uniformly control information analysis tasks, computing power and data resources, realize collaborative scheduling according to information support requirements, and jointly complete information analysis tasks. Using the service-oriented architecture, the core intelligence analysis function, image intelligence analysis service, message intelligence service, open source intelligence analysis service and intelligence data service carries out unified classification management based on the service registration mechanism to form service resource directory. Realize the sharing of intelligence analysis function among nodes in the system.

Real time aggregation of trajectory data. At present, the terminal perceives the real-time access of collected data and comprehensively obtains all kinds of travel data. Established a special analysis model, it masters the trajectory of key areas, and realizes the real-time analysis, research and judgment of intelligence information. The platform includes visual intelligent track analysis and query, research and judgment analysis of abnormal activities, intelligent statistical analysis, dynamic monitoring, analysis and early warning, intelligent information retrieval and other functions which can produce obvious results in a short time.

Closed loop operation of early warning information. Early warning information is synchronously pushed to the public security organs in the control and early warning places, realizing information sharing, breaking the information barrier, and realizing the closed-loop operation of early warning, research and judgment, verification, feedback and other links. Focused on gathering and integrating all kinds of social data,it can play an important role in operations, intelligence research and judgment, carefully study the conversion and processing of all kinds of data, gives full play to the cross secondary comparison of data, and improves the effective utilization of data.

Early warning synchronous mining analysis. Analyze and mine the key tracks and key personnel in the same category and region, and provide stability control suggestions for intelligence work at all levels. The platform has realized the downward extension of system construction and the upward aggregation of data resources, forming a four-level information platform linkage application system; At the same time, it provides platform support for joint operations and cooperation. It provides a strong guarantee for synthetic operations.

5 Summary

This paper proposes an information system architecture based on edge computing. It introduces the advantages of each layer of the system. The system can better complete the cloud edge end collaborative network computing and solve the flow control layer by layer. Because the node location and end-to-node delay are divided into different levels, the traffic volume to be carried by nodes at different levels is different. The capabilities and technical points to be provided are also different. Edge computing needs to solve the following key problems: Resource management and protocol analysis: 1. provide the connection and communication between local devices, realize the local exchange of massive data, provide the ability to adapt and normalize different devices, shield the differentiation of industrial protocols. Storage and forwarding device can provide relatively complete functions of data acquisition, processing, analysis and alarm when the real-time requirements are high, the amount of data transmission is too large or the network connected to the platform is unavailable. At the same time, the local provides a certain storage capacity, which can forward the data to the platform during network recovery. Platform integration realizes comprehensive collaboration with the platform end, flexible data acquisition and distributed computing functions for the decision center at the platform end. It can support seamless running of applications and can be uniformly configured rather than manual compiling and developed programs.

References

1. Kang, Y., et al.: Neurosurgeon: collaborative intelligence between the cloud and mobile edge. In: ACM SIGARCH Computer Architecture News, vol. 45, no. 1, pp. 615–629 (2017)
2. Li, E., Zhou, Z., Chen, X.: Edge intelligence: on-demand deep learning model co-inference with device-edge synergy. In: Proceedings of the 2018 Workshop on Mobile Edge Communications, pp. 31–36 (2018)
3. Machen, A., Wang, S., Leung, K.K., Ko, B.J., Salonidis, T.: Live service migration in mobile edge clouds. Wirel. Commun. 25(1), 140–147 (2018)
4. Mahmud, R., Buyya, R.: Fog and Edge Computing: Principles and Paradigms, pp. 1–35 (2019)

5. Song, C., et al.: Hierarchical edge cloud enabling network slicing for 5G optical fronthaul. J. Opt. Commun. Netw. **11**(4), B60–B70 (2019)
6. El-Sayed, H., Sankar, S., Prasad, M., et al.: Edge of things: the big picture on the integration of edge IoT and the cloud in a distributed computing environment. IEEE Access **6**, 1706–1717 (2018)

A Multi-modal Seq2seq Chatbot Framework

Zhi Ji[(✉)]

The High School Affiliated to Renmin University of China, Beijing, China
`Matrixstroma@163.com`

Abstract. The pandemic has forced young people to stay away from school and friends, complete online learning at home and live at home. Therefore, various mental illnesses such as anxiety and depression occur more frequently. Chatbot is a communication method that is more acceptable to young people. This paper proposes a multi-modal chatbot seq2seq framework, which divides the mental state of young people into different types through multi-modal information such as text and images entered by users in the chatbot. This model combines image description and text summarization modules with the attention mechanism in a multi-modal model to control related content in different modalities. Experiments on multi-modal data sets show that this method has 70% average accuracy and real users who use this system also believe that this method has good judgment ability.

Keywords: Chatbot · Multi-modal · Seq2seq · Machine learning

1 Introduction

Before the outbreak of COVID-19, there were already many online psychotherapeutic applications, and these psychotherapeutic applications were initially consistent with the level of off-line therapy. And it also provides convenience, patients can use it at any time; at the same time, the protection of privacy makes more users willing to actively participate. But relevant doctors are still relatively slow in adopting these tools on a large scale. With the outbreak of COVID-19, medical departments around the world are under tremendous pressure for medical consultations. In fact, COVID-19 not only damages the health of patients, but also the mental health of others by the pandemic [1]. Not only patients and the elderly, many young people and even children also suffer from conditions such as fear, sadness and depression. Psychological trauma. As COVID-19 has caused quarantine and lockdowns in various places, people cannot meet with family and friends, further increasing the possibility of psychological trauma, making it possible for people who were originally normal and healthy to fall into mental illness, and at the same time they cannot realize that this is. A disease and not just an emotion.

These phenomena have led to a huge demand for online psychiatric outpatient systems, whose role is to relieve the pressure on outpatient clinics of medical institutions and provide contactless medical services. The online medical inquiry chatbot system based on artificial intelligence technology can provide online mental medical inquiry.

© The Author(s) 2022
Z. Qian et al. (Eds.): WCNA 2021, LNEE 942, pp. 225–233, 2022.
https://doi.org/10.1007/978-981-19-2456-9_24

Its key technology is the knowledge graph of the medical field. The system relies on entities in one or more fields and performs reasoning or deduction based on the spectrum. Answer the user's question.

The impact of the pandemic on the mental health of children and adolescents was showed in [2], particularly depression and anxiety. It first revealed that countries paid less attention to adolescents' mental health during the pandemic, listing an example of the reduction in beds in hospitals. It also illustrates the COVID-19 pandemic makes it harder to detect adolescent's abnormal behaviors by recommending a reduction in contacts and outdoor activities, leading to a decrease in the number of appointments. The level of anxiety becomes harder to assess, and adolescents get anxious more readily. And a solution is: to help patients with anxiety and depression online. It's proved to be useful to have internet-based care by randomized controlled trials, which provides a strategy for healthcare workers and patient's parents. Online resources like recorded courses, group treatments, and mental health apps provide direct access to instructions for children, which is better suited to the current situation than appointments. Parents far away from their children can have increase care for them and report the abnormalities to doctors, which is helpful to make a diagnosis. Finally, the paper suggests healthcare counselors demonstrate altruism in front of their patients, and stresses the importance of an optimistic mood in the treatment.

In these mentally ill groups, because they have to study online at home, they have broken away from the original traditional teaching mode and cannot have face-to-face communication with teachers and classmates. This has further increased the pressure on young people to study; in addition, due to the fact that they are in the family with their families. The time spent living together has increased, and the relationship between some teenagers and their families has become more tense, which has led to an increasingly serious problem of teenagers' psychological anxiety. At present, scales are commonly used in the evaluation of mental illness in hospitals, which is to evaluate patients through questionnaires. This method may be flawed in the evaluation of mental illness of young people, because compared to adults, young people may be more rebellious. When they are unwilling to undergo psychological tests, they may falsify answers or know how to get high scores based on experience, and avoid being judged. For mental illness.

This paper proposes one kind of chatbot method for the diagnosis of adolescent psychological anxiety. The chatbot model is based on a multi-modal seq2seq model, which is used to analyze the multi-modal interaction data such as text and image when the teenagers were using their chatbot. Experiments show that this structure could reach 71% training accuracy and 63% test accuracy on the existing multi-modal dataset. Preliminary real user tests show that it is correct on the psychological anxiety judging of 15–18 year-old teenagers.

2   Chatbot for Teenager's Depression

A study showed that the physical environments of house settings are more proximal to adolescents, and they have impacts on children's prefrontal cortex (PFC) growth which extends well in children's lives. SES (socioeconomic status) may be correlated to the physical environment of families. A hypothesis that a less-resourced environment leads to a thinner PFC has been made by the author's group [3]. The group conducted in-home interviews with testers, meanwhile scored items in their houses as environmental scores (PHYS test). Hazards, space, noise, cleanliness, interior and external environment are factors assessed. They also collected their brain scan images at UCLA. To make appropriate control of testers, the group tested the basic nurturance and stimulation of the child based on a scale from 0 to 10(SHIF test). All scores ranged from 7 to 10, which provided control of developmental contexts. Ethnicity, educational context, gender, and age are also tested. Cognitive test WRAT revealed the tester's reading, understanding, and math computation skills. The scores were reported as reading scores and mathematical computation scores. To test the relationship between SES and physical environment, the group asked for reports from families about their total income and family household sizes. Testers were divided into five groups based on their data of depth of poverty, and the group used income-to-needs ratios (INR) to report their economic status. The group finally compared MRI surface area maps of testers with standardized size maps to get the effect size value (standard deviation difference), and then used the value to get the conclusion of the thickness of PFC. The group finally used mediation analysis of PHYS, SHIF, WRAT scores, and INR to test their relationships. The comparison showed that adolescents whose parents had more incomes tended to have a better physical environment at home, and they had higher cognitive skills in math. PHYS and SHIF scores were directly proportional to the thickness of the left lateral occipital gyrus, and the WRAT score was positively associated with the thickness of the left frontal gyrus. After mediation analysis of whether PHYS can predict WRAT reading scores, the left superior frontal gyrus was the area associated with PHYS and WRAT reading scores. To sum up, the group concluded that the physical home environment determined the adolescent's reading achievement, and the thicknesses of middle and superior frontal gyri were negatively related to the number of physical problems in the home environments.

The mental health problems from six groups [4]: General population, healthcare personnel, college students, schoolchildren, Hospitality, Sport, and Entertainment industry employees, and others. A series of concerns lead to the abnormal mental health of the general population: Possible disease spread, fearless of ill, financial loss due to unemployment, the uncertainty of test results, and death of family members are all factors that lead to mental health problems in the general population. The healthcare personnel (front-line healthcare workers) experienced the highest level of anxiety and depression. Close contact with patients may make them the source of infection to family members. Intensive works and the possibility of an emergency made them nervous all the time. As a result, they were more likely to have developmental disorders. College students had concerns for their safety and the safety of their families during the pandemic, which led them to have mild anxiety. Lots of part-time jobs and the obstacle to have remote online classes also caused mental stress. The closure was the biggest problem for school children (primarily adolescents). Due to the pandemic, students were needed to stay at home to have online classes, and this led to a lack of activities, disrupted sleeping habits, and loss of resources. Students were struggled to study at home and developed lockdown situations, which were hard to adjust back to normal. For employees in the hospitality, sport, and entertainment industry, the economic strain was the primary reason that led to their stress. A ban on gathering would be a part of modern life after the pandemic, and this led employees to lose their jobs permanently. As a result, they would have mental health problems. As for vulnerable groups (Elderly people, homeless individuals, care homes residents), they already had some chronic diseases (mental disorders like bipolar and diseases like asthma) which made them more likely to get infected.

One research aims to find the reciprocal relationships between excessive internet use and school burnout [5]. The research first shows a school burnout that the engagement of students in Finland decreases because the classroom is in lacks digital devices. Students who used digital technologies felt bored. The school burnout was comprised of exhaustion, cynicism, and a sense of inadequacy. Compared to engagement, it predicts depressive symptoms. School engagement is defined by energy, dedication, and absorption. The research showed a method to increase engagement: Fulfill adolescent's socio-cognitive and emotional needs. School climates and motivation from others are also factors that lead to positive engagement. To start the research, 1702 elementary students were asked to answer a questionnaire about engagement, burnout, internet use, and depression at two different times. EDA, SBI, and DEPE depression Scales were tests that correspond to engagement, burnout, and depression respectively. SES and gender were additional measures. The results of the questionnaire showed that internet use and school burnout are reciprocal positive cross-lagged related. School burnout leads to excessive internet use and depressive symptoms. In components of school burnout, cynicism predicted later inadequacy and inadequacy predicted later cynicism. Exhaustion increased excessive internet uses. Study 2 focused on high school students instead. Using the same method in study 1, researchers found that girls suffered more from depression and school burnout, while boys were suspected of excess internet use. And, exhaustion was found to lead to an increase in internet use. The research showed that the negative attitudes of students may be formed at elementary school, which transformed into school burnout and thus led to excess internet use. About the solution, researchers ask people to promote students to have positive attitudes when they were young.

Fig. 1. Tess chatbot of a participant interacting [7].

An overview of the neurobehavioral changes during adolescence and the impacts of stressful environmental stimuli had on maturation was proposed in [6]. In the first category study, the researcher found that rodents had a higher level of anxiety-like behavior. Rodent's social abilities dropped and their aggression increased. In rats, researchers also observed significant depressive-like behavior, which included high immobility. A specific rodent, mice, formed a depressive-like phenotype when exposed to stress for 10 consecutive days, accompanied by anxiety and lower body weight gains. The social instability stress (1 h isolation per day and then live with a new roommate which PD value was 35 to 40) exerting on mice found that they were more sensitive to drugs like nicotine when they were adults. Additionally, the paper shows that social experiences influence drug-seeking behavior. The paper showed that stress reactivity, mineralocorticoid receptor expression, and glucocorticoid receptor expression changed significantly. In adulthood, HPA activity rose, and the reactivity to stressor increased. Above is the growth of the HPA axis in adolescence. Then, it discussed the impact of stress on the HPA axis growth. Social isolation caused lower corticosterone responses level to stress in adulthood-males, and females had more corticosterone responses. The study showed that adolescents were risk-takers at this time due to the imbalance in the growth of limbic and conical compartments. Immaturity of the cortical region led to novelty-seeking behaviors. And, adolescents were sensitive to rewards, which promote risk-seeking.

Chatbot is an application that can conduct text or voice conversations [7]. Studies have shown that the communication between users and chatbot is also very effective in providing psychological or emotional problems. Woebot is a chat bot that can conduct automatic conversations. While communicating emotions, it also tracks changes in emotions. Tess is an intelligent emotional chatbot, which is shown in Fig. 1, whose method is to find the user's emotion and provide solutions through dialogue with the user. In a study Tess provided emotional support to 26 medical staff, most of these users reported that Tess had a positive effect on their emotions. At the same time, Tess can also reduce the anxiety of many college student volunteers, and can even manage adolescents' depression-related physiological phenomena. The KokoBot platform is an interactive platform for evaluating cognitive abilities. The main feature is that it can conduct point-to-point interaction, and users on the platform can also communicate with other users. Wysa is an emotional intelligent mobile chatbot based on artificial intelligence. The goal is to assist mental health and relieve psychological stress through human-computer interaction. Vivibot's chatbot serves the mental reconstruction of terminally ill teenagers who are undergoing treatment. Pocket Skills is a conversational mobile phone chatbot, mainly responsible for behavior therapy.

3   Multi-modal Seq2seq Model

The information sources that humans interact with the outside world include tactile, auditory, visual, etc., and the resulting media used to carry information includes voice, image, video, text, etc., microphones, cameras, infrared, etc. are sensors responsible for collecting information. The combination of these diverse information can be called multi-modal information. A single modality often only carries the information of its own modality, which has certain limitations. The relationship between each modality can be fully studied through machine learning and other means. Multi-modal is also one of the current research hotspots. Multi-modal methods mainly include Joint Representations and Coordinated Representations.

Multi-modal methods mainly include Joint Representations and Coordinated Representations. As shown in Fig. 2, in the multi-modality, the text processing can use sentence summaries, the purpose is to use the seq2seq model to form short sentence content. In machine translation applications, multi-modality can also be used, and its effect is better than simply using a single text input, which means that images and text sentences need to be input at the same time, and the image needs to be able to describe the text sentence [8].

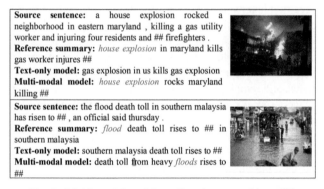

Source sentence: a house explosion rocked a neighborhood in eastern maryland , killing a gas utility worker and injuring four residents and ## firefighters .
Reference summary: *house explosion* in maryland kills gas worker injures ##
Text-only model: gas explosion in us kills gas explosion
Multi-modal model: *house explosion* rocks maryland killing ##

Source sentence: the flood death toll in southern malaysia has risen to ## , an official said thursday .
Reference summary: *flood* death toll rises to ## in southern malaysia
Text-only model: southern malaysia death toll rises to ##
Multi-modal model: death toll from heavy *floods* rises to ##

Fig. 2. Multi-modal model predicts the event objects [8].

The current multi-modal learning is generally based on the deep learning framework. The latest technology is mainly based on the BERT architecture. After pre-training by means of pre-train and transfer, it is applied to other tasks, such as image subtitles, etc. These tasks only require Minor changes [9].

This paper proposes a chatbot method for diagnosing the psychological anxiety of adolescents. The chatbot model is based on the multi-modal seq2seq model. The specific structure is shown in Fig. 3, where the image caption technology is used to extract the text description of the image at the front end of the model, and the attention mechanism is used in the multi-modal model to control the associated part of the image and text,

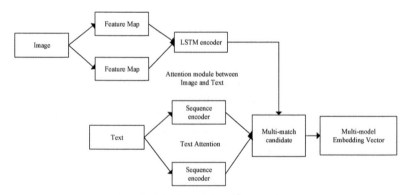

Fig. 3. Multi-modal seq2seq chatbot.

which is used to analyze the use of chat by teenager's multi-modal data such as text and images during the chatbot.

4 Experimental Result

In order to find the effectiveness of the structure proposed in this article, we selected part of the Microsoft COCO Caption data set [10] and LCSTS data set [11], which are merged with own chatbot image and text dataset and conducted training and testing. The user fills in the standard psychological scale as the ground truth of data. In evaluating the degree of user anxiety, we divide the degree of anxiety into 0–5 levels, which correspond to 0%, 0%–20%, 20%–40%, 40%–60%, 60%–80% and above 80% anxiety level of the user in the overall ranking.

Table 1. Comparison of the indicators on training dataset

Heading level	Precision	Recall	F1
TF-IDF decision tree	0.63	0.39	0.24
LSTM	0.69	0.30	0.21
Multi-modal Seq2seq	0.71	0.36	0.23

Table 2. Comparison of the indicators on testing dataset

Heading level	Precision	Recall	F1
TF-IDF decision tree	0.58	0.40	0.24
LSTM	0.61	0.38	0.23
Multi-modal Seq2seq	0.63	0.47	0.27

The experimental results are shown in Table 1. The results on the training set have an average accuracy of 71%; in the test, k-fold cross-validation is used for verification, and an average accuracy of 63% is obtained. In comparison, the results of TF-IDF Decision Tree on the training set are 63% average accuracy, and the results on the test set are 58% average accuracy; the results of LSTM are 69% training set average accuracy and 61% respectively. Average accuracy of the test set (Table 2).

Finally, five teenagers aged 15–18 years old were invited to test the chatbot. 3 of the 5 teenagers had a more anxious mental state. Using this chatbot, they obtained results consistent with their own cognition.

5 Conclusion

With the outbreak of COVID19, teenagers who study and live at home are more likely to suffer from mental illness and anxiety symptoms. This paper proposes a multi-modal chatbot scheme, which analyzes and judges the mental state of teenagers when they use chatbot through multi-modal information such as text and images. The model is a seq2seq model, which combines image text description extraction and text summarization modules, and uses an attention mechanism in a multi-modal model to control related content in different modalities, and is used to analyze text and images when teenagers use chat bots and other multi-modal data. Experiments show that this structure can achieve better accuracy on the existing multi-modal data set, and it has also received better feedback from real users.

References

1. Feijt, M., de Kort, Y., Bongers, I., Bierbooms, J., Westerink, J., IJsselsteijn, W.: Cyberpsychol. Behav. Soc. Netw. 860–864 (2020)
2. Courtney, D., Watson, P., Battaglia, M., et al.: COVID-19 impacts on child and youth anxiety and depression: challenges and opportunities. Can. J. Psychiatry **65**(10), 688–691 (2020)
3. Uy, J.P., Goldenberg, D., Tashjian, S.M., et al.: Physical home environment is associated with prefrontal cortical thickness in adolescents. Dev. Sci. **22**(6), e12834 (2019)
4. Khan, K.S., Mamun, M.A., Griffiths, M.D., et al.: The mental health impact of the COVID-19 pandemic across different cohorts. Int. J. Mental Health Addict. 1–7 (2020)
5. Salmela-Aro, K., Upadyaya, K., Hakkarainen, K., et al.: The dark side of internet use: two longitudinal studies of excessive internet use, depressive symptoms, school burnout and engagement among Finnish early and late adolescents. J. Youth Adolesc. **46**(2), 343–357 (2017)
6. Iacono, L.L., Carola, V.: The impact of adolescent stress experiences on neurobiological development. In: Seminars in Cell and Developmental Biology, vol. 77, pp. 93–103. Academic Press (2018)
7. Dosovitsky, G., Pineda, B.S., Jacobson, N.C., et al.: Artificial intelligence chatbot for depression: descriptive study of usage. JMIR Formative Res. **4**(11), e17065 (2020)
8. Li, H., Zhu, J., Liu, T., et al.: Multi-modal sentence summarization with modality attention and image filtering. In: IJCAI, pp. 4152–4158 (2018)
9. Moon, J.H., Lee, H., Shin, W., et al.: Multi-modal understanding and generation for medical images and text via vision-language pre-training. arXiv preprint arXiv:2105.11333 (2021)
10. Chen, X., Fang, H., Lin, T.Y., et al.: Microsoft coco captions: data collection and evaluation server. arXiv preprint arXiv:1504.00325 (2015)
11. Hu, B., Chen, Q., Zhu, F.: LCSTS: a large scale Chinese short text summarization dataset. arXiv preprint

The Research on Fishery Metadata in Bohai Sea Based on Semantic Web

Meifang Du[(✉)]

Shandong Technology and Business University, Yantai, China
8049870@qq.com

Abstract. In this paper, a data sharing and management mechanism suitable for the characteristics of fishery industry was established to clarify the phenomenon of heterogeneous Web data and Information Island based on Semantic Web technology, and unified interface specification information platform was established. Form the specification of metadata from the physical and chemical database, developing and publishing the corresponding metadata management tool, assisting, assisting and guiding a specialized database centre, completing the construction of metadata from the professional database.

Keywords: Metadata annotations · Web semantics · Fishing industry

1 Introduction

The Bohai Sea area is the key area of social and economic development in China. Development and use of fishery information resources in the Bohai Sea directly affects the social and economic development of the area. Currently, with the rapid development of fisheries economics, the investigation and scientific research of environmental resources in the surrounding waters of the Bohai Sea and it have accumulated rich basic data of various marine environment. These professional resources for fisheries information are distributed in maritime administrative departments, marine institutions at all levels, scientific research institutes and other services.

However, there are still many defects and deficiencies in the integration of marine fishery resources in China, such as the lack of a unified definition of basic information of fishery management; For the equipment used in construction, data resources cause fragmentation of data storage management at different levels of information technology development, and there are too many redundant data and inconsistencies. The level of data sharing cannot meet the requirements of the unit for the overall development and use of Information Resources. A large number of data does not provide a unified data interface, does not use general standards and specifications, cannot obtain A shared public data source, and is responsible for a large number of information islands.

The existence of these problems causes the management and value of Bohai Sea fisheries to be reduced, the quality of the use of increased costs, management cannot obtain effective support for decision-making data. Although the collection of maritime fishing information and statistical work have constituted an enormous database, MAS

© The Author(s) 2022
Z. Qian et al. (Eds.): WCNA 2021, LNEE 942, pp. 234–240, 2022.
https://doi.org/10.1007/978-981-19-2456-9_25

due to poor processing and analysis of information, not directly from the database system at various levels and from the collection of data and wide use.All this leads to the Marine Fishing Database system producing large amounts of data can not extract sublimated information in useful information to meet the needs of the managers, eventually making the level of use of the information resources is low, caused large amounts of waste.

2 The Research Contents

2.1 Metadata Annotations

Metadata is data about data, i.e. information on content, quality, status and other characteristics in the database (data attribute, data set or data warehouse, etc.). A semantic continuum is formed by the above classification.

2.2 Metadata Framework

Metadata can be one of two ways. One way is direct access to metadata, one type is to capture all types of database operation process of metadata. Set of metadata standards and specifications. In the process of database system operation can capture the metadata.

(1) Design is the designer and developer used to define metadata requirements, metadata requirements, and includes data model, business transformation work design.
(2) Physical metadata: use of tools to run establishment, management and access to metadata.
(3) Operational metadata: When carrying out data integration activities, operational metadata will tell users what will happen to change, especially about YOUR influence on how the Data Integration Source works.
(4) Project metadata: used to produce documents, audit development efforts, assign accountability and process change management issues. Guided: persons, responsible, tools, users and management operation.

The metadata database system can realize the following functions:

(1) Data entry: 1) Direct Input keypad. 2) included existing text files. 3) including the original scanning image.
(2) Preview and output: the information from the database, the results from the recovery of the query and the statistical analysis can be directly via the browser screen for a given form to the form, statement or graph of statistical analysis, Users allowed to Show changes in content can be submitted to the database server. At the same time, data can be directly through the printer output.
(3) Edit and modify: Authorised users can edit and change the information in the database. Modifying the general process is to extract MS according to the information state, information editing, Outcome of the presentation.

(4) Data recovery query, data consultation and recovery query refer to own metadata. Include a simple query, query consultation, merge query, Query and recovery results fulfil the conditions that will be shown As a query and data recovery. Details of the query data, respectively, using the format of the corresponding navigation. In no leak and under the premise of protection of intellectual property rights, For some data You can provide direct online download services. Data and information download will adopt the corresponding file download directly.

(5) User access control and monitoring of user information: according to different types of users, determine user permissions for the operation of the network database, the FIM to ensure the safe operation of the system. To record user information for tracking.

(6) System management functions: system administrators and data managers to maintain, update, system to manage the user, database registry can be added, deleted, modified, edited operation, etc.

2.3 Topic-Oriented Meta Database Building

An important step in building the Subject-oriented Meta Database is to establish the Bohai Fishery Theme Management model and obtain the modeling of metadata according to the subject model.

The modeling process of the subject is shown in Fig. 1. The Subject model is obtained from the existing business model by Specialist Persons, which can be divided into fisheries management specialists, data analysts and software developers.

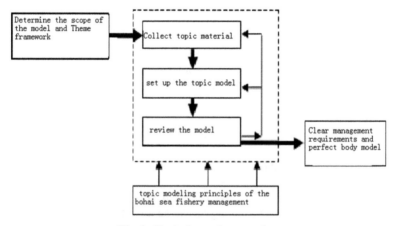

Fig. 1. Topical metadata template

2.4 Metadata Resource Query Algebra System

Logical calculation and query algebra is the basis of the query of data. In a relational database theory, the expression of relational calculation of security is an important problem. If a query expression cannot be evaluated within finite steps, and obtain a finite set the results of this expression are referred to as the safe. Otherwise, the expression is safe. In Metadata Resources management, there are such problems.

Logical calculation and query algebra are the basis of the query of data. In the traditional theory of the relational database, relational calculus expression is an important safety problem. If the query expression cannot be evaluated in limited steps and the result set is limited, the expression is called safe expression. Otherwise, the expression is safe. There are such problems in the management of metadata resources.

Research of question algebra plays an important role in the field of data management. Common operational semantics are used to compare query definitions, query optimizations, and query capabilities for query languages. In relational databases, Codd has proposed a relational algebra that has constructed a theoretical foundation for the success of relational algebra. In the data model research, query algebra has become a part of the data model for the past decade. Whether there is a corresponding algebraic system, whether the data model that studied the XML data model of the object oriented query algebra model and the query algebra system is an important symbol of maturity.

3   Research Methods

3.1   Metadata Standards Set

Standard procedures are divided into nine stages: preliminary stage, project stage, draft form, opinion, review, approval, release stage, examination stage, and abolition stage.

Standard specification description elements:

Standard No. of China

Standard Title in China

Standard Title in English

……

Governer Code

Drafting Committee

3.2   Research on Metadata Semantic Model

Characteristics of metadata are analyzed generally. The metadata format is complex. In addition to the simple format of the data dictionary, there are many complex levels, and the metadata format is changeable. In general, read only is used during system operation. Metadata is usually used scatter with cross platform and cross process characteristics.

In order to share data and resources, it is more complicated to organize model data and fields, and to simplify the model of relational data resources provided by different organizations and to model metadata models. Obviously, conventional object oriented models cannot achieve this goal. Figure 2 shows mapping relationships between metadata and domain ontology.

The semi-automatic semantic association framework between heterogeneous data sources is shown in Fig. 3. The framework takes as input semi-structured documents in the database (Web page XML documents, etc.) and unstructured documents such as this document. Through the shallow natural language processing, such as (Chinese word

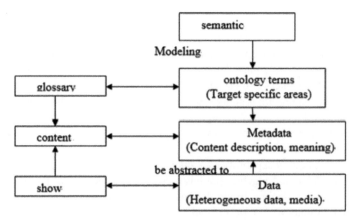

Fig. 2. Demand of fishery management resource modeling in Bohai sea

segmentation except stop words, part of speech tagging, key phrase identification, entity noun identification, etc.), vectorization is carried out. Then machine learning and data mining methods are used to analyze the semantic relationship of the implied concepts.

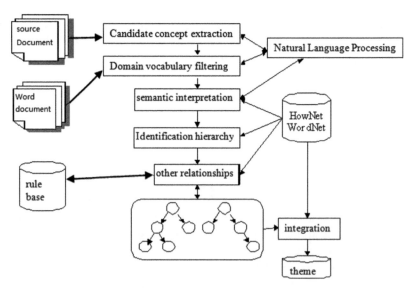

Fig. 3. Semi-automatic semantic association framework

3.3   Establishment of Meta Database System for Fishery Management in Bohai Sea

Includes the use and management of metadata, the metadata database system must record the following information:

(1) The type of it ?
(2) Where is it?
(3) From where?
(4) What is it related to?
(5) Who is responsible for it?
(6) What terms, vocabularies, and business domains are associated with it?
(7) What will be the impact of any changes to it?
(8) What will be their properties and relationships when they are exported to another tool?

3.4 Constructing the Maintenance and Renewal Management Mechanism of Fishery Management Metadata Resources in the Bohai Sea

Database development technology includes database management technology and database online publishing technology.

There are many database management systems to choose from, such as Sybase SQL Server, Informix SQL Server, Oracle SQL Server and so on. The system development can use Microsoft SQL Server as the database management software on the Server, because the advantages of Microsoft SQL Server can be reflected in the following aspects: Perfect combination with the operating system, the use of Windows security mechanism and their own security mechanism combined, with safe and reliable performance; Large data volume support; Concurrency control, automatic backup; With the good combination of development tools, using VC, VB, InterDev, PowerBuilder and so on can be very convenient in SQL Server platform for database application development.

4   Conclusions

This paper can provide a unified standard for different Marine fisheries departments to add established fishery databases to the information platform. Formulate metadata specifications of physical and chemical databases, develop and release corresponding metadata management tools, assist and guide all professional data centers to complete the metadata construction of their professional databases.

Developed a high availability and high efficiency data application service system platform based on meta-directory, established data input, collection, management, inquiry, and the corresponding authority management mechanism. Realize unified management and service provision of existing scattered data through advanced metadata directory technology.

Through the metadata management control, the database management system to achieve dynamic database loading, when the structure of the data changes, can be achieved by modifying and maintaining the metadata directory library, and the corresponding data application system without reconstruction. Therefore, the system has good versatility and is easy for scientific and technical personnel to master and use. It provides an ideal soft environment for the retrieval and management of various scientific and technical data. Its application has important theoretical and practical significance.

References

1. Arenas, M., Gottlob, G., Pieris, A.: Expressive languages for querying the semantic web. ACM Trans. Database Syst. **43**(3), 1–45 (2018)
2. Siddiqui, I.F., Lee, S.U.-J.: Access control as a service for information protection in semantic web based smart environment. J. Korean Soc. Internet Inf. **17**(5), 9–16 (2016)
3. Augusto, L., Carvalho, M.C., Garijo, D., Medeiros, C.B., Gil, Y.: Semantic software metadata for workflow exploration and evolution. In: 2018 IEEE 14th International Conference on e-Science (e-Science), vol. 1, pp. 431–441 (2018)
4. Shirgahi, H., Mohsenzadeh, M., Haj Seyyed Javadi, H.: Trust estimation of the semantic web using semantic webclustering. J. Exp. Theor. Artif. Intell. **29**(3), 537–556 (2017)
5. Strobin, L., Niewiadomski, A.: Linguistic summaries of graph datasets using ontologies: an application to semantic web. J. Intell. Fuzzy Syst. **32**(2), 1193–1202 (2017)

Design of Portable Intelligent Traffic Light Alarm System for the Blind

Lili Tang[(⊠)]

College of Computer and Information Engineering, Zhixing College of Hubei University,
Wuhan 430011, People's Republic of China
toney2001@126.com

Abstract. The system is composed of STC single chip microcomputer, color signal recognizer and sensor control module, wireless communication control module, voice and video synthesizer and broadcast control module. STC MCU adopts STC89C52; color recognition sensor module uses gy-33 color recognition sensor, which can identify the current traffic light conditions; wireless communication module uses nRF24L01 made by Nordic company, which needs to be installed at the sending end and the receiving end to send the current traffic light information; the speech synthesis broadcasting module uses the TTS speech synthesis broadcasting module xfs5152ce of iFLYTEK, after data recognition and analysis, it finally sends voice alarm about traffic lights to the blind, so as to effectively guide the blind whether it can pass through, so as to ensure the safety of the blind. This design combines artificial intelligence with daily life, which not only meets the development trend of the information age, but also meets the needs of the current society. It has a broad market prospect in the application of intelligent travel.

Keywords: Hand held · Intelligent alarm · Real time remote monitoring · Travel of the blind · Artificial intelligence

1 Introduction

Nowadays, the number of blind people in China is the largest in the world, with more than 6 million blind people. Visual barriers seriously affect the blind people's access to information and perception of the environment, making it impossible for them to travel normally, even in places they often visit and familiar environment, There are also all kinds of stumbling, let alone never set foot in the place, so if you want to go to a completely strange, never crossed street, but because you can't get real-time road conditions, then their travel safety is difficult to achieve even the lowest guarantee, it's just like this, many blind people don't want to go out of the house, so they have no way to better integrate into the society and achieve their goals The value of life, which is a pity for the blind, is the loss of national and social resources, so it is urgent to effectively help the blind travel safely and normally [1].

The intelligent traffic light alarm system for the blind is designed to solve the problem of blind travel. It takes the single-chip microcomputer as the central controller, as the

© The Author(s) 2022
Z. Qian et al. (Eds.): WCNA 2021, LNEE 942, pp. 241–249, 2022.
https://doi.org/10.1007/978-981-19-2456-9_26

data collection terminal, identifies the traffic lights through the color recognition sensor, monitors the status of the traffic lights in real time, and transmits the information to the single-chip microcomputer. After data recognition and analysis, it finally identifies the blind with hardware modules such as voice synthesis broadcast module Voice warning [2].

2 Overall Design Scheme

The main body of this design is composed of two parts: the sender and the receiver. STC MCU module and wireless communication module are common at both ends of the transceiver. MCU module is used to collect data, and wireless communication module makes the sender and receiver communicate. The color recognition sensor module is unique to the transmitter, through this module to identify the traffic lights, the data information will be transmitted to the MCU. The receiving end analyzes and synthesizes the data received by MCU through its unique voice synthesis broadcast module, and finally completes the voice alarm for the blind. The general design scheme is shown in the figure below (Fig. 1).

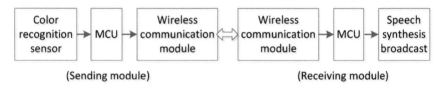

Fig. 1. The overall scheme design

Among the above two terminals, the transmitter needs at least one single chip micro-computer to collect and monitor the traffic light information in real time; one red, one yellow and one green LED light and three buttons to correspond with each other one by one to simulate the operation of road traffic lights; a wireless communication module [3] as the communication transmitter; at least one color recognition sensor to identify the color of LED lights, So as to judge the current traffic light situation. The receiver needs at least one single chip computer to receive and monitor the traffic light information; it needs a wireless communication module as the receiver to communicate; it needs a voice synthesis broadcast module [4] to process the received traffic light information, and finally broadcast it through voice synthesis.

2.1 Software Design of Transmitter

The function of the sender is to identify the traffic lights at the intersection through the color recognition sensor, and transmit the traffic light information to the MCU. When the judgment data is received, the information is transmitted to the receiver through the wireless communication module. The software design of the transmitter is as follows (Fig. 2).

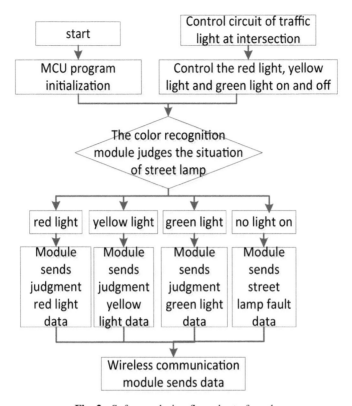

Fig. 2. Software design flow chart of sender

The function realization in the figure above is mainly completed by two processes, which complement each other. The first core task of the process is to complete the identification of the traffic lights at the intersection, mainly through the three primary colors principle in gy-33 module [5, 6]; the second core task of the process is to complete the judgment of the traffic lights at the intersection (red light, yellow light, green light or street light fault), select the current working mode of the street light, and complete the wireless communication with the receiver module.

2.2 Software Design of Receiver

The function of the receiver is to receive the traffic information transmitted by the sender through the wireless communication module, and send the traffic information to the MCU. After the MCU judges whether the data is red, yellow, green or no light, it sends the information to the speech synthesis broadcast module, and finally broadcasts the situation of the intersection to the blind, telling them whether they can pass at this time. The software design of the receiver is shown in the figure below (Fig. 3).

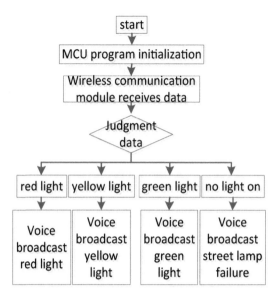

Fig. 3. Software design flow chart of receiver

3 Design Features and Extension Description

3.1 Feature Introduction

This design is based on color recognition sensor, voice synthesis broadcast, wireless communication and MCU technology, combined with social phenomenon and demand, as well as new concept innovation. Whether from the selection of single chip micro-computer, different module selection and communication protocol scheme, or from the sender to the receiver, it is very different from the existing blind products in the market. This design uses today's most common processor to complete an unusual design. Its characteristics are summarized as follows:

(1) The color recognition module identifies the current traffic lights.
(2) The sending end can collect and monitor the current traffic lights in real time through MCU.
(3) The communication between transmitter and receiver can be completed by wireless communication module.
(4) The receiver can receive the current traffic light information.
(5) The receiving end can transmit the current traffic light information to the speech synthesis broadcast module through the single chip microcomputer.
(6) The current traffic light information can be intelligently broadcast to the blind through the speech synthesis broadcast module.

Among them, the communication mode of this design uses the enhanced short burns protocol [7–9] of n0rdic company, as shown in the following Table 1.

Table 1. Enhanced short burns protocol form

Classification data	Sender data type(uchar)	Receiver data type(uchar)	Explanation
	0xAA	0xAA	Received data 0xAA, indicating that the current status is red
	0xBB	0xBB	Received data 0xBB, indicating that the current status is green
	0xCC	0xCC	Received data 0xCC, indicating that the current status is yellow
	0xDD	0xDD	Received data 0xDD, indicating street lamp maintenance failure

3.2 Extended Description

The intelligent traffic light alarm system for the blind can not only complete the functions described above, but also expand the following functions:

(1) Real time monitoring the current traffic light information and the location of the blind through the mobile App.
(2) It can be used together with relevant map navigation software to intelligently broadcast traffic lights during navigation.
(3) The color recognition sensor can recognize traffic lights accurately and quickly.
(4) It can realize long distance wireless communication.

4   Scheme Difficulties and Key Technologies

The difficulties of this design are as follows:

(1) When the sender identifies the traffic lights at the intersection, it is easy to be affected by the surrounding environment, which leads to the recognition of the traffic light color is not fast and accurate enough.
(2) The wireless communication module has a certain distance limit. If the transmission distance exceeds a certain range, wireless communication can not be realized, and the wireless communication module is installed at every traffic light intersection, which costs a lot of manpower and material resources in the early stage.
(3) The circuit diagram and program design of receiver and transmitter.

The key technologies are as follows:

(1) Gy-33 program modularization writing.

(2) The sender software is written.
(3) The software of receiver is written.
(4) Enhanced short burns communication protocol setting

5 System Simulation and Result Analysis

5.1 Overall Appearance of Intelligent Traffic Light Alarm System

The appearance design of the intelligent traffic light alarm system for the blind is shown in the figure. The whole design is divided into two parts: the sender and the receiver. The transmitter includes STC89C52 MCU, gy-33 color recognition sensor and nRF24L01 wireless communication module. The receiver includes nRF24L01 wireless communication module, STC89C52 MCU and xfs5152ce voice synthesis broadcast module (Fig. 4).

Fig. 4. Physical picture of intelligent traffic light alarm system for the blind

5.2 Overall System Debugging

The debugging of the blind intelligent traffic light alarm system includes the debugging of the sender and the receiver. Among them, the overall debugging of this design also includes: traffic lights, color recognition sensor module, wireless communication module, intelligent recognition street lights, voice report debugging, etc.

Speech Synthesis Debugging. Install the USB to TTL driver "ch340_341_32-bit.rar" or "ch340_64.rar" according to whether the computer system is 32-bit or 64 bit. After installing the driver, insert the USB-TTL module into the computer, open "my computer", find the "device manager" in the "device" option, click "com and LPI port", and then compare it with ch340. Open the "xfs5152ce PC demonstration tool" software, select the required port, write the required Chinese characters in the sent text, and then click "start synthesis" to synthesize the voice.

Wireless Communication Debugging. If the functions of interrupt request (IRQ) and acknowledgement character (ACK) can be realized at the same time, after the communication is successfully completed: for the receiving node, the effective data that can be recognized as successfully received through the enhanced ShockBurst protocol is IRQ = 0; For the transmitting node, the received ACK = IRQ = 0 is returned by the receiving node (Fig. 5).

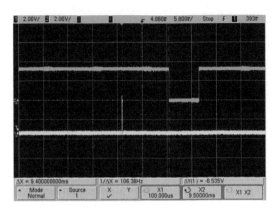

Fig. 5. Configuration process of CE and IRQ signals

In the figure, after CE (yellow signal) = 1, about 10ms, that is, after the number of transmissions reaches the maximum upper limit, IRQ (green signal) = 0. There are two possibilities for this situation: the configuration of the transmitting node is inconsistent with that of the receiving node (the bytes or frequencies transmitted and received are different); There is no receiving node (Fig. 6).

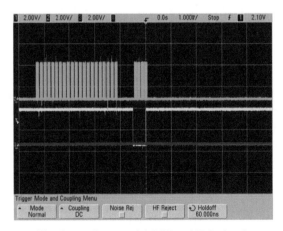

Fig. 6. Send successful SCK and IRQ signals

It can be seen from the figure that after sending the last SCK (green signal) signal of the first batch, IRQ (yellow signal) = 0 after 1ms at most (Fig. 7).

The logic shown in the figure above is as follows: Ce (purple signal) = 1. At this time, the transmitting node just completes the signal configuration process. Under different

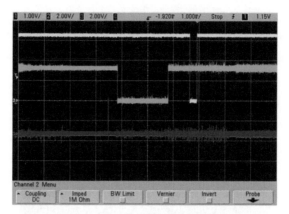

Fig. 7. SCK, IRQ, CE signal configuration process

communication conditions, the phase of IRQ (green signal) of receiving node and IRQ (yellow signal) of transmitting node will also be different. For the above reasons, the ACK signal needs to be sent by the transmitting end for many times before the receiving end can receive it successfully.

Intelligent Broadcast Traffic Light Test. Connect the power supply of sending end and receiving end, turn on the red light, yellow light and green light in turn, and place the color recognition sensor module above the LED. If the voice broadcast information is consistent with the street light, the system works normally.

6 Conclusion

After many times of program modification and system debugging, the design of the intelligent traffic light alarm system for the blind is completed, and all the expected functions can be achieved. The color recognition module, wireless communication module and voice broadcast module are all normal. The recognition accuracy of traffic lights, the agility of wireless communication and the accuracy of voice broadcast all meet the expected requirements. The significance of this design is to integrate the intelligent traffic light alarm system into the actual situation of social life, which can effectively solve the problem of blind travel. It is a major trend of social development, and also the aspiration of the people.

Acknowledgments. In this paper, the research was sponsored by the Science and Technology Research Program of 2021 Hubei Provincial Education Department (Project No: B2021410).

References

1. Lu, H.L.: Exploring the best way to guide the blind to travel. China Disabled **1**, 46–47 (2019)
2. Zhao, N., Luo, S.S.: Application status and key technologies of artificial intelligence. J. China Acad. Electron. Sci. **12**, 590–592 (2017)
3. Chen, C., Li, R.X., Liu, T.T.: Research on wireless data transmission system based on nRF24L01. Electron. Sci. Technol.**29**, 22–24, 27 (2016)
4. Ren, S.Y.: Research on speech reminder based on speech synthesis. Commun. World **9**, 258–259 (2018)
5. Long, J.P., Han, L.: An online led detection method based on color sensor. Mach. Tool Hydraulics **11**, 30–35 (2016)
6. Stiglitz, R., Mikhailova, E., Post, C., et al.: Soil color sensor data collection using a GPS-enabled smartphone application. Geoderma **296**, 108–114 (2017)
7. Li, J.D., Xiao, W.J., Liu, W.S.: Design of microgrid communication architecture based on nRF24L01 and Ethernet. Electron. World **11**, 182–183 (2017)
8. Izumi, S., Yamashita, K., Nakano, M., et al.: Normally off ECG SoC with non-volatile MCU and noise tolerant heartbeat detector. IEEE Trans. Biomed. Circ. Syst. **9**, 641–651 (2017)
9. Heriansyah, H., Nopriansyah, A.R., Istiqphara, S.: Evaluasi Kinerja testbed routing protocol berbasis node MCU ESP8266 pada Perangkat IoT. MIND J. **5**, 135–148 (2021)

Multi-objective Reliability Optimization of a Pharmaceutical Plant by NSGA-II

Billal Nazim Chebouba, Mohamed Arezki Mellal$^{(\boxtimes)}$, and Smail Adjerid

LMSS, Faculty of Technology, M'Hamed Bougara University, Boumerdes, Algeria
mellal.mohamed@gmail.com

Abstract. This work addresses the use of a MO optimization algorithm to deal with the reliability optimization problem in order to determine the redundancy and reliability of each component in the system. Often, these problems are formulated as a single-objective problem with mixed variables (real-integer) and is subject to various design constraints. Classical solution approaches were limited to deal with these problems and most recent solution approaches are based on nature-inspired optimization algorithms which belong to artificial intelligence (AI). In the present paper, the problem is solved as a MO optimization problem through the Non-dominated Sorting Genetic Algorithm II (NSGA-II) to generate the set of optimal solutions, also called Pareto. The latter helps the decision-maker. The case studied consists of a pharmaceutical plant.

Keywords: Reliability · MO optimization · Genetic algorithms · NSGA-II

1 Introduction

Industry 4.0 involves high-tech systems and requires reliable subsystems to meet the requirements of the companies. Reliability of systems belongs to dependability studies. By definition, the reliability is the ability of an item to perform given functions during a given period time and under given conditions. A system with high-level reliability should be investigated at the design stage by resorting to various methods, notably adding identical and/or different redundant components that perform the same functions, increasing the component reliability, or both options a mixture. The problem is described by a non-linear optimization problem [1]. These problems are hard to solve due to the complexity, nonlinearity, high computational time, and finding the optimal solutions. Therefore, various methods of artificial intelligence (IA), notably nature-inspired algorithms, have been proposed to solve these problems. During the last decades these algorithms have been widely used and proven their effectiveness in solving various problems.

The paper aims to implement a MO optimization algorithm (namely the NSGA-II) to deal with the reliability optimization problem to reach the highest reliability level at the lowest cost under the design constraints of space, weight, and cost.

2 Problem Description

The MO reliability optimization problems are mainly described as [2, 3]:

© The Author(s) 2022
Z. Qian et al. (Eds.): WCNA 2021, LNEE 942, pp. 250–256, 2022.
https://doi.org/10.1007/978-981-19-2456-9_27

2.1 Reliability Allocation

$$\text{Maximize } R_S(r) = R_S(r_1 r_2, \ldots, r_m)$$
$$\text{Minimize } C_S(r) = C_S(r_1 r_2, \ldots, r_m) \tag{1}$$

Subject to

$$g_j(r_1, r_2, \ldots, r_m) \le b$$
$$0 \le r_i \le 1; \; i = 1, 2, \ldots, m \tag{2}$$
$$r \subset \mathbb{R}^+$$

where $R_S(\cdot)$ and $C_S(\cdot)$ are the system reliability and cost, $g(\cdot)$ is the set of constraints, r_i is the component reliability, m is the number of subsystems, and b is the vector of limitations. This problem involves real design variables only.

2.2 Redundancy Allocation

$$\text{Maximize } R_S(n) = R_S(n_1 n_2, \ldots, n_m)$$
$$\text{Minimize } C_S(n) = C_S(n_1 n_2, \ldots, n_m) \tag{3}$$

Subject to

$$g_j(n_1, n_2, \ldots, n_m) \le b$$
$$0 \le n_i \le n_{i\,max}; \; i = 1, 2, \ldots, m \tag{4}$$
$$n_i \in \mathbb{Z}^+$$

where n_i is the number of redundant components. This problem involves integer design variables only.

2.3 Reliability-Redundancy Allocation (RRAP)

$$\text{Maximize } R_S(r, n) = R_S(r_1 r_2, \ldots, r_m; \; n_1 n_2, \ldots, n_m)$$
$$\text{Minimize } C_S(r, n) = C_S(r_1 r_2, \ldots, r_m; \; n_1 n_2, \ldots, n_m) \tag{5}$$

Subject to

$$g_j(r_1, r_2, \ldots, r_m; \; n_1, n_2, \ldots, n_m) \le b$$
$$0 \le r_i \le 1; \; 0 \le n_i \le n_{i\,max}; \; i = 1, 2, \ldots, m \tag{6}$$
$$r \subset \mathbb{R}^+, n \in \mathbb{Z}^+$$

The values of R_s and C_s are given in the Pareto front [4].

3 NSGA-II

The NSGA-II has been proposed in [4]. It is the MO version of the genetic algorithms which is inspired by nature evolution. It has been successfully implemented to solve many problems, such as design optimization, energy management, and layout problems.

Algorithm 1 illustrates the pseudo-code of the NSGA-II implemented in the present paper.

Algorithm 1. Pseudo-code of NSGA-II [4].

- M: population size
- N: archive size
- t_{max}: max number of generations
- **Begin**
- Initialize P_A^0 randomly, set $P^0 = \emptyset$, $t=0$.
- **While** t $< t_{max}$
- $P^t = P^t + P_A^t$
- Assignment of adaptation to P^t
- $P_A^{t+1} = \{N$ best individuals from $P^t\}$
- MP (mating pool) $= \{$select M individuals randomly from P_A^{t+1} by appliying a binary tournament$\}$
- $P^{t+1} = \{$generate M new individuals$\}$
- $t=t+1$
- **Output**
- Generate non-dominated solutions from P_A^t

Constraint Handling

In the literature, many techniques were developed to deal with the constraints. To handle the design constraints (resource limitation), the penalty function method is adopted in the present paper [5]. The constraints are introduced to the objective function using penalty terms. Therefore, the MO RRAP becomes as follows:

$$Fitness_1 = -R_S(r_1, r_2, \ldots, r_m, n_1, n_2, \ldots, n_m) + \psi(r_1, r_2, \ldots, r_m, n_1, n_2, \ldots, n_m) \tag{7}$$

$$Fitness\_2 = C_S(r_1, r_2, \ldots, r_m, n_1, n_2, \ldots, n_m) + \psi(r_1, r_2, \ldots, r_m, n_1, n_2, \ldots, n_m) \tag{8}$$

where $\psi(r_1, r_2, \ldots, r_m, n_1, n_2, \ldots, n_m)$ is the penalty function, calculated as follows:

$$\psi(r_1, r_2, \ldots, r_m, n_1, n_2, \ldots, n_m) = \sum_{j=1}^{M} \phi_j \cdot \max(0, g_j(r_1, r_2, \ldots, r_m, n_1, n_2, \ldots, n_m))^2 \tag{9}$$

where ϕ_j are the penalty factors (constant values). The values of these factors are fixed after several tests.

4 Numerical Case Study

The investigated case study consists of a pharmaceutical plant (see Fig. 1). The NSGA-II including the constraint handling described in Sect. 3 is used to solve this problem.

This pharmaceutical plant involves ten subsystems connected in series [6]. The raw material is transferred from a subsystem to another one till the end of the production line, chronologically.

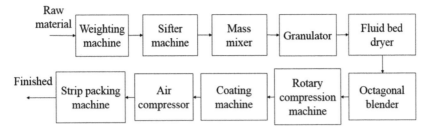

Fig. 1. Pharmaceutical plant

The MO RRAP of this pharmaceutical plant is given as follows:

$$\text{Maximize } R_S = \prod_{i=1}^{10} [1 - (1 - r_i)^{n_i}]$$

$$\text{Minimize } C_S = \sum_{i=1}^{10} C(r_i)(n_i + exp(\tfrac{n_i}{4}))$$

(10)

Subject to

$$g_1(r, n) = \sum_{i=1}^{5} C(r_i)(n_i + exp(\tfrac{n_i}{4})) \leq C$$

$$g_2(r, n) = \sum_{i=1}^{10} v_i n_i^2 \leq V$$

(11)

$$g_3(r, n) = \sum_{i=1}^{10} w_i(n_i * exp(\tfrac{n_i}{4})) \leq W$$

$$0.5 \leq r_i \leq 1 - 10^{-6}, \ r \subset \mathbb{R}^+$$

$$1 \leq n_i \leq 10, \ n \subset \mathbb{Z}^+$$

$$0.5 \leq R_S \leq 1 - 10^{-6}$$

where $C(r_i) = \alpha_i(-T/\ln r_i)^{\beta_i}$ is the cost of the component at subsystem i, T is the mission time, w_i is the weight of the component at subsystem i. C, V, and W are the limits of cost, volume, and weight, respectively.

In [5, 7], the problem has been investigated as a single-objective problem by taking the overall reliability as a target. Data of this system are given in Table 1.

Table 1. Data of the system [5, 7].

Subsystem i	$10^5\,\alpha_i$	β_i	v_i	w_i	V	C	W	$T(h)$
1	0.611360	1.5	4	9	289	553	483	1000
2	4.032464	1.5	5	7				
3	3.578225	1.5	3	5				
4	3.654303	1.5	2	9				
5	1.163718	1.5	3	9				
6	2.966955	1.5	4	10				
7	2.045865	1.5	1	6				
8	2.649522	1.5	1	5				
9	1.982908	1.5	4	8				
10	3.516724	1.5	4	6				

5   Results and Discussion

The implemented NSGA-II with the constraint handling was implemented using MAT-LAB and run on a PC with Intel Core I7 (6 GB of RAM and 2.20 GHz) under Windows 7 of 64 bits. The parameters of the implemented NSGA-II are given in Table 2. These parameters were carefully fixed after several simulations.

Table 2. Parameters of the implemented NSGA-II.

Parameters	Values
Population	100
Crossover	0.7
Offspring	2 * round(pCrossover * nPop/2)
Mutation	0.4
Mutants	round(pMutation * nPop)
Mutation	0.02
Mutation step	0.1 * (VarMax - VarMin)

Figure 2 shows the obtained Pareto front for the tradeoff between the system reliability and system cost. It can be observed that the redundancy and reliability of the components which give high reliability increases the cost, i.e., highest system reliability is more expensive. Each point corresponds to an optimal number of redundant components and the corresponding reliabilities. The solutions of the Pareto front are optimal and the decision-maker can choose a specific solution after deep further investigations based on the main target.

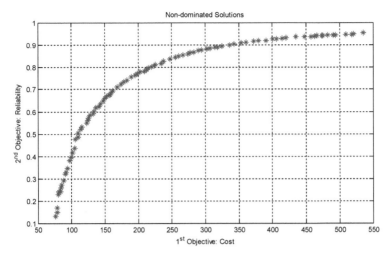

Fig. 2. Pareto front

6 Conclusions

MO optimization problems are complex problems that need strong solution approaches. Artificial intelligence has contributed by proposing nature-inspired optimization algorithms which can tackle these problems. This paper addressed the MO RRAP through a pharmaceutical plant as a case study. The NSGA-II has been implemented to deal with the problem and the penalty function has been used to handle the constraints. The results obtained have been given in a Pareto front that helps the decision-maker choosing an adequate solution. Future works will focus on an approach allowing to consider the constraints as other objectives.

References

1. Kuo, W.: Optimal Reliability Design: Fundamentals and Applications. Cambridge University Press, Cambridge (2001)
2. Hsieh, Y.-C., Chen, T.-C., Bricker, D.L.: Genetic algorithms for reliability design problems. Microelectron. Reliab. **38**, 1599–1605 (1998)
3. Xu, Z., Kuo, W., Lin, H.H.: Optimization limits in improving system reliability. IEEE Trans. Reliab. **39**, 51–60 (1990)
4. Deb, K., Pratap, A., Agarwal, S., Meyarivan, T.: A fast and elitist multiobjective genetic algorithm: NSGA-II. IEEE Trans. Evol. Comput. **6**, 182–197 (2002)
5. Mellal, M.A., Zio, E.: A penalty guided stochastic fractal search approach for system reliability optimization. Reliab. Eng. Syst. Saf. **152**, 213–227 (2016)

6. Garg, H., Sharma, S.P.: Multi-objective reliability-redundancy allocation problem using particle swarm optimization. Comput. Ind. Eng. **64**, 247–255 (2013)
7. Garg, H., Sharma, S.P.: Reliability-redundancy allocation problem of pharmaceutical plant. J. Eng. Sci. Technol. **8**, 190–198 (2013)

Construction of SDN Network Management Model Based on Virtual Technology Application

Zhong Shu[1], Boer Deng[1], Luo Tian[1], Fen Duan[1], Xinyu Sun[1], Liangzhe Chen[1(✉)], and Yue Luo[2]

[1] Jingchu University of Technology, Jingmen 448000, Hubei, China
chen_lz1991@jcut.edu.cn
[2] Jingmen Mobile Media Co. Ltd, Jingmen 448000, Hubei, China

Abstract. This paper designs a virtual SDN network management model constrained by fair and equal network management information access mechanisms by analyzing the problems existing in the universality of existing SDN network management models. Starting with the three-tier structure of the SDN network management system, the main parameters involved in the network management service function, information processing and transmission channel construction in the system were strictly and normatively defined. The design of virtual nodes is regarded as the core element of the network management system, and the information transmission inside it adopts logical operation; The network management service function and the channel for realizing the network management service function are isolated, and the iterative search, analysis and update mechanism is enabled in the network management information transmission channel. By constructing the experimental verification platform and setting the evaluation parameters of the system performance objectives, the scalability and timeliness of the model were evaluated from two aspects: the deployment of network virtual nodes and the dynamic control of network management information channels. The collected experimental core evaluation parameters, the realization time of the network management service function, can show that the dynamic distribution mechanism of network management information can be cross-applied to each virtual node, and the channel update mechanism of network management information can adjust the information processing queue in real-time. The network management system model that has been built realizes the separation of management and control of the network management system and has the characteristics of independent operation, autonomous function, self-matching, rapid deployment and dynamic expansion.

Keywords: Network function virtualization · OpenFlow communication protocol · Virtual node of a network · Channel iterative update · Separation of network management and control

1 Introduction

Because of the current heterogeneous network environment, building an SDN-based network management model by applying the above research results does not have strong

© The Author(s) 2022
Z. Qian et al. (Eds.): WCNA 2021, LNEE 942, pp. 257–268, 2022.
https://doi.org/10.1007/978-981-19-2456-9_28

universality [1–7]. The main reasons are: the research and application of computer network technology are developing rapidly, new networking technologies are emerging one after another, the research and application of network management technologies supporting it are given priority, and faults in the technical application are inevitable results, which is also a common phenomenon in heterogeneous network systems; The main body of research and development and practical application of network management technology is numerous network technology developers, and developers are used to modelling network management system based on their own rules and products, and there will inevitably be deviations in the implementation of unified modelling standards. Based on this main factor, according to the Network Functions Virtualization (NFV) standard put forward by ETSI Standardization Organization, this paper firstly determines the three-tier structure of network management, namely, user layer, service layer and device layer, and realizes the virtualization of network management functions and resources in the three-tier structure, and then designs the virtual network management node structure. Applying the virtual network dynamic management and control mechanism, introducing the concept of fair and equal network management to control information access, a general SDN network management model is constructed, and its performance is evaluated.

2   The Construction of Virtual Network Management Framework

2.1   Application Layer Construction

Constructing a decentralized and distributed network management system can realize the high integration of network management information transmission, control and management. Among the three elements of the application layer, the network communication lines can be extended to the Internet system, and the network operation management service and network security management service can be extended to the cloud management platform. Figure 1 shows the component set and information transmission process of the application layer of virtualized network management services.

According to the structure diagram of the application layer and the diagram of information transmission process shown in Fig. 1, to build the application layer of network management service based on OpenFlow communication mechanism, firstly, it is necessary to define the system configuration service (Network Management Services1, Abbreviated as NMS1), system control service (NMS2), system performance detection service (NMS3), information flow collection service (NMS4), information flow control service (NMS5), safety detection service (NMS6), fault alarm service (NMS7), data detection service (NMS8) and data analysis service (NMS9). Then, these nine service types are identified and their attributes are marked, and the service functions of NMS1–NMS9 are identified by P1–P9, and their service function attributes can be defined by themselves according to certain programming rules (not listed here). Then, according to all the element sets of the application layer, all the service functions provided by this set are defined, which is called Virtual Network Function Element Collection, abbreviated as VNFEC. Finally, the specific service contents of all the element sets are defined, The main element sets of will be all service function subsets (define this subset as S), attribute subsets of all service functions (define this subset as F), input parameter subsets between service functions and service function attributes (define this subset as I), output parameter subsets between service functions and service function attributes (define this subset as O), The time subset of network management service function realization (this subset is defined as T), the subset of information exchange channel established between every two network management service functions (this subset is defined as L), and the subset of information exchange channel connection state (that is, the channel can be started) (this subset is defined as Q) are composed of six subsets, of which seven subsets are S, F, I, O, T, L and Q. Figure 3 shows the information exchange process of an application layer network function element set to complete a network management event.

$$VNFEC = [S; F; I; O; T; L; Q] \tag{1}$$

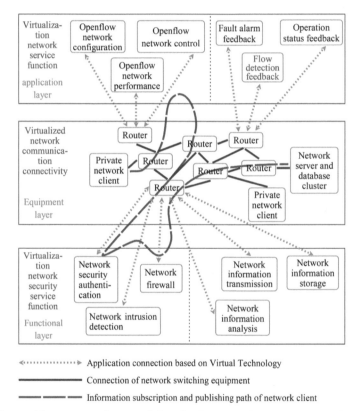

Fig. 1. Composition structure diagram of virtualization network management service application layer.

2.2 Functional Layer Construction

The functional layer construction of virtualized network management system mainly solves two problems: one is to provide network management service functions, and the other is to provide network management service channels. To construct a functional layer, it is necessary to define three-element sets of service function, service channel and the connection state of channel respectively. The formation of the three-element sets mainly depends on the determination of various parameters.

In Formula 1, the set of service function elements is defined as S, and the network management service function identifier is defined as: $(S \rightarrow S_p \in (S_{p+1} \sim S_{p+n}))$; The period for realizing network management service functions can be defined as T, and the time for completing one or more network management service functions can be defined as: $(T \rightarrow T_p \in (T_{p+1} \sim T_{p+n}))$; the collection of software and hardware resources managed by the network management service function can be defined as $(R \rightarrow R_p \in (R_{p+1} \sim R_{p+n}))$. In the whole S_p, T_p and R_p, and one-to-one relationship, the processing process is a single channel, and when multiple events occur simultaneously, you can selectively choose the processing process to build the channel according to the need. when multiple events occur, each event is shared in T_p, and, due to the one-to-one correspondence for T_p, S_p and R_p, the S_p identification and R_p resources occupied by each event processing are shared. The benefit of this is to discard the complexity of multi-parameter definition through design running time limit, time interval, time cyCle adopted in many systems, reduce the parameters of the system when programming, and ensure that the hierarchy of the system is clear.

According to the above analysis, the service function set S can be defined by formula (2), where, S_p, T and R must be described by vectors.

$$S = [S_p \in (S_{p+1} \sim S_{p+n}); T(T_p \in (T_{p+1} \sim T_{p+n})); \\ R(R_p \in (R_{p+1} \sim R_{p+n}))] \tag{2}$$

Formula (3) is the definition of the service channel element set L, among them, S_{p+i} for the output identification after the completion of the previous service function, S_{p+j} is the received input identification for the latter service function, $O(S_{p+i})$ is the corresponding attribute for the output identification, and $I(S_{p+j})$ is the corresponding attribute for the input identification. The attribute here represents the data information processed by the corresponding service function. Formula (4) is the definition of the input and output data information D, where E is the collection of network management events, F is the collection of network service function attributes, and k is the definition rules for the VNFEC set of all service functions of the virtual network management system. The parameters in the above formula are all vector representations.

$$L \in (S_{p+i}, S_{p+j}), L_F = O(S_{p+i}) \cap I(S_{p+j}) \tag{3}$$

$$D = [E; F; k; L_F] \tag{4}$$

Only when the service channel is opened can all kinds of service functions play a role in sequence. In formula (1), the channel connection state element set is defined as Q(a dynamic collection). Q_0 for the initial channel connection state, Q_e and Q_{e-1} is the channel connection state for the first and previous event, then Q_e can be defined as:

$$Q_e = G * (Q_{e-1}) + H * G(s \times s) \tag{5}$$

In formula (5), G represents a vector matrix consisting of the number of all service channels and the number of all service functions in the designed functional layer; H represents a vector matrix consisting of the number of actually needed service channels (1) and the number of actually needed service functions (s) in the event; the vector $G(s \times s)$ represents the matrix of $s \times s$ dimension.

Whether the service channel is on or off can be defined by $G(i, j)$ definition, which $L(i, j)$ represents the connection state of the previous service function with the subsequent service function, $G(L(i, j))$ describes a certain connection, and the connection state $L(i, j)$ is represented by the vector-matrix, with only two values: either connected or disconnected.

2.3 Equipment Layer Construction

In Fig. 1, the devices in the device layer are mainly divided into two categories: network switching devices and network analysis and operation devices. These two types of devices will be virtually applied in the network management system, so they need to be described abstractly. Therefore, these devices first need to be defined by multi-angle configuration parameters like the set elements in the application layer and the functional layer. Then define the resource allocation mechanism for service functions and the resource allocation mechanism for service channels.

The number of functional processes that devices can accept can be defined as C. The entire content of network management resources can be completely defined by formula (6), in which c is the mapping function of t (the time of network management service function realization) and r (the network management resource set), which can be expressed by (c: T → R), and the constituent elements in T and R sets have been defined in the previous functional layer construction. It should be noted that the specific information of these devices, such as the model, function and performance of the devices, should not be defined here.

$$R = [R_p \in (R_{p+1} \sim R_{p+n}); c(c_p \in (c_{p+1} \sim c_{p+n}))] \tag{6}$$

The resource allocation of service channels also needs to be defined by constructing element sets. Its main components include four-element sets: service function set S, service channel set L, priority of service function operation X and allocation process function Y of network management resources. If the service channel resource allocation set is defined as V, then formula (7) can describe the network resource requirements.

$$V = [S; L; x; y] \tag{7}$$

3 Application of Fair Peer-to-Peer Access Mechanism

3.1 Design Fair Peer-to-Peer Access Mechanism

The application of peer-to-peer information access mechanisms in network management and control is the basis of a dynamic combination of network management service functions. The key part of fair and peer-to-peer information access mechanism application lies in the virtual network management node in the network management service channel. That is to say, the key to the application of a fair and equal information access mechanism is to design virtual network management nodes, and to realize the interconnection between virtual network service nodes in a fair and equal way is the main goal.

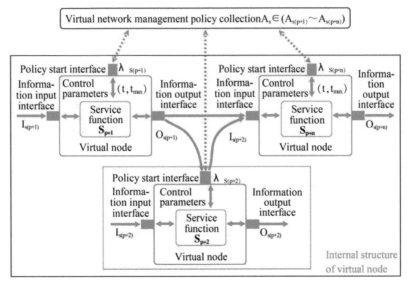

Fig. 2. Network management service virtual channel structure diagram.

To design a virtual network management node, firstly, the functional attributes of network management services need to be uniformly encapsulated. The premise that the functional attributes of network management services can be encapsulated is that it is a kind of data information. Under the platform of big data and cloud computing, the best way to uniformly encapsulate information is to express information in the form of granularity, and Granular Computing (GRC) must be carried out before information encapsulation [8].

The application of granularity and the definition of data I/O interface are the key strategies for the construction of virtual network management nodes and the connection of virtual network management nodes. Figure 2 shows the virtual channel structure diagram of network management service based on a fair peer-to-peer access mechanism and the internal structure diagram of a single virtual node.

3.2 The Internal Structure Design of Virtual Nodes

In the internal structure of the virtual node, the information flow representation and the operation process of input and output all adopt logical operation mode, which is completely different from the coding operation mode commonly used in other software

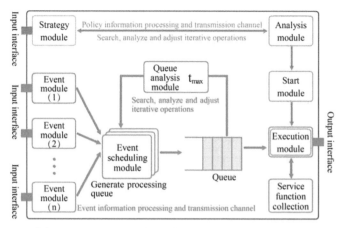

Fig. 3. Diagram of the process of dynamic device connection and information processing and transmission by virtual nodes.

system designs. The realization of its network management service function is mainly based on the unified planning operation strategy, which is the key operation organizer of the network management service function operation strategy. Running policies are planned for different service functions, connection modes between virtual nodes, descriptions of input and output information, etc. They are also a set, which can be expressed by using A_s, and a collection of running policies for a service feature can be defined as $A_s = [A_{sp} \in (A_{s(p+1)} \sim A_{s(p+n)})]$.

The input and output flow of network management information flow in a virtual node mainly consists composed of four elements, single operation policy A_{sp}, the information transmission channel L_{sp}, policy execution part and network management service function S_p; the control parameters to be defined are mainly t and t_{max}; the main logical operation data information includes operation policy start instruction $\lambda_{sp} \in (\lambda_{s(p+1)} \sim \lambda_{s(p+n)})$, input information $I_{s(p+1)} \sim I_{s(p+n)}$, and output information $O_{s(p+1)} \sim O_{s(p+n)}$, Fig. 3 shows an information operation transmission process of network management virtual nodes, wherein the information transmission channel $I_{s(p+1)} \sim I_{s(p+n)}$ provides processing information to the policy execution part through logical operation, the operation execution process $A_{s(p+1)} \sim A_{s(p+n)}$ defined by the policy execution part and the processed output interface. The policy execution component also needs to complete the data information packaging, the operation and processing rule setting of the service function, the establishment of the connection channel of each virtual node, and the construction of the internal communication mechanism.

3.3 Dynamic Control Strategy of Network Management Information Channel

The operation strategy of the whole network management system and the processing and transmission of network management event information not only need to provide the information transmission channel but also need to introduce the management and control mechanism of the channel, which can be realized through the overall deployment of the network management channel. For the deployment of network management channels, first of all, it is necessary to formulate the deployment rules of transmission channels for network management function information and operation strategy information and apply the corresponding scheduling update rules to realize the overall dynamic management and control, so that it can have limited intelligent management. Figure 4 shows the dynamic deployment plan of the whole information transmission channel of the network management system.

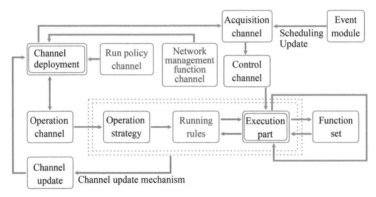

Fig. 4. Dynamic deployment planning strategy of the whole channel of information transmission in a network management system.

The management and control of the network management event information transmission channel mainly depend on the realization of the network management event processing scheduling update mechanism shown in Fig. 5, and its dynamic performance is mainly reflected in the t(max) judgment conditions of the queue task analysis module. The management and control of the operation policy information transmission channel mainly depend on the operation policy set and the policy execution component shown in Fig. 6. By formulating the operation policy rules, the dynamic update instructions of the operation policy channel are analyzed and calculated, and the construction of the policy channel update set is completed, thus realizing the redeployment of the entire network management channel. This is also the key to the dynamic deployment of network management information transmission channels. Here, the information transmission channels of the whole network management system can be defined in detail, in which the network management event transmission channel can be defined as $L_{sp} \in (L_{s(p+1)} \sim L_{s(p+n)})]$, the running policy channel set can be defined as $L_{ap} \in (L_{a(p+1)} \sim L_{a(p+n)})]$ when the two channels are dynamically updated, their range of values adjusts dynamically.

4 System Performance Verification

4.1 System Scalability Verification

The system scalability experiment mainly verifies the deployment mechanism of network virtual nodes. On the premise that 200 network management service events happen at the same time, the experiment sets these 200 network management service events as five parallel processing sets (five parallel processing sets match five network management servers and five virtual nodes at most; Each set handles 40 combined network management service events, aiming at the simultaneous parallel processing capability of the system and the combined capability of network management function services), and configures multiple network management data information processing servers (actually, the network management information processing nodes corresponding to multiple virtual nodes are the combination of virtual nodes and network management processing nodes; The purpose is to provide the information processing and operation ability suitable for large-scale network system management, essentially providing multiple CPUs).

The experimental results show that the number of virtual nodes and corresponding servers is small, and the time from the occurrence of network management events to the start of network management event processing is the longest. Because the network management service events are divided into many single events, the advantages of fully opening the processing queue are not fully reflected, and it takes the longest time from the start of network management event processing to the completion of network management event processing. With the increasing number of virtual nodes and the corresponding servers, the corresponding network management events are dynamically distributed to the corresponding processing units, and the factors of uncertain time consumption for different network management functions are counted, realization in case of change of network management event handling and scheduling mode and the dynamic distribution mechanism is cross-applied to different virtual nodes. Therefore, the time from the occurrence of network management events to the start of network management event processing and the completion of network management event processing shows a steady downward trend. The network management system model designed in this paper fully embodies the centralized management of network management functions and the distributed control of network management information transmission channels. The mechanism of combining and publishing network management events can be successfully realized. The virtualized network management information processing units are closely connected, the dynamic association increases or decreases the deployment of network management information processing units is flexible, and the expansion performance of the whole system is superior.

4.2 System Timeliness Verification

The system timeliness experiment is mainly aimed at verifying the dynamic control mechanism of the network management information channel. The experiment is also based on the premise that 200 network management service events occur at the same time, and five parallel processing sets are set, and multiple network management data information processing servers are configured to record the running time of the network management system and the realization time of network management service functions.

The experimental results show that, under the condition that the policy channel update mechanism is not enabled, because five virtual nodes and five network management servers are started to operate, the experimental results are the same as those in the timeliness verification experiment. In the role of the policy channel update mechanism, more network management events will be added to the information processing queue in time, and it will have the ability to deal with some network management emergencies.

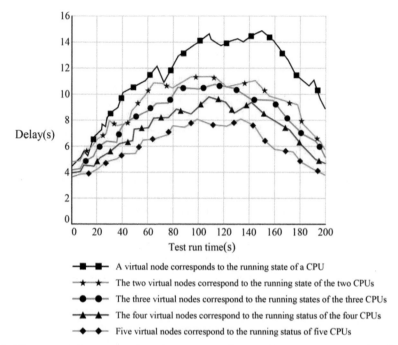

Fig. 5. Diagram of time-consuming change state of network management service function realization under the condition of virtual node setting change.

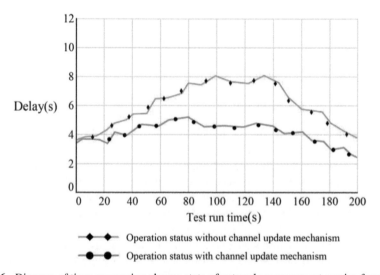

Fig. 6. Diagram of time-consuming change state of network management service function.

5 Conclusion

According to NFV standard, taking the internal structure design of virtual nodes as a breakthrough, this paper constructs a universal virtual SDN network management model by introducing logical operations to control information transmission, classify and dynamically control network management information channels. The main achievements of the research work include:

(1) The network management functions and resources of the network management system are virtualized; All network management functions are centralized management, and network management information transmission channels are distributed applications.
(2) The network management service channel based on a fair peer-to-peer access mechanism, which encapsulates network management data information in a container virtual way, can flexibly handle multiple network management service functions.
(3) The number of processing functions and processing time of virtual network management nodes are relatively fixed, which can better analyze the network state information and network operation state in real-time; Extensible interfaces for managing network service functions, service channels and resources can realize the construction of flexible network management system.
(4) The virtual network management node adopts logical operation to construct the input and output channels of internal information, which simplifies the structural complexity of the mathematical model and improves the running efficiency of the system.
(5) The independent, dynamic, and combined construction of the two information transmission channels of operation strategy and network management events is also the key to the construction of a virtual network management system.

Acknowledgements. The authors are grateful for the financial support of the Scientific Research Project of Hubei Education Department (Grant No. Q20204306 and B2020195), Jingmen Science and Technology Project (Grant No. 2021YFZD076, 2020YFYB049 and 2021YFYB119), Scientific Research Project and Team of Jingchu University of Technology (Grant No. YY202102 and TD202101).

References

1. Bari, M.F., Roy, A.R., Chowdhur, S.R., et al.: Dynamic controller provisioning in software defined networks. In: Proceedings of 9th International Conference on Network and Service Management (CNSM), pp. 18–25, IEEE (2013)
2. Mogul, J.C., Au, Y.A., et al.: Corybantic: towards the modular composition of SDN control programs. In: Proceedings of the Twelfth ACM Workshop on Hot Topics in Networks. ACM (2013)
3. Shin, S., Porras, P.A., et al.: FRESCO: modular composable security services for software-defined networks. In: Proceedings of NDSS (2013)

4. Blendin, J., Ruckert, J., et al.: Software-defined network service chaining. In: Proceedings of 2014 Third European Workshop on Software Defined Networks (2014)
5. Csoma, A., Sonkoly, B.A.Z., et al.: Multi-layered service orchestration in a multi-domain network environment. In: Proceedings of 2014 Third European Workshop on Software Defined Networks (EWSDN). IEEE (2014)
6. Sonkoly, B.A.Z., Czentye, J.A.N., et al.: Multi domain service orchestration over networks and clouds: a unified approach. In: Proceedings of the 2015 ACM Conference on Special Interest Group on Data Communication. ACM (2015)
7. Lee, G., Kim, M., et al.: Optimal flow distribution in service function chaining. In: Proceedings of the 10th International Conference on Future Internet. ACM (2015)
8. Wang, T.T., Rong, C.T., et al.: Survey on technologies of distributed graph processing systems. J. Softw. **29**(03), 569–586 (2018)

Research on Interdomain Routing Control in SDN Architecture

Liangzhe Chen[1], Yin Zhu[1], Xinyu Sun[1], Yinlin Zhang[1], Gang Min[2], Yang Zou[2], and Zhong Shu[1,2(✉)]

[1] Jingchu University of Technology, Jingmen 448000, Hubei, China
421934337@qq.com
[2] Jingmen Mobile Media Co. Ltd, Jingmen 448000, Hubei, China

Abstract. Aiming at the difficulty of network management due to the coexistence of traditional BGP network and new SDN network, this paper proposes a routing update algorithm with clear interdomain structure and network exception handling ability. By defining SDN, BGP-SDN fusion and BGP three network domains, a packet transmission path with route discovery and update capability were formed through the three network domains in sequence. On the premise of reducing the communication delay range, the route update delay is set, and the exception handling mechanism is introduced. Specify the master controller to make the Interzone routing control rules and make routing updates a key parameter in the data flow table of border switches. The algorithm firstly ensures the absolute unimpeded communication between network domains, provides a reliable time guarantee for network exception handling, enhances the connection between control servers between network domains and between control servers and boundary switches, and improves the synchronization of multiple links in network communication. By using Mininet to build a simulation experiment platform, the reliability and feasibility of the proposed algorithm are verified from the perspectives of data packet loss and Interzone route update delay, and it is suitable for application and implementation in the current Internet environment.

Keywords: Software defined network · Border gateway protocol · OpenFlow · Computer network domain · Domain controller

1 Introduction

The main application of the computer network domain is the BGP network protocol. The BGP network mainly works out discovering the next routing node independently and following consistent communication rules within a defined domain and among constituent domains. In the BGP inter-domain boundary routing protocol, the routing control is mainly based on the IP address from the communication destination, and the selection of routing path is derived from adjacent routers. In addition, the transparency and intuition of the routing algorithm are not strong [1].

In the SDN framework mode, the main existing problems are reflected in the updating process of routing paths between network domains. The information loss of data stream

© The Author(s) 2022
Z. Qian et al. (Eds.): WCNA 2021, LNEE 942, pp. 269–278, 2022.
https://doi.org/10.1007/978-981-19-2456-9_29

(namely packets) transmitted between network domains occurs from time to time. The most fundamental reason is the mixed application of traditional network and SDN network technology. The key point of the problem is that the two network management modes have different setting strategies for packet transmission control parameters. In fact, there is no relatively consistent standard for constraint [2–4].

Herein, based on the premise of a clear definition of traditional network domain (Route discovery technology which mainly refers to route discovery technology with BGP border Gateway protocol as the core), SDN network domain and BGP-SDN fusion network domain, and based on the application of route update mechanism in SDN network domain, according to the principle of collaborative and consistent interdomain route discovery, The consistency of routing update policies of three types of Interzone's is constrained to achieve the goal of no packet information loss from the whole mechanism. At the same time, by constructing a standard SDN architecture model and introducing the improved algorithm under its model framework, simulation experiments are carried out from the perspectives of packet loss in data transmission and routing update delay between network domains to verify the reliability and feasibility of the algorithm proposed in this paper.

2 Implementation of the Algorithm in this Paper

To prevent packet loss or network communication interrupt, two key problems need to be solved. One is relatively independent of each domain control server, data packets in asynchronous problem, the other is that the SDN network communication mechanism and BGP do not match the network communication mechanism, which temporarily interrupts the network communication problems (Fig. 1).

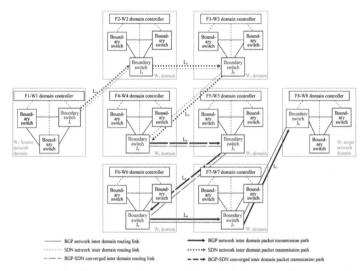

Fig. 1. Interzone routing discovery and update policy and packet transmission path design process proposed in this paper.

As for the effective fusion of SDN network communication mechanism and BGP network communication mechanism, the main solution is to design a highly matched control algorithm between SDN and BGP Network management control policy, focusing on the design of a master control server that can coordinate the control of all inter-domain control servers and synchronize the same task. The main process of the algorithm includes: (1) master control server to send all the network domain "after submit inter-domain routing updates available path" information, the information is changed after routing updates main path information, the information sent by the included in the "all requirements of the network domain has confirmed to receive offers available transmission path to apply for" information, only in the network domain feedback after all complete information, The master server will initiate the next command. In this step, the master controller collects statistics on the SDN network domain, BGP and SDN fusion network domain, and BGP network domain components involved in route updating. (2) The master control server first sends the request of "Enabling interdomain routing to update available paths" to all SDN domains. After receiving the request, the control server in the SDN domain sends the enable instruction to the boundary switch in the domain. The boundary switch completes the parameter update in the data flow table. The in-domain control server feedback the received and completed instructions to the master control server. (3) In the same way as the second step, the master controller sends the request of "Enabling interdomain routing to update available paths" to all BGP-SDN fusion domains, and completes all corresponding instructions with the cooperation of the intra-domain control server and the intra-domain boundary switch. (4) In the same way as in the second step, the master controller sends the request of "Enabling Interzone routing to update available paths" to all BGP network domains, and directs the intra-domain control server and intra-domain boundary switches to complete corresponding instructions.

In the above process, performed by the master control server to set an information exchange round-trip time limit (defined as routing update delay), make sure that the network domain control server and boundary switch when performing routing update instruction, will be the last time the configuration parameter, reset all the data in ensuring accurate routing updates instruction execution at the same time, To a certain extent, it can also improve the synchronization of each operation process. T_{F-J} are mentioned in the algorithm of the concept of routing updates, its exact meaning is according to the prescribed three kinds of a network domain, must first be connected from the source to the target network between domains, independent SDN network domain, BGP - domain SDN fusion, BGP network domain three packet transmission path (so that we can ensure that network system without a cross-domain communication no difference to the target domain), Then, according to the actual status of the network environment, a transmission path that can complete the packet transmission process is constructed according to the sequence of "SDN network domain → BGP-SDN fusion domain → BGP network domain". In the definition of process flow of source network domain → SDN network domain → BGP-SDN fusion domain → BGP network domain → target network domain must be considered complete. Figure 2 shows the algorithm flow in this paper.

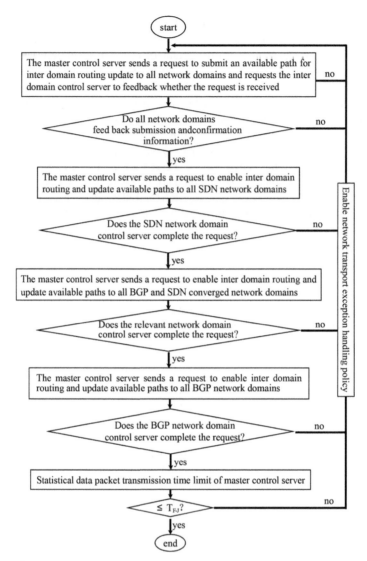

Fig. 2. Flow chart of Interzone routing update algorithm proposed in this paper.

To realize the above algorithm, the control server set corresponding to the defined network domain set W_{sdn}, W_{bgp}, SW_{sdn}, BW_{bgp} and W_{s-b} needs to be defined first, which can be defined as F_{sdn}, F_{bgp}, SF_{sdn}, BF_{bgp} and F_{s-b} according to the sequence of the above network domain set SW_{sdn}, BW_{bgp}, W_{s-b}, SF_{sdn}, BF_{bgp} and F_{s-b}. When the algorithm is implemented, it only needs to define the updated parameters. The parameters before the update can be defined by initializing the updated parameters.

In the master control server, it is also necessary to define some control information of interdomain routing updates. According to the algorithm flow mentioned above, the control information updated for the four main Interzone routes can be defined as I_{F-J}

(from the master control server), I_{sdn} (from the control server in the SDN domain), I_{s-b} (from the control server in the BGP-SDN fusion domain), and I_{bgp} (from the control server in the BGP domain). In the whole network system, a packet transmission task to be completed can be defined as $K_D(W_A \rightarrow W_B)$, and the master control server can be defined as F_{KD}. If the control server in the information source domain is specified as the master server, it needs to be defined $F_A \in F_{KD}$. Based on the above analysis, the algorithm needs to be designated as the master control server, SF_{sdn} is the SDN intra-domain control server, BF_{bgp} is the BGP-SDN intra-domain control server, and the BGP intra-domain control server as F_{KD}. In addition to the function of formulating routing update policies, other intra-domain control servers can implement routing update policies.

To ensure that the entire network system $W_A \rightarrow W_B$. In the control part, the algorithm proposed in this paper emphasizes that on the premise of clearly defining three types of network domains, transmission parameters must be set strictly in the order of "SDN network domain $\rightarrow$ BGP-SDN fusion domain $\rightarrow$ BGP network domain", and the execution of instruction tasks must be completed in order. The transmission parameters of SDN network domain must be set first based on: In the current application of the network system, the processing leading network management status of SDN is network communication mechanism, and BGP network communication mechanism is mainly applied to traditional network system (the mixture, heterogeneous network system is the current network management must face the situation), because the SDN is obviously better than the BGP network management system of network management mechanism, therefore, As long as SDN technology is used in a network domain, it should be used preferentially, and the priority level of pure SDN network domain should be set to the highest.

How do you implement $W_A \rightarrow W_B$ The goal of normal network communication to W can be verified by elimination. If the communication between W_A and W_B cannot be achieved. There must be at least one boundary switch in the whole network. At a certain point in time (or period), it is impossible to ensure smooth communication between the $W_A \rightarrow SW_{sdn} \rightarrow W_{s-b} \rightarrow BW_{bgp} \rightarrow W_B$ network domain sets in sequence. If the boundary switch with possible problems is defined F_e, the reliability of network communication can be finally concluded by finding out the relationship with some key network domain sets and judging whether the inter-domain transmission of packets can be realized. The verification and analysis process is as follows:

(1) F_e does not belong to SW_{sdn} Y BW_{bgp} the domain set, which means F_e is not in the network system to which the study belongs. During routing updates, F_e is impossible to receive any data packets, and F_e the possibility of forwarding incorrect data packets does not exist. Even if F_e packets forwarded are irrelevant to this task, inter-domain data packet transmission can be realized $W_A \rightarrow W_B$.

(2) F_e belongs to the domain set but does not belong to BW_{bgp} the domain set. Before routing update, F_e is impossible to receive any data packets; When the second step process of the algorithm in this paper is started, F_e the second step process algorithm cannot be directly enabled. However, after adjustment through the feedback mechanism, data packet transmission can be realized. Theoretically, at least, we can know how to realize $W_A \rightarrow W_B$ inter-domain data packet transmission.

(3) F_e belongs to the SW_{sdn} I BW_{bgp} domain set, which means that the F_e routing update path provided by one of the domain sets is adopted in two stages SW_{sdn} or BW_{bgp} to realize $W_A \rightarrow W_B$ inter-domain packet transmission.

(4) F_e does not belong to the SW_{sdn} domain set but belongs to the BW_{bgp} domain set. Before routing update, it means that F_e interdomain packet transmission is realized $W_A \rightarrow W_B$ through the routing transmission path provided by the domain set in two stages. Before routing update, BW_{bgp} domain set starts the packet transmission path; SW_{sdn} domain collection starts the packet transport path before routing updates.

Through the above assumption F_e and the network state, the relation of four domain sets of packets can be seen from $W_A \rightarrow W_B$ an analysis of the network communication results, F_e impossible to interrupt transmission in the network communication between $W_A \rightarrow W_B$ domain problems, also proved the reliability of the algorithm in this paper, at the same time, also verified F_e must belong to the above definition of one of the four control server, there is no possibility of a problematic boundary switch.

3   Experiment and Discussion

3.1   Experimental Platform

In the constructed experimental platform, TCP communication protocol (UDP communication protocol can also be used) is the main communication mode between the four types of intra-domain control servers. Some literature points out that the high efficiency of data communication can be guaranteed by using distributed technology [5–7]. The virtual network simulation tool Mininet was used to configure the OpenFlow boundary switch [8], and the algorithm in this paper was planted in the master control server and three types of domains. Data packet transmission required in the experiment was completed by network performance testing tool Iperf [9].

3.2   Data Packet Loss Verification

The experiment sends data packets (mainly image files and video files) to the target network domain from the information source through the master control server, and the transmission of data packets is successively through L_1–L_2–L_3–L_4–L_5–L_6–L_7 Channel, packets are encapsulated through UDP communication mechanism, and packets are set in two encapsulation modes of 1400 bytes and 20 bytes (at the same time, the purpose of using 20 bytes to encapsulate packets is to provide higher data transmission rate for the experiment). The sending data rate is divided into eight levels and increases successively. The main verification parameters are the number of data packets lost and packet loss rate, and the number of abnormal processing packets and exception processing rate. Main evaluation index parameters, experimental condition parameters and experimental result parameter values are shown in Table 1.

Table 1. Statistical table of simulation results under successively increasing data transmission rates using two packet encapsulation methods.

Packet encapsulation	Data send rate	Data packets received	Number of lost packets	Packet loss rate (%)	Number of abnormal processing packets	Abnormal packet processing rate (%)
1400-byte	100 Mbps	9723	0	0	0	0.000
	200 Mbps	18412	0	0	3	0.016
	300 Mbps	26436	0	0	12	0.056
	400 Mbps	35218	0	0	15	0.043
	500 Mbps	44929	0	0	27	0.060
	600 Mbps	52194	0	0	42	0.080
	700 Mbps	61875	0	0	51	0.082
	800 Mbps	69157	0	0	68	0.098
20-byte	1.0 Gbps	34517	0	0	102	0.30
	1.2 Gbps	36254	0	0	157	0.43
	1.3 Gbps	36572	0	0	136	0.37
	1.4 Gbps	36925	0	0	118	0.32
	1.5 Gbps	38073	0	0	103	0.27
	1.6 Gbps	41128	0	0	98	0.24
	1.7 Gbps	42576	0	0	92	0.21
	1.8 Gbps	43914	0	0	84	0.19

3.3 Verifying Interzone Route Update Delay

In the experiment, data packets (with different sizes of transmission files) were sent from the source network domain to the target network domain through the master control server, and the data packets were transmitted through L_1–L_2–L_3–L_4–L_5–L_6–L_7 Channel, packets are encapsulated by THE CTP communication mechanism. The average rate of sending data is divided into sixteen levels and increases successively. The main statistical verification parameter is the average transmission rate (S_S), transfer file size (D_S), routing update times (R_S), the normal transmission delay of data packets (T_S), route update delay (T_{F-J}), the delay increased by routing update ($T_\triangle$). Main evaluation index parameters, experimental condition parameters and experimental result parameter values are shown in Table 2.

Table 2. Statistical table of simulation results for different incoming files with successively increasing data transmission rates.

S_S(Gbps)	D_S(G)	R_S(b)	T_S(s)	$T_{F\text{-}J}$(s)	T_Δ(s)
0.80	5.69	11	58.94	59.01	0.07
0.90	6.87	8	59.16	59.28	0.12
1.00	7.56	10	56.73	56.84	0.11
1.10	8.43	9	61.37	61.55	0.18
1.20	9.20	10	60.54	60.74	0.20
1.30	10.14	9	62.05	62.20	0.15
1.40	11.75	8	59.66	59.84	0.18
1.50	12.69	11	60.98	61.19	0.21
1.60	13.51	9	58.61	58.74	0.13
1.70	14.18	10	61.32	61.48	0.16
1.80	21.32	8	62.81	63.03	0.22
1.90	25.43	11	60.17	60.36	0.19
2.00	28.97	9	59.93	60.17	0.24
4.00	36.16	9	61.54	61.90	0.36
8.00	49.71	11	60.12	60.43	0.31
12.00	69.36	10	60.96	61.61	0.65

3.4   Discussion of Experimental Results

The data packet loss detection experiment is mainly to verify whether the proposed algorithm can transfer files from the information source to the target network domain under the co-existence of multiple network domains. With the continuous increase of packet transmission rate, all the packet loss rates shown in the experimental results are 0%, which further verifies the correctness of the theoretical analysis mentioned above. Of exception handling the number of packets, is, in fact, this algorithm's ability to perform routing update test, under the condition of giving a large amount of data transmission, almost under the different data transmission rate, all collected complete exception handling the total number of packets, illustrates the proposed routing update policy has played a role; If the number of exception processing packets is not too large, it indicates that the algorithm in this paper can independently find idle transmission paths, and the design requirements of Interzone route discovery can be realized.

In the experiment of interdomain routing delay detection, the delay length increased by routing update can verify the efficiency of the proposed algorithm. Experimental results show that in the process of packet transmission, the number of route updates is not high and remains relatively stable, indicating that the algorithm is very accurate and fast to find the switch on the idle boundary of the transmission path. Most of the implementation of route updates does not enable the exception handling strategy, which makes the route update delay is not long. In the process of a routing update, the added delay is not long, and with the continuous increase of packet transmission rate, The value increase of T_Δ is not significant and has little impact on the normal transmission of packets. The results not only further verify the effectiveness of the routing update strategy but also apply to the current constantly developing Internet environment.

In the routing update, the parameters of the data flow table of the boundary switch are mainly set for the part that generates the update, which reduces the space occupation of the internal register in the switch. In the process of data transmission, only the communication delay between the master control server and each network domain server and the communication delay between each network domain server and the corresponding domain boundary switch are defined, which improves the synchronization of multiple processing links in packet transmission. The implementation of the algorithm is mainly accomplished through the master control server, which does not add too many application functions in the SDN module, BGP module, other control servers in the domain and boundary switch, greatly simplifying the subsequent development and application complexity.

4 Conclusion

Based on the definition of network domain and the relationship between domains, this paper takes route discovery and key problem solving as the main breakthrough direction and proposes an algorithm to control routing updates between network domains under the Framework of SDN.

(1) In the BGP and SDN converged network domain, the communication protocols are inconsistent, and the control server configuration is relatively independent, which is the main cause of data transmission packet loss and network communication interruption between network domains.
(2) Whether in the BGP network domain or the SDN fusion network domain, the configured interdomain routing discovery control server must be based on the SDN network communication rules.
(3) In the whole network system, deploy an Interzone routing update master control server, which can effectively prevent the abnormal phenomenon caused by the abnormal processing process of network communication.
(4) Define clearly the network domain and set the transmission parameters in the data flow table of the boundary switch according to the correct sequence of a pure SDN network domain, BGP-SDN fusion network domain, and pure BGP network domain, which is the key to realize the non-loss transmission of packets between network domains.

Based on the coexisting simulation experiment platform of the SDN and the BGP-SDN technology, the above conclusions are verified and show that the set between the source domain and target network domain packet transmission path, control of the transmission in the process of data packet loss, the phenomena of routing updates time delay is short, the algorithm has high reliability and feasibility.

Acknowledgements. The authors are grateful for the financial support of the Scientific Research Project of Hubei Education Department (Grant No. Q20204306 and B2020195), Jingmen Science and Technology Project (Grant No. 2021YFZD076, 2020YFYB049 and 2021YFYB119), Scientific Research Project and Team of Jingchu University of Technology (Grant No. YY202102 and TD202101).

References

1. Gupat, A., Vanbever, L., Shahbaz, M., et al.: SDX: a software defined internet exchange. ACM SIGCOMM Comput. Commun. Rev. **44**(4), 551–562 (2015)
2. Jain, S., Kumar, A., Mandal, S., et al.: B4: experience with a globally-deployed software defined WAN. In: Proceedings of the ACM SIGCOMM 2013 Conference on SIGCOMM, pp. 3–14. ACM (2013)
3. Mizrahi, T., Saat, E., Moses, Y.: Timed consistent network updates in software-defined networks. IEEE/ACM Trans. Netw. **24**(6), 1–14 (2016)
4. Alimi, R., Wang, Y., Yang, Y.R.: Shadow configuration as a network management primitive. ACM SIGCOMM Comput. Commun. Rev. **38**(4), 111–122 (2008)
5. Xu, Y.H., Sun, Z.X.: Research development of abnormal traffic detection in software defined networking. J. Softw. **31**(01), 183–207 (2020)
6. Xiao, Y., Fan, Z.-J., Nayak, A., Tan, C.-X.: Discovery method for distributed denial-of-service attack behavior in SDNs using a feature-pattern graph model. Front. Inf. Technol. Electron. Eng. **20**(9), 1195–1208 (2019). https://doi.org/10.1631/FITEE.1800436
7. Hu, T., Zhang, J.H., Wu, J., et al.: Controller load balancing mechanism based on distributed policy in SDN. Acta Electron. Sin. **46**(10), 2316–2324 (2018)
8. Ulf, N.: Investigating the possibility of speeding up Mininet by using Netmap, an alternative Linux packet I/O framework. Procedia Comput. Sci. **8**(126), 1885–1894 (2018)
9. Lei, M.: Research on Traffic Scheduling Algorithm Based on SDN Data Center. Technological University, Xi'an (2018)

Human Action Recognition Based on Attention Mechanism and HRNet

Siqi Liu, Nan Wu, and Haifeng Jin[✉]

Department of Cyberspace Security, Changchun University, Changchun, China
200701178@mails.ccu.edu.cn

Abstract. A human action recognition network (AE-HRNet) based on high-resolution network (HRNet) and attention mechanism is proposed for the problem that the semantic and location information of human action features are not sufficiently extracted by convolutional networks. Firstly, the channel attention (ECA) module and spatial attention (ESA) module are introduced; on this basis, new base (EABasic) and bottleneck (EANeck) modules are constructed to reduce the computational complexity while obtaining more accurate semantic and location information on the feature map. Experimental results on the MPII and COCO validation sets in the same environment configuration show that AE-HRNet reduces the computational complexity and improves the action recognition accuracy compared to the high-resolution network.

Keywords: Deep convolutional network · Human motion recognition · High resolution network · Attention mechanism

1 Introduction

Human action recognition is an important factor and key research object for the development of artificial intelligence. The purpose of human action recognition is to predict the type of action visually. And it had important applications in security monitoring, intelligent video analysis, group behavior recognition and other fields, such as the detection abnormal behavior in ship navigation and the identification of dangerous people in the transportation environment of subway stations. Other scholars had applied action recognition technology to smart home, where daily behavior detection, fall detection, and dangerous behavior recognition were getting more and more concentrate from researchers.

Literature [1] proposed an improved dense trajectories (referred to as iDT), which is currently widely used. The advantage of this algorithm is that it is stable and reliable, but the recognition speed was slow. With the innovation and development of deep learning technology, the method of image recognition had been further developed. Literature [2] had designed a new CNN (Convolutional Neural Network) action recognition network-3D Convolutional Network, This net extracted features from both temporal and spatial dimensions and performs 3D convolution to capture motion information in multiple

© The Author(s) 2022
Z. Qian et al. (Eds.): WCNA 2021, LNEE 942, pp. 279–291, 2022.
https://doi.org/10.1007/978-981-19-2456-9_30

adjacent frames for human action recognition. In the literature [3], a two-stream expansion 3D convolutional network (referred to as TwoStream-I3D) was used for feature extraction. And in literature [4], Long Short-Term Memory (referred to as LSTM) had been used.

In papers that use the two-stream network structure, researchers have further improved the two-stream network. The literature [4] used a two-stream network structure based on the proposed temporal segmentation network (TSN) for human action recognition, literature [5] used a deep network based on learning weight values to recognize action types, literature [6] uses a ResNet network structure. As the connection method of dual-stream network, and the literature [7] used a new two-stream that is three-dimensional convolutional neural network (I3D) based on a two-dimensional convolutional neural network to recognize human actions. These types of deep learning methods lead to a significant increase in the accuracy of action recognition.

All the above improvements were based on convolutional neural networks, and the spatially and temporally based self-attentive convolution-free action classification methods had been proposed in the literature [8], which could learn features directly from frame-level patch sequence data. This type of method directly assigned weight values through the attention mechanism, which increases the complexity of model processing and ignores the structural information of the picture itself during pre-processing and feature extraction.

For human behaviour action recognition in video data or image data, both need to transform the data carrier into sequence images, then recognizing human actions in static images can be transformed into an image classification problem. The advantage of the convolution method applied to the action classification in the image is that it could learn through hierarchical transfer, save the reasoning and perform new learning on subsequent levels, and feature extraction had been performed when training the model. There was no need to repeat this operation. However, on those data which was not pre-processed, it was not possible to rotate and scale images with different scales, and the human features extracted using convolution operations do not reflect the overall image description (e.g., "biking" and "repairing" may be divided into one category), so it is necessary to use attention network to recognize local attribute features.

Based on the above research, the action recognition in this paper uses high-resolution network HRNet as the basic network framework, at the same time, making improvements to the basic modules of HRNet, and improving the HRNet base module by using Channel Attention and Spatial Attention to further increase the local feature information extracted from the feature maps, besides, allowing the feature maps exchange with each other in terms of spatial information. At the same time, the fusion output of HRNet has been improved. We have designed a fusion module to perform gradual fusion operations on the output feature maps, and finally output the feature maps after multiple fusions. The main work of this paper is as follows:

(1) Designed the basic modules AEBasic and AENeck which integrate the attention mechanism. While extracting image features with high resolution, it improves the weight of local key point information in image features, reduces the loss caused by key point positioning, and has better performance than the HRNet network model.

(2) Compared with the original three outputs of HRNet, we designed a new fusion output method, fused the feature maps layer by layer to obtain more sufficient semantic information in the feature maps.

2 Overview of HRNet and Attention Mechanism

2.1 HRNet Network Structure

The HRNet network structure started with the Stem layer, as shown in Fig. 1. the Stem layer consists of two stride-2 3 * 3 convolutions. After the Stem layer, the image resolution reduced from R to R/4, while the number of channels changed from RGB three channels to C. as shown in Fig. 1. The main body of the structure was divided into four stages, while containing four parallel convolutional streams, the resolution R in the convolutional stream is R/4, R/8, R/16 and R/32, respectively, and the resolution is kept constant in the same branch. The first stage contains four residual units consisting of a bottleneck layer of width 64, followed by a 3 * 3 convolution that changes the number of channels of the feature map to R. The second, third and fourth stages contain 1, 4 and 3 of the above modules, respectively.

In the modular multi-resolution parallel convolution, each branch contains 4 residual units, each residual unit contains two 3 * 3 convolutions of the same resolution with batch normalization and nonlinear activation function ReLu. the number of channels in the parallel convolution stream is C, 2C, 4C, 8C, respectively.

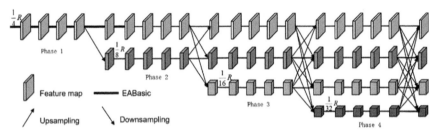

Fig. 1. HRNet structure.

2.2 Attentional Mechanisms

The attention mechanism plays an important role in human perception. For what is observed, the human visual system does not process the entire scene at once, but selectively focuses on a certain part so that we can better understand the scene. Also in the field of Machine-vision, using the attention mechanism can make the computer better understand the content of the picture. The following describes the channel attention and spatial attention used in this paper.

Channel Attention. For a channel feature map $F(X) \in R^{C*H*W}$,the feature map has height H and width W and contains C channels. In some learning tasks, not all channels contribute equally to the learning task, some channels are less important for this task, while others are very important for this task. Therefore, computer needs to assign channel weights according to different learning tasks.

Literature [9] proposed SENet, a channel-based attention model, as shown in Fig. 2. Through compression (F_{sq}) and excitation (F_{ex}) operations, the weight ω of each feature channel was calculated. The weight ω of the feature channel is used to indicate the importance of the feature channel. and the learned feature channel weights ω vary for different learning tasks. Subsequently, the corresponding channel in the original feature map F is weighted using the feature channel weight ω, that is, each element of the corresponding channel in the original feature map F is multiplied by the weight to obtain the channel attention feature map ($\tilde{X}$). In short, channel attention is focused on "what" is a meaningful input image. The larger the feature channel weight ω, the more meaningful the current channel; conversely, if the feature channel weight ω is smaller, the current channel is meaningless.

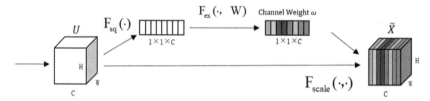

Fig. 2. SENet structure.

Spatial Attention. In the literature [3–7], researchers had used a model based on a two-stream convolutional network and made improvements on the original, using the improved model for image feature extraction, which had improved the accuracy of action recognition but still essentially uses convolution to extract image features.

When performing the convolution operation, the computer divides the whole image into regions of equal size and treats the contribution made by each region to the learning task equally. In fact, each region of the image contributes differently to the task, thus each region cannot be treated equally. Moreover, the convolution kernel is designed to capture only the local spatial information, but not the global spatial information. Although the stacking of convolutions can increase the receptive field, it still does not fundamentally change the situation, which leads to some global information being ignored.

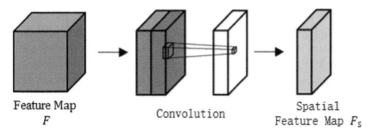

Fig. 3. CBAM spatial attention module.

Therefore, some researchers have proposed the CBAM (Convolutional Block Attention Module) model [10], which uses the spatial attention module to focus on the location information of the target, and the area with prominent significance for the task increases the attention, while the area with less significance is Reduce attention, as shown in Fig. 3.

3   Action Recognition Model Based on Attention Mechanism

The performance improved by modifications on the convolutional network only can no longer meet the needs of the study, inspired by the literature [10–12], we choose to fuse the convolutional neural network and the attention mechanism to improve the network performance. A high-resolution network, HRNet, is used in the literature [13] to maintain the original resolution of the image during convolution and reduce the loss of location information, so we add attention mechanisms to the selected HRNet network model and propose an action recognition model based on channel attention and spatial attention mechanisms, AE-HRNet (Attention Enhance High Resolution Net).

3.1   AE-HRNet

AE-HRNet inherits the original network structure of HRNet, which contains four stages, as shown in Fig. 4. The reason for using four stages is to let the resolution of the feature map decrease gradually. Due to the adoption of a substantial downsampling operation, which leads to the rapid loss of details such as location information and human action information in the feature map, it is difficult to guarantee the accuracy of the prediction

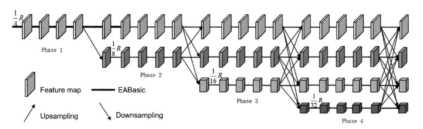

Fig. 4. AE-HRNet structure.

even if the feature information is learned from the blurred image and then restored by upsampling the image. Therefore, in each stage, parallel branches with 1, 2, 3, and 4 different resolutions and number of channels are used to maintain the high resolution of the image while performing the downsampling operation, which allows the location information to be retained.

The specific processing of the AE-HRNet network model is as follows.

(1) In the pre-processing stage, the resolution of the image is unified to 256 * 256, and two standard stride-2 3 * 3 convolutions are used, so that the input resolution is 1/4 of the original resolution, at the same time, number of channels becomes C.
(2) Take the pre-processed feature map as the input of stage 1, and extract the feature map through 4 EABasic modules.
(3) In the following three stages, EANeck pair features with different resolutions (1/4, 1/8, 1/16, 1/32) and channel numbers (C, 2C, 4C, 8C) are used respectively Figure for feature extraction.

The basic network architecture used in our experiment is HRNet-w32. The resolution and the number of channels will be adjusted between each stage. At the same time, the feature maps between the resolutions will also be exchanged and merged to form a feature map with richer semantic information.

3.2 ECA (Enhance Channel Attention) Module

The structure of ECA (Enhance Channel Attention) module is shown in Fig. 5, firstly, the convolved feature maps are pooling by Max Pooling and Avg Pooling respectively. In order to maximize the retention of image features, we use both Max Pooling and Avg Pooling; then we use two 1 * 1 convolutions on the pooling feature maps; next, we add those two feature maps and use the Sigmoid activation function to obtain the channel attention feature map with dimension C * 1 * 1. Finally, multiply the channel attention feature map F_c with the original feature map F, and reduce the output dimension to C * H * W to get the new feature map.

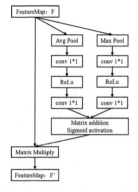

Fig. 5. ECAttention module.

3.3   ESA (Enhance Spatial Attention) Module

The ESA (Enhance Spatial Attention) module is shown in Fig. 6. The original feature map is also subjected to Max Pooling and Avg Pooling, then we concatenate the two parts of the feature map to get a tensor which dimension is 2 * H * W, use a convolution operation with a convolution kernel size of 7 or 3 to make the number of channels 1 and keep H and W unchanged. Then use the Sigmoid function to get a dimension of 1 * H * W. Finally, matrix multiplication is used to multiply the spatial attention feature map with the feature map output by the ECA module, and the output dimension is restored to C * H * W, and the final feature map is obtained.

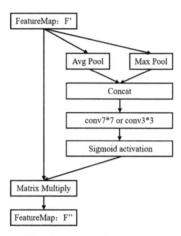

Fig. 6. ESAttention module.

3.4   EABasic and EANeck Modules

The EABlock module consists of ECA module and ESA module. The main modules of HRNet network model are Bottle neck module and Basic block module. In order to integrate with the attention mechanism, we designed EABlock (Enhance Attention block) module to add it to the Bottle neck module and Basic block module, as shown in Fig. 7, called EABasic (Enhance Attention Basic) module and EANeck (Enhance Attention Neck) module, as shown in Fig. 7.

In EABasic, the image of dimension C * H * W input from the Stem layer is convolved by two consecutive 3 * 3 convolutions to obtain a feature map F of dimension 2C * H * W. The number of channels is increased from C to 2C in the first convolution, and the number of channels does not change in the second convolution. The feature map F is then input to the EABlock, and the feature map weights are weighted using the ECA module as well as the ESA module, and the final output feature map.

In EANeck, the image of dimension C * H * W input from the Stem layer is first convolved by 1 * 1 and the number of channels is changed from C to 2C, and then the feature map width and height are maintained unchanged using 3 * 3 convolution with padding of 1. Finally, the feature map of image dimension 2C * H * W is obtained using 1 * 1 convolution F. Subsequently, the feature map F is input to EABlock, and the feature map weights are weighted using ECA module and ESA module, and finally the feature map is output.

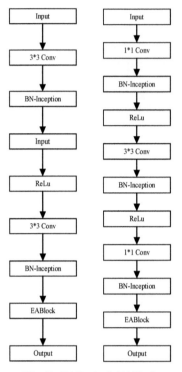

Fig. 7. EABasic & EANeck

3.5   Aggregation Module

When outputting fused features, the output of the aggregation module is redesigned to gradually fuse the extracted feature maps with a view to obtaining richer semantic information, as shown in Fig. 8. That is, the output of branch 4 is first subjected to up-sampling operation, and then feature fusion is performed after unifying with the dimensionality of the output of branch 3 to form a new output 3, and so on, and finally the fused features with the highest resolution are output.

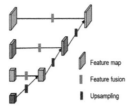

Fig. 8. Aggregation module.

4   Experiment

4.1   MPII Data Set

Description of MPII Data Set. The MPII data set contains 24,987 images, a total of 40,000 different instances of human action, of which 28,000 are used as training samples and 11,000 as testing samples. The label contain 16 key points, which are 0-right ankle, 1-right knee, 2-right hip, 3-left hip, 4-left knee, 5-left ankle, 6- pelvis, 7- chest, 8- upper neck, 9- top of head, 10-right wrist, 11-right elbow, 12-right shoulder, 13- left shoulder, 14- left elbow, 15- left wrist.

Evaluation Criteria. The experiments were trained on the MPII training set, and verified using the MPII validation set. The calibration criteria are accuracy top@1 and top@5. We divide the MPII data set into 20 categories based on behavior, and output 1 and 5 image feature labels respectively after training using the model. if the output labels are consistent with the real labels, then the prediction is correct, and vice versa, the prediction is wrong.

The accuracy top@1 refers to the percentage of the predicted labels that match the true labels in the same batch of data with 1 label output; the accuracy top@5 is the percentage of the predicted labels that contain the true labels in the same batch of data with 5 labels output.

Training Details. The experimental environment in this paper is configured as follows: Ubuntu 20.04 64−bit system, 3 GeForce RTX 2080ti graphics cards, and pytoch1.8.1 deep learning framework is used for training.

The training was performed on the MPII training set with a uniform image scaling crop of 256 * 256. The initial learning rate of the model is 1e−2, which is reduced to 1e−3 in the 60th round, 1e−4 in the 120th round, and 1e−5 in the 180th round. each GPU batch training is 32, and the data are enhanced using random horizontal rotation (p = 0.5) and random vertical rotation (p = 0.5) during the training process.

Experimental Validation Analysis. The data results of this paper on the MPII validation set are shown in Table 1. The results show that our AE-HRNet model compared with the improved HRNet, although increased spatial attention and channel attention, the amount of calculation of the model has increased, from the original 8.03 GFLOPs to 8.32 GFLOPs, but the amount of parameters 41.2 * 107 drops to 40.0 * 107, and the parameter amount is 3% less than HRNet. Compared with HRNet-w32 network,

Table 1. Experimental results of MPII data set.

Network	Parameters(10^7)	Computing power(GFLOPs)	Top@1(%)	Top@5(%)
ResNet-50	34.0	8.92	75.24	–
ResNet-101	54.0	12.41	75.78	–
HRNet-w32	41.2	8.03	73.90	94.06
AE-HRNet[Ours]	40.0	8.32	74.62	**95.03**

AE-HRNet network has an accuracy rate of top@1 increased by 0.72%, and an accuracy rate of top@5 increased by 0.97%.

Since both ResNet50 and ResNet101 in Simple Baseline use pre-trained models, and neither HRNet nor our model use pre-trained models, compared to ResNet50 in Simple Baseline, the accuracy of HRNet-w32 is Top@1 lower than Simple Baseline By 1.34%, our AE-HRNet accuracy rate only dropped by 0.62%.

4.2   COCO Data Set

Description of COCO Data Set. The COCO dataset contains 118287 images, and the validation set contains 5000 images. The COCO dataset contains 17 key points in the whole body in the COCO data set annotation, which are 0-nose, 1-left eye, 2-right eye, 3-left ear, 4-right ear, 5-left shoulder, 6-right shoulder, 7-left elbow, 8-right elbow, 9-left wrist, 10-right wrist, 11-left hip, 12-right hip, 13-left knee, 14-right knee, 15-left ankle, 16-right ankle.

In this paper, we use part of the COCO data set, the training set contains 93,049 images and the validation set contains 3,846 images. It is divided into 11 action categories according to labels, which are baseball bat, baseball glove, frisbee, kite, person, skateboard, skis, snowboard, sports ball, surfboard and tennis racket.

Evaluation Criteria. The tests used for our evaluation criteria on the COCO data set are accuracy top@1 and top@5, and the details are described in MPII Data Set Evaluation Criteria.

Experimental Details. When training on the COCO data set, the images were first uniformly cropped to a size of 256 * 256, and the other experimental details used the same parameter configuration and experimental environment as the MPII data set, as detailed MPII Data Set Experimental Details.

Experimental Validation Analysis. The data results of this paper on the COCO validation set are shown in Table 2. The AE-HRNet model operation volume rises to 8.32 GFLOPs compared with that of HRNet. The number of parameters in the AE-HRNet network is reduced by 3% compared with that of HRNet. At the same time, the accuracy of the AE-HRNet network is 0.87% higher than that of HRNet-w32 on top@1 and 0.46% higher than HRNet-w32.

Compared with ResNet50 in Simple Baseline, the accuracy rate of AE-HRNet has increased by 1.09%, and the accuracy rate of ResNet101 has increased by 1.03%.

Table 2. COCO dataset experimental results.

Network	Parameters(10^7)	Computing power(GFLOPs)	Top@1(%)	Top@5(%)
ResNet-50	34.0	8.93	70.03	–
ResNet-101	54.0	12.42	70.09	–
HRNet-w32	41.2	8.31	70.25	98.27
AE-HRNet[Ours]	39.9	8.32	71.12	**98.73**

5 Ablation Experiment

In order to verify the degree of influence of the ECA module and ESA module on the feature extraction ability of AE-HRNet, AE-HRNet containing only ECA module and ESA module were constructed respectively.

It was trained and validated on the COCO data set and MPII data set respectively, and both were not loaded with pre-trained models, and the experimental results are shown in Table 3.

Table 3. Results of ablation experiments.

Datasets and models	Top@1(%)	Top@5(%)
MPII	74.62	95.03
MPII-WithoutESA	69.19	93.71
MPII- WithoutECA	73.44	94.66
COCO	71.12	98.73
COCO-WithoutESA	69.79	98.26
COCO-WithoutECA	70.10	98.60

On the MPII data set, the accuracy rates of AE-HRNet top@1 and top@5 are 74.62% and 95.03%, respectively. After using only the ECA module, top@1 drops by 5.43%, and top@5 drops by 1.32%; only use After the ESA module, top@1 dropped by 1.18%, and top@5 dropped by 0.37%.

On the COCO data set, the accuracy rate of AE-HRNet top@1 is 71.12%, and the accuracy rate of top@5 is 98.73%. After only using the ECA module, top@1 drops by 1.33%, and top@5 drops by 0.47%; After using the ESA module, the accuracy rate of top@1 dropped by 1.02%, and top@5 dropped by 0.13%.

6 Conclusion

In this paper, we introduced ECA module and ESA module to improve the basic module of HRNet, and built EABasic module and EANeck module to form an efficient human action recognition network AE-HRNet based on high-resolution network and attention mechanism, which can obtain more accurate semantic feature information on the feature map while reducing the complexity of operation and retaining the key spatial location information. The spatial location information, which plays a key role, is retained. This paper improves the accuracy of human action recognition, but further improvement is needed in the parametric number of models.

In addition, this paper is validated on the MPII validation set and the COCO validation set, and a larger data set can be used for action recognition validation if conditions permit; on the premise of ensuring the accuracy of the network model for action recognition, how to perform real-time human action recognition in the video data set is the main direction of future research.

References

1. Wang, H., Cordelia, S.: Action recognition with improved trajectories. In: Proceedings of the IEEE International Conference on Computer Vision (2013)
2. Ji, S., Xu, W., Yang, M., et al.: 3D convolutional neural networks for human action recognition. IEEE Trans. Pattern Anal. Mach. Intell. **35**(1), 221–231 (2012)
3. Liu, L.X., Lin, M.F., Zhong, L.Q., et al.: Two-stream inflated 3D CNN for abnormal behaviour detection. Comput. Syst. Appl. **30**(05), 120–127 (2021)
4. Zeng, M.R., Luo, Z.S., Luo, S.: Human behaviour recognition combining two-stream CNN with LSTM. Mod. Electron. Technol. **42**(19), 37–40 (2019)
5. Wang, L., Xiong, Y., Wang, Z., et al.: Temporal segment networks: towards good practices for deep action recognition. In: European Conference on Computer Vision. Springer, Cham, pp. 20–36 (2016). https://doi.org/10.1007/978-3-319-46484-8_2
6. Lan, Z., Zhu, Y., Hauptmann, A.G., et al.: Deep local video feature for action recognition. In: Proceedings of the IEEE Conference on Computer Vision and Pattern Recognition Workshops, pp. 1–7 (2017)
7. Zhao, L., Wang, J., Li, X., et al.: Deep convolutional neural networks with merge-and-run mappings. arXiv preprint arXiv:1611.07718 (2016)
8. Carreira, J., Zisserman, A.: Quo vadis, action recognition? A new model and the kinetics dataset. In: Proceedings of the IEEE Conference on Computer Vision and Pattern Recognition, pp. 6299–6308 (2017)
9. Jie, H., Li, S., Gang, S., et al.: Squeeze-and-excitation networks. IEEE Trans. Patt. Anal. Mach. Intell. **PP**(99) (2017)
10. Woo, S., Park, J., Lee, J.Y., et al.: CBAM: Convolutional Block Attention Module. arXiv preprint arXiv:1807.06521v. Springer, Cham (2018). https://doi.org/10.1007/978-3-030-012 34-2_1
11. Guo, H.T., Long, J.J.: High efficient action recognition algorithm based on deep neural network and projection tree. Comput. Appl. Softw. **37**(4), 8 (2020)
12. Li, K., Hou, Q.: Lightweight human pose estimation based on attention mechanism[J/OL]. J. Comput. Appl. 1–9 (2021). http://kns.cnki.net/kcms/detail/51.1307.tp.20211014.1419.016. html
13. Sun, K., Xiao, B., Liu, D., et al.: Deep High-Resolution Representation Learning for Human Pose Estimation. arXiv e-prints (2019)

Recognition Model Based on BP Neural Network and Its Application

Yingxiong Nong, Zhibin Chen, Cong Huang, Jian Pan, Dong Liang, and Ying Lu[✉]

Information Center of China Tobacco Guangxi Industrial CO. LTD., Nanning, Guangxi, China
03429@gxzy.cn

Abstract. The BP neural network model used in data classification can change the traditional manual classification, which has the disadvantages of low efficiency and subjective interference. According to the principle of BP, this paper determines the relevant parameters of network structure, and establishes an optimized BP. The BP model is used to analyze the chemical composition data of tobacco leaves to determine the grade of tobacco leaves. Experiments show that this model has better recognition accuracy than KNN and random forest model. It effectively improves the efficiency of classification and reduces the interference of subjective factors in classification.

Keywords: BP neural network · Classification · Data normalization · Tobacco grade

1 Introduction

Tobacco leaf is an important raw material of the tobacco industry. Its grade purity will directly affect the quality and taste of cigarettes produced by the tobacco industry. Therefore, the classification of tobacco leaf grade is of great significance [1]. In the traditional tobacco grading process, it mainly depends on relevant professionals to comprehensively evaluate the tobacco grade, and identify the tobacco grade through vision, touch, smell and other senses. The classification method of artificial tobacco leaf has strong subjectivity and is closely related to the experience of professionals. Different experts may classify tobacco leaves into different grades, which is inefficient, difficult to guarantee the accuracy, and consumes a lot of human and material resources [2]. In view of the limitations of manual classification of tobacco leaves, some technical schemes have been put forward in relevant literature. Literature [3] proposed to use band light source and light intensity to classify the grade of tobacco leaves. Literature [4] proposed tobacco classification based on clustering and weighted k-nearest neighbor, and classified tobacco classification according to infrared spectroscopy. Reference [5] used entropy method to weight the features of samples, introduced the weight of features in the calculation of sample distance, and used KNN algorithm to classify tobacco leaf chemical composition data. If there is a lot of noise in tobacco data, KNN classification cannot eliminate the interference of noise, so the accuracy will be affected. Literature [6] applies random forest algorithm to tobacco grade classification, which can achieve good results when

Z. Qian et al. (Eds.): WCNA 2021, LNEE 942, pp. 292–302, 2022.
https://doi.org/10.1007/978-981-19-2456-9_31

there are many samples in the data set. However, the random forest algorithm cannot show its advantages on the small sample data set in this paper. Literature [7] proposed an automatic classification method of tobacco leaves based on machine vision, which realizes the classification of tobacco leaves according to the feature extraction and recognition of tobacco images. However, in the process of tobacco leaf image recognition, the actual situations such as folding of tobacco leaf images and mixing of front and back sides of tobacco leaves are not considered. Literature [8] proposed to classify tobacco grades by near-infrared spectroscopy and use partial least squares discrimination method to classify tobacco grades. However, infrared spectroscopy equipment is expensive and cannot be used on a large scale. Aiming at the above problems, this paper studies the tobacco grade recognition technology based on BP model. BP has strong nonlinear mapping ability and associative memory for external stimuli and input information, so it has strong recognition and classification ability for input samples [9]. BP has high accuracy in tobacco leaf chemical composition data set classification and solve the disadvantages of low efficiency and strong subjectivity.

2   Data Acquisition and Analysis of Tobacco Grade

The chemical composition of tobacco leaf is one of the important factors affecting the taste and quality of cigarette [10], which includes reducing sugar, total alkaloids, total sugar, potassium, total nitrogen, starch and other components. The experimental data of this paper come from different flue-cured tobacco bases in Guangxi, Yunnan, Chongqing and Hunan of China. Flue-cured tobacco leaves are mainly divided into four grades: B2F, C2F, C3F and X2F. The BP model is introduced to identify the tobacco chemical composition data set. When the tobacco grade needs to be divided, the predicted tobacco grade information can be obtained by inputting the tobacco chemical composition information. Table 1 is partial records in the database about the chemical composition data and grades of tobacco leaves.

Table 1. Chemical composition and grades of tobacco leaves.

Total sugar (%)	Reducing sugar (%)	Total alkaloids (%)	K(%)	Cl (%)	Total N (%)	Starch (%)	Tobacco Leaf Grade
27.2	23.1	0.78	3.39	2.18	2.18	5.26	B2F
32.0	27.5	0.56	2.33	2.48	1.49	6.25	C2F
30.6	25.2	0.59	2.94	2.35	1.86	5.65	C3F
30.1	27.8	0.53	2.49	2.57	1.70	5.78	C2F
29.8	28.2	0.31	2.61	2.94	1.82	6.60	C3F
20.2	18.7	0.15	3.83	2.43	2.10	4.54	B2F
32.4	27.0	0.11	2.00	2.96	1.43	3.79	X2F

Table 2 summarize the proportion of chemical components contained in B2F tobacco grade, C2F tobacco grade, C3F tobacco grade and X2F tobacco grade.

Table 2. Chemical proportion of tobacco grades.

Composition	B2F Proportion	C2F Proportion	C3F Proportion	X2F Proportion
Total sugar	15.60%–44.6%	24.9%–42%	16.2%–45.2%	22.8%–45.8%
Cl	0.08%–1.21%	0.2%–0.62%	0.03%–1.11%	0.02%–1.08%
Total N	1.36%–2.85%	1.4%–2.27%	1.07%–2.76%	1.03%–2.67%
Starch	1.31%–9.77%	1.72%–9.22%	1.25%–13.18%	1.39%–8.95%
K	1.00%–3.79%	1.51%–2.93%	1.37%–4.97%	1.59%–5.22%
Reducing sugar	11.5%–35.6%	20%–31.42%	13.2%–35.5%	18.9%–33.58%
Total alkaloids	0.72%–5.08%	1.55%–4.5%	0.96%–4.1%	0.81%–4.73%

It can be seen from Table 2 that in the proportion of chemical components of B2F tobacco grade, total sugar accounts for the highest proportion of all chemical components and chlorine accounts for the lowest proportion. The fluctuation range of total sugar and reducing sugar is the largest. The total sugar can reach 15.6% at the lowest time and 44.6% at the highest time. Reducing sugar accounted for 11.5% at the lowest time and 35.6% at the highest time.

It can be seen from Table 2 that in the proportion of chemical composition of C2F tobacco grade, the overall change trend of chemical composition of tobacco leaf is consistent with that of other grades, the proportion of total sugar is the highest, followed by reducing sugar. But the difference is that the lowest proportion of total sugar is 24.9%, and the lowest proportion of reducing sugar is 20%, which is higher than other grades. In the proportion of chlorine, the lowest is 0.2% and the highest is 0.62%, which is much higher than other grades.

It can be seen from Table 2 that in the proportion of chemical composition of C3F tobacco grade, the proportion trend of chemical composition of tobacco leaf is generally consistent with that of other grades. However, compared with B2F, the proportion of potassium in C2F can reach 4.97%, which is higher than that of 3.79% and 2.93% in B2F and C2F. The highest proportion of starch was 13.18%, which was also higher than the other three grades.

According to Table 2, in the proportion of chemical composition of X2F tobacco grade, the proportion of total sugar and reducing sugar is much higher than that of B2F tobacco grade and C2F tobacco grade, second only to that of C2F. However, the change trend of overall component proportion is similar to that of B2F grade.

From Table 2, it can be found that the chemical composition information of Different Tobacco Grades changes greatly, and the chemical composition proportion between each tobacco grade also has great similarity. If identified by professionals, when the chemical composition proportions of two different grades of tobacco leaves are relatively similar, it is difficult for professionals to determine what grade the two kinds of tobacco leaves belong to. Because the proportion of chemical components between different grades is not stable in a small range, on the contrary, it will fluctuate in a large range, which may also lead to overlap between different tobacco grades. Therefore, if professionals only rely on experience and personal subjectivity to judge the grade of tobacco leaves, there are defects.

3 Establishment of Tobacco Grade Recognition Model Based on BP

The chemical composition of tobacco leaves is analysed to judge the grade of the tobacco leaves. This problem belongs to the classification problem of machine learning. To realize multi-dimensional data classification, BP is hierarchical, which is composed of input layer, middle layer and output layer. All neurons in adjacent layers are fully connected. Each neuron obtains the input response of the BP network and generates the connection weight. From the output layer to each intermediate layer, the connection weight is corrected layer by layer by reducing the error between the desired output and the actual output, and returned to the input layer. The process is repeated, and it is completed when the global error of the network tends to the given minimum value [11].

3.1 Input Data Preprocessing

The main factor affecting the grade is the chemical composition. The total sugar, reducing sugar, total alkaloids, potassium, chlorine, total nitrogen and starch in the tobacco chemical composition data set are determined as seven characteristics, which are set as the BP input layer data and expressed by x1, X2,…, X7 respectively. Take the tobacco grade as the BP output layer data, expressed by Y.

The tobacco data were normalized. The normalization of data sets can effectively raise the prediction accuracy and accelerate the convergence speed of the model. The input data X1, X2,…, X7 of the network are linearly normalized and processed according to Formula (1).

$$x = \frac{x - x_{min}}{x_{max} - x_{min}} \tag{1}$$

Encode the BP output layer data: 1 represents B2F tobacco grade, 2 represents C2F tobacco grade, 3 represents C3F tobacco grade, and 4 represents X2F tobacco grade.

3.2 BP Network Structure Design

(1) Input and output layer design

The input index of BP model is the chemical composition of tobacco leaves, and the output is the grade of tobacco leaves. So, the input layer has 7 nodes and the output layer has 1 nodes.

(2) Hidden layer design

When BP has enough hidden layer nodes, it can approximate the nonlinear function with arbitrary accuracy [12]. Therefore, a three-layer BP model is adopted in this paper. But too many hidden layer neurons will not only increase the computational complexity, but also produce the problem of over fitting [13]. Too few hidden layer neurons will affect the accuracy of output results. Generally, the number of hidden layer nodes is determined by Formula (2).

$$h = \sqrt{m+n} + a \tag{2}$$

The parameters h, m and n in Formula (2) are the number of hidden layer nodes, the number of input layer nodes and the number of output layer nodes respectively. And a is a constant between [1, 10]. According to Formula (2), the number of neurons in the hidden layer is calculated to be between 3 and 13. In this paper, the number of BP hidden layer neurons is set as 6. The BP design is shown in Fig. 1.

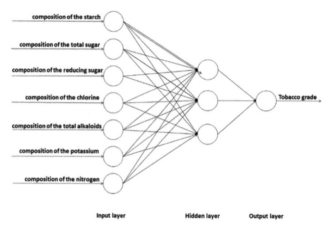

Fig. 1. BP design drawing.

(3) Activate function selection

The activation function of the hidden layer in the BP is a nonlinear function [14], because the combination of linear functions is a linear function itself. Increasing the number of network layers can not calculate more complex functions, so the nonlinear function must be introduced. Types of activation functions: ReLU, Sigmoid, Tanh, etc. The ReLU, Sigmoid and Tanh are shown in Formulas (3), (4) and (5) respectively.

$$f(x) = \frac{A}{1 + e^{-x}} \tag{3}$$

$$f(x) = \frac{e^{2x} - 1}{e^{2x} + 1} \tag{4}$$

$$f(x) = \max(0, x) \tag{5}$$

The research shows that the ReLU activation function is generally used for hidden layers. For the output layer, if it is classified and split, the Sigmoid function is used

[14]. Sigmoid function represent output probability. The prediction of tobacco grade is realized by inputting relevant attribute values through the joint action of input layer, hidden layer and output layer.

3.3 BP Network Training

The training of BP model includes the forward propagation process of data set and the back propagation process of error. Forward propagation of data set: represent the chemical composition data and tobacco grade information contained in tobacco leaves with (x, y), and input the sample data into BP model. At the same time, set the weight of the network model and the threshold of the last iteration, and the output of neurons is calculated layer by layer. Error back propagation: determine the influence gradient of the weight and threshold of the last layer and the previous layers on the total error, and then modify the weight and threshold to minimize the target error. The following steps are the network training process.

(1) Initialize the network model. The data set includes the chemical composition of tobacco leaves and the corresponding grade of tobacco leaves. The input data is the chemical composition X of tobacco leaves, and the number of input features is expressed by P. The number of hidden layers is expressed in M. The output layer is tobacco grade y, because there is only one output, and the number of output layers is 1.

(2) Get hidden layer data R. Input x_i according to the characteristics of tobacco chemical information x_i. The weights of input layer and hidden layer are ω_{ij}, hidden layer threshold a_j. Calculate the hidden layer output as R. As shown in Formula (6).

$$R_j = f\left(\sum_{i=1}^{p} \omega_{ij}x_i - a_j\right), j = 1, 2, \ldots, m \qquad (6)$$

(3) According to the hidden layer output R, the weight between the hidden layer and the output layer ω_j, and the output layer threshold b to calculate the tobacco grade prediction L.

$$L = g\left(\sum_{j=1}^{m} R_j w_j - b\right) \qquad (7)$$

Where f represents the hidden layer activation function ReLU and g represents the output layer activation function Sigmoid. After obtaining the prediction output L, BP prediction error E is calculated from the expected output Y using Formula (8). The smaller the value of MSE, the better the accuracy of the prediction model.

$$e = \frac{1}{2}(L - Y)^2 \qquad (8)$$

According to the error E, the weight ω_{ij} and threshold a_j between the network input layer and the hidden layer is updated. And the weight ω_j and Threshold b between the hidden layer and the output layer is updated. η indicates the learning rate.

$$\omega_{ij} = \omega_{ij} + \eta(1 - R_j)x_i\omega_j e, i = 1, 2, \ldots, p; j = 1, 2, \ldots, m \qquad (9)$$

$$\omega_j = \omega_j + \eta H_j e, j = 1, 2, \ldots, m \qquad (10)$$

$$a_j = a_j + \eta H_j \left(1 - R_j\right)\omega_j e, j = 1, 2, \ldots, m \qquad (11)$$

$$b = b + e \qquad (12)$$

(4) Finally, the end of training is judged according to whether the target error is reached or the number of iterations. If satisfied, it ends. Otherwise, return to step 2.

4 Simulation Experiment

4.1 Experimental Setup

Set the relevant parameters of BP. Set the excitation functions of the BP hidden layer and output layer as ReLU and Sigmoid respectively, the BP training function Traingdx and BP performance is evaluated by MSE. The characteristic numbers of input layer, hidden layer and output layer are 7, 6 and 1 respectively. Number of iterations Epochs, expected error e, learning rate η are set to 6000, 0.000001 and 0.02 respectively.

4.2 Analysis of Experimental Results

Figures 2, 3, 4 and 5 show the prediction results of the system for different tobacco grades.

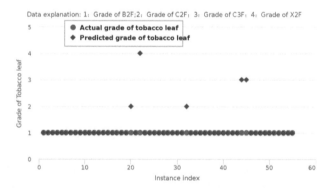

Fig. 2. Comparison of actual and predicted tobacco leaf grade of B2F.

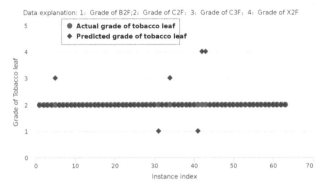

Fig. 3. Comparison of actual and predicted tobacco leaf grade of C2F.

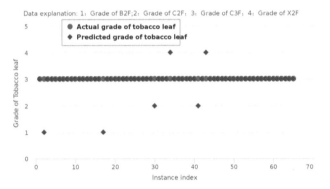

Fig. 4. Comparison of actual and predicted tobacco leaf grade of C3F.

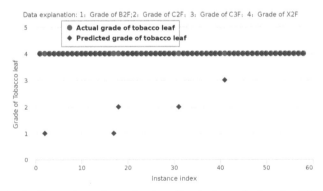

Fig. 5. Comparison of actual and predicted tobacco leaf grade of X2F.

Figures 2, 3, 4 and 5 show the prediction results of four tobacco grades. The ordinate in the figure represents the tobacco grade, including 1: B2F grade, 2: C2F grade, 3: C3F grade and 4: X2F grade. The orange dot indicates the actual tobacco grade, and the blue dot indicates the predicted tobacco grade. When the actual tobacco grade is consistent with the predicted tobacco grade, two points will coincide, that is, when all points are on the line corresponding to the grade, the prediction result is the best. It can be observed that in the test set data, the predicted grade of most tobacco sample data can well coincide with the actual grade, which shows that the model can correctly predict the tobacco grade of most tobacco sample data. However, there are still a few data that cannot be correctly identified, which may be related to the tobacco data itself. The proportion of chemical components of different grades of tobacco leaves is the most highly similar. In addition, it may also be related to the model itself. The selection of the number of hidden layer neurons and hidden layer layers of the BP model and the selection of activation function will have a certain impact on the prediction accuracy of the model.

In the data set, 70% is set as the training set, and the training model is established by BP neural network algorithm. The remaining 30% data were used as a test set to predict 30% tobacco grade. Finally, the predicted grade is compared with the actual grade of 30% tobacco leaves and displayed at the front of the web page. The effect is shown in Fig. 2, 3, 4 and 5, and the prediction results are shown in Table 3. The recognition rate of B2F grade of tobacco leaves reached 90.09%, C2F grade of tobacco leaves reached 90.47%, C3F grade of tobacco leaves reached 90.77%, X2F grade of tobacco leaves reached 91.38%, and the overall average recognition rate was 90.67%.

Table 3. Tobacco leaf grade prediction results under BP model.

Tobacco grade name	Number of test samples	The number of Correct identification	Recognition rate
B2F	55	50	90.90%
C2F	63	57	90.47%
C3F	65	59	90.77%
X2F	58	53	91.38%

The above literature mentioned that KNN and random forest are applied to tobacco grade recognition. Now these two algorithms are compared with BP. See Table 4 for comparison results. The data set in this paper belongs to small samples and data with noise. BP has nonlinear characteristics. By fitting the change law of input data through multi-layer neurons, it can denoise and fit small sample data, so it can obtain higher classification accuracy.

Table 4. Comparison of tobacco leaf grade recognition rate.

Tobacco grade name	Random Forest	KNN	BP
B2F	87.27%	85.45%	90.90%
C2F	88.89%	87.30%	90.47%
C3F	87.69%	84.62%	90.77%
X2F	91.38%	86.21%	91.38%

5 Conclusion

With the higher and higher requirements of customers for the quality of tobacco leaves, the current manual grading of tobacco leaves has some limitations, such as strong subjectivity, consuming human and material resources and so on. In this paper, the chemical composition data of tobacco leaves are used as the training set, the BP model is established, and the tobacco grade classification technology based on BP is developed. The purpose is to solve the disadvantages of low efficiency and high subjectivity of artificial tobacco grading. Experiments show that the proposed algorithm achieves better recognition accuracy than KNN and random forest. Deep neural network has better performance than traditional neural network and has been widely used [15]. In the next step, we will use deep neural network to predict tobacco grade.

Acknowledgments. This research is funded by the Guangxi Science and Technology Planning Project (GX[2016] No. 380), and the Science and Technology Planning Project of Guangxi China Tobacco Industry Co., Ltd. (No. GXZYCX2019E007).

References

1. Tan, X., Yunlan, T., Yingwu, C.: Intelligent classification method of flue-cured tobacco based on rough set. J. Agric. Mach. **06**, 169–174 (2009)
2. Shuangyan, Y., Zigang, Y., Siwei, Z., et al.: Automatic tobacco classification method based on near infrared spectroscopy and PSO-SVM algorithm. Guizhou Agric. Sci. **46**(12), 141–144 (2018)
3. Zhiqian, Q.: Effects of different light sources and light intensity on tobacco classification. Guizhou university, China Guiyang (2020)
4. Hang, L.: The research on tobacco classification based on clustering and weighted KNN . China Zhengzhou: Zhengzhou university (2017)
5. Hui, Z., Kaihu, H., Zhou, Z.: Application of EM-KNN algorithm in classification of re-dried tobacco leaves. Software **39**(06), 96–100 (2018)
6. Hari, S., Maria, P.A.: Prediction of tobacco leave Grades with ensemble machine learning methods. In: International Congress on Applied Information Technology, pp. 1–6 (2019)
7. Zhenzhen, Z.: Method for automatic grading of tobacco based on machine vision. China Chongqing: Southwest university (2016)

8. Guo, T., Kuangda, T., Zuhong, L., et al.: Classification of tobacco grades by near-infrared spectroscopy and PLS-DA. Tobacco Sci. Technol. **309**(04), 60–62 (2013)
9. Qing, C., Wei, L., Kejun, Z.: A neural network recognition model based on aroma components in tobacco. J. Hunan Univ. **33**(02), 103–105 (2006)
10. Guiting, H., Chengchao, Z., Weijun, Z., Zhengjiang, Z.: Application of BP neural network based on model identification in photovoltaic system MPPT. Comput. Meas. Control **25**(10), 213–216 (2017)
11. Lin, W., Zhihong, L., Zicheng, X.: Study on relationship between acid aroma with polyphenol content, chemical composition and taste characteristics of flue-cured tobacco. J. Agric. Sci. Technol. **21**(05), 159–169 (2019)
12. Qiyi, Q., Chengxiang, G., Shuai, W., Xuyi, Y., Ningjiang, C.: On BP neural network optimization based on particle swarm optimization and cuckoo search fusion. J. Guangxi University (Nat Sci Ed) **45**(04), 898–905 (2020)
13. Runa, A.: Research on text classification based on improved convolutional neural network. Inner Mongolia University for Nationalities, China Tongliao (2020)
14. Xiao, Q., Chengcheng, H., Shi, Y., et al.: Research progress of image classification based on convolutional neural network. Guangxi Sci. **27**(6), 587–599 (2020)
15. Konovalenko, I., Maruschak, P., Brezinová, J., et al.: Steel surface defect classification using deep residual neural network. Metals **846**(10), 1–15 (2020)

Multidimensional Data Analysis Based on LOGIT Model

Jiahua Gan[1,2(✉)], Meng Zhang[3], and Yun Xiao[3]

[1] Transport Planning and Research Institute, Ministry of Transport, Beijing 100028, China
ganjh@tpri.org.cn
[2] Laboratory for Traffic and Transport Planning Digitalization, Beijing 100028, China
[3] School of Urban Construction and Transportation, Hefei University, Hefei, Anhui, China

Abstract. Logit Model is an important method for empirical analysis of multi-source data. In order to explore the traffic safety mechanism, The Paper taked traffic behavior data as an example, researched personal characteristics of truck drivers, Analyzed the influence of the driver's personal traits on traffic violations. Based on the binary logistics regression model, the analysis model of traffic violations was established. The results show that personality, driver's license level, daily driving time, transportation route, vehicle ownership, and occupational disease are important factors that affect drivers' violations. Further data analysis shows that truck drivers with bile personalities, driving for more than 12 h per day, no fixed transportation routes, and vehicles with loans have the highest probability of violations. The data analysis conclusion provides data basis for truck driver management and improving truck traffic safety.

Keywords: Truck transportation · Traffic violations · Logistics regression model · Behavior analysis · Data mining

1 Introduction

People, vehicles, roads, and the environment are the four elements of traffic safety, among which people have a significant impact on safe driving. According to the traffic accident statistics of various countries in the world, road traffic accidents caused by human factors are as high as 80% to 90%, and road traffic accidents caused by drivers themselves account for more than 70% [1]. By analyzing the psychological factors of drivers and combining them with questionnaire surveys, Yang Yu et al. proposed improving the psychological quality of drivers in order to achieve driving safety [2]. Wu Di et al. analyzed the traffic accidents in Anhui Province in 2019. Among the 22 large road traffic accidents with more than 3 deaths, those caused by the illegal behavior of drivers accounted for the majority [3].

Driving behavior has a significant impact on traffic safety [4]. Yan Ge et al. studied the association between impulsive behavior and violations using data from 299 Chinese drivers. The results show that the driver's impulsivity is positively correlated with the driver's positive behavior and some common violations. The other three dimensions

© The Author(s) 2022
Z. Qian et al. (Eds.): WCNA 2021, LNEE 942, pp. 303–315, 2022.
https://doi.org/10.1007/978-981-19-2456-9_32

of dysfunction are negatively correlated with positive driving behavior, and positively correlated with abnormal driving behavior and fines [5]. Zhang Mengge et al. established an association model between road conditions and abnormal driving behavior based on current research status of driving behavior at home and abroad, combined with data of abnormal driving behavior from the Internet of Vehicles OBD, thereby establishing a research idea for identifying road traffic safety risks.

Many scholars have paid attention to the correlation between the driver's personal characteristics and driving behavior [6]. Lourens et al. deduced from the Dutch database that there is a relationship between violations and traffic accidents in different types of annual mileage and that there is no difference in the degree of involvement of male and female drivers in accidents. The rate of accidents among young drivers is the highest [7]. Wang et al. employed the Eysenck Personality Questionnaire (EPQ) and the Symptom Self-Rating Scale (SCL-90-R) to assess the personality and mental health of truck drivers, as well as investigate the link between mental health and personal traits. These findings provide a theoretical foundation for truck driver selection and intervention strategies for high-risk drivers, which will help to better manage road traffic safety construction and reduce road traffic injuries.

The Logistic regression model has been used by many researchers to investigate the association between a driver's personal characteristics and traffic safety behavior. Lin Qingfeng et al. built a Logistic regression model to analyze the relationship between motor vehicle driver attributes, non-motor vehicle driver attributes, motor vehicles, non-motor vehicles, roads, and the environment, and the relationship between the driver's fault and the severity of the accident. The results show that the severity of motor vehicle accidents is significantly related to seven variables, including the motor vehicle driver's driving age, motor vehicle safety status, road alignment, and the alignment and motor vehicle driver's fault [8]. Tian Sheng et al. utilized Pearson correlation analysis and multiple regression model analysis to survey 1,800 primary and middle school children in Guangzhou, and the results showed that education, awareness, attitude, and personal variables influence young people's traffic safety practices [9].

The current study has conducted a pretty extensive investigation into the relationship between driver behavior and traffic safety. However, its concentration is primarily on ordinary drivers, with little investigation into the features of truck drivers. This article investigates the impact of truck drivers' personal characteristics on violations, investigates the relationship between the two, and searches for appropriate personal characteristics for truck drivers in order to provide a theoretical foundation and reference for professional truck driver selection.

2   Research Methods

2.1   LOGISTIC Regression Model

The Logistics regression model is a classification model that investigates the link between classification outcomes and affecting factors. It can be defined as the likelihood of influencing factors on a specific outcome. The Logistic regression model is an important model for assessing personal traffic behavior in the field of road traffic. It can analyze the impact of one or more influencing factors on a non-numerical classification result, and more accurately and comprehensively describe the decision-making behavior of individuals or groups, has achieved relatively rich research results. This paper applies it to the field of truck transportation safety analysis, employing a binary logistic regression model and a truck driver's driving behavior selection model based on the model theory. The model is constructed and calibrated using personal information collected from truck drivers via online questionnaire surveys.

The driving dependent variable y of the model is a binary variable with values of 1 and 0, and x is a risk factor that affects y. Let the probability of $y = 1$ under the condition of x be:

$$P = P(y = 1|x) = \frac{e^{\alpha+\beta x}}{1 + e^{\alpha+\beta x}} = \frac{\exp(\alpha + \beta x)}{1 + \exp(\alpha + \beta x)} \tag{1}$$

This article mainly adopts the binary logistic regression model, and its mathematical model is:

$$P = P(y = 1|x) = \frac{e^{\alpha+\beta x}}{1 + e^{\alpha+\beta x}} = \frac{\exp(\alpha + \beta_1 x_1 + \beta_2 x_2 + \beta_2 x_2 + \cdots \beta_K x_K)}{1 + \exp(\alpha + \beta_1 x_1 + \beta_2 x_2 + \cdots \beta_K x_K)} \tag{2}$$

2.2 Questionnaire Design and Survey

Questionnaire Design. The author designs a questionnaire based on some phenomena existing in reality and combines them with existing related research. According to Song Xiaolin et al.'s examination of connected accidents, men were responsible for a higher proportion of road accidents caused by speeding than women [10]. Lourens et al. found that age is related to drivers' violations [7]. Chuang and Wu found that sleep problems can cause stress in professional drivers [6]. Salar Sadeghi Gilandeh found that driving behavior is related to road conditions [5]. Gender, age, education level, years of employment, personality, household registration, driver's license level, and other factors are combined in this article to create a questionnaire with a total of 18 factors, including the truck driver's gender, age, education level, years of employment, personality, household registration, driver's license level, and so on.

From a psychological point of view, the driver's personality is divided into depressive qualities (sensitive, frustrated, withdrawn, indecisive, slow recovery from fatigue, slow response), and bloody (calm, tolerant, focused and hardworking, patient and hardworking. But inflexibility, lack of enthusiasm, conservatives), mucus quality (enthusiasm, ability, adaptability, wit, lack of focus, changeable emotions, lack of patience), bile quality (excited, short-tempered, straightforward, enthusiastic, But the mood is lower when the energy is exhausted).

Data Acquisition and Processing. In order to improve the accuracy of the data, this survey uses the real-name system to fill in the blanks. In order to meet the universality, we chose to put the questionnaire online and send the link to the truck driver through the truck company in Anhui Province to collect the questionnaire. Truck drivers are required to fill out the questionnaire objectively and impartially. The business managers will answer the questions that the driver has. Finally, a total of 1354 papers have been filled out. There is no invalid questionnaire due to the driver's personal reasons, and the effective questionnaire is 100%.

Table 1. Driver's statistical information.

Category	Frequency	Percentage	Category	Frequency	Percentage
Gender			Age		
Male	1331	98.30%	≤25	18	1.33%
Female	23	1.70%	26-35	285	21.05%
Education level			36-45	661	48.82%
Junior high school and below	880	64.99%	46-55	368	27.18%
Senior middle school	350	25.85%	≥56	22	1.62%
Junior college	100	7.39%	Years of employment		
Bachelor and above	24	1.77%	1-2 years	66	4.87%
Personality			3-5 years	161	11.89%
Depressive qualities	77	5.69%	6-10 years	325	24.00%
Bloody	643	47.49%	More than 10 years	802	59.24%
Mucous quality	326	24.08%	Driver's license level		
Bile	308	22.75%	A2	867	64.03%
Household registration			A1	48	3.55%
Rural	994	73.41%	B2	380	28.06%
Urban	360	26.59%	B1	14	1.03%
Monthly mileage			C1	45	3.32%
Below 5000KM	266	19.65%	Daily driving time		
5000-10000KM	584	43.13%	Less than 8 hours	632	46.68%
10000-15000KM	337	24.89%	8-10 hours	397	29.32%
15000-20000KM	85	6.28%	10-12 hours	190	14.03%
Above 20000KM	82	6.06%	12 hours or more	135	9.97%
Drive for four consecutive hours			Whether there is a fixed transportation route		
Yes	1284	94.83%	Yes	730	53.91%
No	70	5.17%	No	624	46.09%
Several days off each month			Number of drivers in the car		
1-2 days	141	10.41%	1 people	921	68.02%
3-4 days	329	24.30%	2 people	422	31.17%
5-8 days	301	22.23%	2 people or more	11	0.81%
More than 8 days	205	15.14%	Monthly income(yuan)		
No rest, wait for the goods to rest	378	27.92%	Below 5000	176	13.00%
Vehicle ownership			5000-8000	426	31.46%
Owned vehicles have no arrears	502	37.07%	8000-10000	347	25.63%
Owned vehicle has arrears	539	39.81%	10000-15000	279	20.61%
Hired to drive	313	23.12%	More than 15000	126	9.31%
Whether there is an occupational disease			Vehicle attachment situation		
No	594	43.87%	Atachment	1132	83.60%
Cervical spondylosis	503	37.15%	Semi-attached	222	16.40%
Hypertension	66	4.87%	Violation of the previous year		
Heart disease	9	0.66%	0 times	416	30.72%
Stomach disease	165	12.19%	1 times	232	17.14%
Other disease	17	1.26%	2 times	278	20.53%
			3 times or more	428	31.61%

3   Establishment and Improvement of Driving Violation Behavior Model

3.1   Descriptive Statistical Analysis

Truck drivers are the subjects of this study. According to statistics, a total of 1354 people were investigated, including 938 people who violated regulations and 416 people who did not. There are 1331 male drivers and 23 female drivers (Table 1).

3.2   Reliability Analysis

In this paper, the Cranbach α coefficient is used to analyze the reliability of the questionnaire through SPSS 23.0 software, and the calculation result is α = 0.143 (Table 2).

Table 2.   Driver's statistical information.

Kronbach Alpha	Kronbach Alpha based on standardized terms	Number of category
0.143	0.120	18

The SPSS 23.0 software was used to analyze the validity of the questionnaire, and the results are shown in Table 3. The KMO coefficient is 0.680, which is greater than 0.50, and the Sig value is 0.00, which is less than 0.05. Therefore, factor analysis can be performed.

Table 3.   Kmo and Bartlett test.

Kmo sampling appropriateness quantity		0.680
Bartlett sphericity test	Approximate chi-square	2610.199
	Degree of freedom	153
	Saliency	0.000

3.3   Logistic Model Analysis

The Choice of Dependent and Independent Variables. Based on whether truck drivers violate the regulations, the total number of people is planned to be classified into two types: violation and non-violation. The value of the dependent variable Y is shown in the table below. As shown in Table 4, according to the questionnaire data, all items are set as independent variables (X).

Table 4. Dependent variable.

Y	0	No violation
	1	Violation

Initially, we used the SPSS 23.0 software to perform binary logistic regression analysis on 18 factors, with a significance level of $= 0.05$ and the forward LR method (forward stepwise regression method based on maximum likelihood estimation). First, use the score test method to screen the independent variables. According to whether the p value corresponding to the score value meets the given significance level, the variables that meet the requirements are initially selected as shown in Table 5.

Table 5. Score test result.

Influencing factors	Score	Degree of freedom	Saliency
Age	0.049	1	0.825
Gender	0.748	1	0.387
Personality	5.315	1	0.021
Years of employment	5.272	1	0.022
Education level	16.525	1	0
Household registration	10.960	1	0.001
Driver's license level	6.087	1	0.014
Monthly mileage	14.535	1	0
Daily driving time	24.613	1	0
Drive for four consecutive hours	1.500	1	0.221
Whether there is a fixed transportation route	3.856	1	0.050
Several days off each month	1.463	1	0.226
Number of drivers in the car	1.442	1	0.230
Monthly income(yuan)	50.403	1	0
Vehicle ownership	24.602	1	0
Whether there is an occupational disease	35.437	1	0
Vehicle attachment situation	12.656	1	0

Table 6. Model (if item is removed).

Step	Variable	Degree of freedom	Saliency
1	Monthly income(yuan)	1	0.000
2	Monthly income(yuan)	1	0.000
	Vehicle attachment situation	1	0.000
3	Education level	1	0.002
	Monthly income(yuan)	1	0.000
	Vehicle attachment situation	1	0.000
4	Education level	1	0.002
	Monthly income(yuan)	1	0.000
	Vehicle attachment situation	1	0.001
	Whether there is an occupational disease	1	0.003
5	Education level	1	0.001
	Monthly income(yuan)	1	0.000
	Vehicle ownership	1	0.017
	Vehicle attachment situation	1	0.026
	Whether there is an occupational disease	1	0.002
6	Education level	1	0.002
	Daily driving time	1	0.024
	Monthly income(yuan)	1	0.000
	Vehicle ownership	1	0.013
	Vehicle attachment situation	1	0.061
	Whether there is an occupational disease	1	0.009

Determine the significance of all the influencing factors according to the preliminary test, and then gradually substitute all the influencing factors into the equation. When the parameter estimation value changes by less than 0.001, the estimation is terminated at the 7th iteration, and the following results are initially obtained, as shown in Table 6.

Model Checking. In this comprehensive test of the binary logistic regression model coefficients, one line of the model outputs the likelihood ratio test results of whether all the parameters in the logistic regression model are 0, as shown in Table 7. Where the significance level is less than 0.05, it means that the OR value of at least one of the included variables in the fitted model is statistically significant, that is, the model is overall meaningful.

Table 7. Comprehensive test of model coefficients.

		Chi-square	Degree of freedom	Saliency
Step 6	Step	5.108	1	0.024
	Block	97.778	6	0.000
	Model	97.778	6	0.000

In this paper, Hosmer and Lemeshow tests are used to test the goodness of fit of the model, and the calculated significance level is 0.781 > 0.005, which indicates that the model fits well, as shown in Table 8.

Table 8. Comprehensive test of model coefficients.

Step	Chi-square	Degree of freedom	Saliency
6	4.775	8	0.781

After preliminary fitting model calculations, six factors including personality, driver's license level, daily driving time, whether there is a fixed transportation route, vehicle ownership, and whether there is an occupational disease are selected from the analysis results, and SPSS 23.0 software is used to target these six factors. Perform binary Logistic regression analysis, select the significance level $\alpha = 0.05$, and use the input method. The final result is consistent with Table 6. In the comprehensive test of model coefficients, the significance level is less than 0.05, indicating that the model is meaningful in general. In the Hosmer and Lemeshow test, the significance level is 0.731 and greater than 0.05, indicating that the model fits well. It can be seen that the truck driver's personality, driver's license level, daily driving time, whether there is a fixed route, the ownership of the vehicle, and whether there is an occupational disease have a significant impact on the driver's traffic violations.

4   Discuss

Based on the data from the questionnaire survey, a binary logistics model for truck drivers is established for comprehensive analysis. In this section, the author will discuss the relevant results of other scholars on the factors that affect drivers' traffic violations, and compare the results of this article to get more information and practical suggestions.

According to previous related research, personality is divided into depressive, bloody, mucous, and bile (easily excited, short-tempered, straightforward, enthusiastic, but depressed when energy is exhausted). According to previous related studies, the driver's personality changes from depression to bile, and the driving speed is getting faster and faster. The number of people with bloody and mucous personalities is the highest among them [12, 13], and this survey confirms this.The situation is roughly the same. The bloody personality has the most people in this article, with 643 people, accounting for 47.49% of the total number of people, 432 of whom have broken the rules, accounting for 67.19%; the mucus personality has 326 people, accounting for 24.08% of the total number of people, and 231 of whom have broken the rules. People accounted for 70.86%; 308 people with biliary personalities accounted for 22.75% of the total, with 225 of them having 73.05% violations; and depressive personalities affected 77 people, or 5.69% of the total, with 48 of them having major depression.Violations made up 62.34% of the total.The significant difference between drivers with bloody and biliary personalities is bigger, implying that drivers with biliary personalities are more prone to committing infractions while driving, and that drivers with biliary personalities require special attention at work. To strengthen their self-control and avoid traffic offenses caused by high-speed driving, such people must be supervised.

The driver's license level is quite different in the model of the truck driver's personal attributes and violation behavior (significance = 0.014). The investigated truck driver obtained primarily A2 driver's licenses, with a total of 867 people, accounting for 64.03% of the total.Among them, 602 people have violated regulations, accounting for 69.43%; the second is the B2 driver's license type, with a total of 380 people, accounting for 28.06% of the total, of which 254 people have violated the regulations, accounting for 66.84%; and the C driver's license type has a total of 45 people, accounting for 3.32% of the total, of which 26 people have violated the rules, accounting for 57.78%. With the trend toward larger vehicles, truck drivers with A2 licenses have increasingly become the mainstream. At present, driving a tractor requires an A2 driver's license, which must be increased on the basis of obtaining a B2 driver's license. It is not possible to directly apply for the test, and a motor vehicle that is driven during the internship period is not allowed to tow a trailer. Due to the high cost of taking photos, it is also one of the reasons why it is difficult to attract young practitioners to enter. At present, some auto manufacturers have introduced automatic tractors, but they have to apply for an A2 driver's license.

In the past, a large number of relevant studies have shown that fatigue driving is one of the important causes of traffic accidents [14]. There are also many reasons for fatigue driving. Among them, the driver's perceptual reaction time and the ability to maintain attention increase with the driver's drowsiness. Sleep is reduced [15], and daily driving time is also one of the important factors that make people fatigued. This article divides the daily driving time into 8 h or less, 8–10 h, 10–12 h, and 12 h or more. There were 632 people under 8 h, accounting for 46.68% of the total, of which 230 offenders accounted for 36.39%; there were 397 people under 8–10 h, accounting for 29.32% of the total, of which 157 offenders accounted for 39.55%; 190 people in 10–12 h, accounting for 14.03% of the total number of people, of which 65 offenders accounted for 34.21%; and 135 people over 12 h, accounting for 9.97% of the total number, accounting for 9.97% of the total number, of which 102 offenders People accounted for 75.56%. The special working environment of truck drivers makes them generally work longer hours and be labor-intensive. 53.32% of truck drivers drive 8 h or more per day, and there is a risk of fatigue driving, which may lead to violations.

Whether there is a fixed transportation route is quite different in the model of a truck driver's personal attributes and violation behaviors (significance $= 0.050$). There are 730 people with fixed transportation routes, accounting for 53.91% of the total, of which 488 people are in violation. It accounted for 66.85%; there were 624 people without fixed transportation routes, accounting for 46.09% of the total number, of which 448 people who violated regulations accounted for 71.79%. There is a higher rate of violations without fixed transportation routes, which may be due to driving on an unfixed road section, leading to traffic accidents due to unfamiliar road conditions when driving. It shows that different driving environments have a greater impact on the driver.

In the model of a truck driver's personal attributes and violation behavior, vehicle ownership and whether there is an occupational disease are very different, and the significance is 0.000.The survey shows that 76.88% of truck drivers report that their vehicles are self-owned vehicles, 39.81% of which are currently in the process of repaying their loans, and only 23.12% of truck drivers drive vehicles that belong to their employer or fleet. Self-employed truck drivers are still more common, with back-loan drivers taking up more space. There are 502 people without arrears in their own vehicles, accounting for 37.07% of the total, of which 357 people are in violation of the rules, accounting for 71.12%; 539 people are in arrears with their vehicles, accounting for 39.81% of the total, and among them, 417 are in violation of the rules. People accounted for 77.37%; there were 313 hired drivers, accounting for 23.12% of the total number, of which 162 offenders accounted for 51.76%. At present, there is a "0" down payment model in the truck sales market. Financial companies use ultra-low threshold "0" down payment or low down payment methods to attract a large number of truck drivers to enter the freight market. Financial companies turn the down payment burden into high monthly payments and high fees (maintenance, etc.), which increases purchase costs. At the same time, drivers are required to attach their vehicles to the anchoring company and charge higher anchorage fees, insurance premiums, and inspection fees, which further increases the driver's burden. Affiliated companies can obtain a large number of vehicle input invoices and transfer them to other markets. At the same time, when the loan expires, the driver asks to transfer the vehicle out, generally facing the problem of a high transfer-out fee. The survey shows that 56.13% of truck drivers suffer from one or more occupational diseases such as stomach disease, cervical spondylosis, and back pain due to long-term driving. A total of 760 people have occupational diseases. Among them, 559 people who violate regulations account for 73.55%. The health problems of truck drivers are worth causing. focus on. 43.87% of truck drivers did not have the above-mentioned health problems because of their low working years or short driving time each day. There were 594 people without occupational diseases, of which 377 people who violated regulations accounted for 63.47%.

People usually think that age and driving age are very related to drivers' violations. Leixing et al. found that as the driver's age changes, his driving behavior will change accordingly, which will affect driving safety [16]. Fang Yuerong believes that drivers between the ages of 40 and 52 have relatively slow driving speeds, more stable driving behaviors, and safer driving [17]. The research in this article found that age and driving age have no obvious relationship with whether truck drivers have traffic violations.

5   Conclusion

This article investigated the personal attributes and violations of truck drivers and obtained 1354 traffic violation data samples. The driver's infraction data was mined and evaluated using the logistics model, and the following findings were drawn:

(1) Whether truck drivers will violate the rules is significantly related to six variables: personality, driver's license level, daily driving time, whether there is a fixed transportation route, vehicle ownership, and whether there is an occupational disease. Among them, personality, daily driving time, whether there is a fixed transportation route, and vehicle ownership are positively related to violations.
(2) Further data analysis shows that this group of people who are bile, drive more than 12 h a day, have no fixed transportation routes and have loans for their own vehicles are most likely to have violations during the driving process, which can be further improved in the future. Investigate and research this part of the group. When hiring drivers, relevant departments can conduct personality tests. They can strengthen management and coaching for this portion of the group among the existing truck drivers.

6   Practical Implications and Directions for Further Research

In this study, there is no guarantee that the data filled in by the surveyed persons when filling out the questionnaire is authentic. Some people have personal subjective emotions when filling out the questionnaire, which leads to a certain deviation in the data filled in. Therefore, in the future research work, it is necessary to adjust the existing survey methods.

The data obtained from the questionnaire survey in this article has certain deficiencies. Among them, there are too few female drivers and they are not representative. The sample data is not enough, it can only represent part of the truck drivers in Anhui, and cannot distinguish the personal attributes of the drivers in the plain area and the mountain forest area. The dependent variables used in this model are divided into two types of violations and non-violations. In future research, violations can be divided into high-risk violations and low-risk violations in more detail, so that truck drivers in different regions can be studied in detail.

Acknowledgment. This work was supported by National Key R&D Program of China Entitled "Multimodal Transportation Intelligent Integration Technology and Equipment Development" (2019YFB1600400). Our thanks also go to those who volunteered in this research.

References

1. National Bureau of Statistics of the People's Republic of China 2019, China Statistical Yearbook-2019. China Statistics Press, Beijing
2. Yu, Y., Shubo, C.: 2021 Psychological factors influencing the driving safety of drivers and countermeasures. Fire Fighting Circle (Electronic Edition) 7(04), 50–52+54 (2021)
3. Di, W., Zhihan, W., Xiaobao, C.: Analysis of the major road traffic accidents in Anhui Province in 2019. Road Traffic Manage. 12, 40–41 (2020)
4. Mokarami, H., Alizadeh, S.S., Pordanjani, T.R., et al.: 2019 The relationship between organizational safety culture and unsafe behaviors, and accidents among public transport bus drivers using structural equation modeling. Transp. Res. 65(Aug.), 46–55 (2019)
5. ASSG, AMHH and BAJA 2018 Examining bus driver behavior as a function of roadway features under daytime and nighttime lighting conditions: driving simulator study-sciencedirect. Safety Sci. 110, 142–151
6. AYSC and BHLW: Stress 2013 strain, and health outcomes of occupational drivers: An application of the effort reward imbalance model on Taiwanese public transport drivers. Transp. Res. Part F: Traffic Psychol. Behav. 19(4), 97–107 (2013)
7. Lourens, P.F., Vissers, J.A.M.M., Jessurun, M.: Annual mileage, driving violations, and accident involvement in relation to drivers' sex, age, and level of education. Accid. Anal. Prev. 31(5), 593–597 (1999)
8. Qingfeng, L., Yuanchang, D., Jihua, H.: Logistic regression analysis of factors affecting driver fault and accident severity in non-mechanical traffic accidents. Safety Environ. Eng. 026(005), 187–193 (2019)
9. Sheng, T., Erhui, L.: Analysis of youth traffic safety behavior based on multiple regression model. Traffic Inf. Safety (002) 98–102 (2015)
10. Jinghong, R., Jun, H., Zhuoqing, Z.: 2020 On the problems and countermeasures of rural road traffic management. Road Traffic Manage. 435(11), 42–43 (2020)
11. Xiaolin, S., et al.: The mediating effect of driver characteristics on risky driving behaviors moderated by gender, and the classification model of driver's driving risk. Acc. Anal. Prev. 153, 106038 (2021)
12. Zhongxiang, F., Huazhi, Y., Jing, L., et al.: The influence of driver's personal characteristics on driving speed. J. Traffic Transp. Eng. 12(006), 89–96 (2012)
13. Dong, Y., Changxi, M., Pengfei, L., et al.: The influence of BRT driver's personal attributes on driving speed. Traffic Inf. Safety 036(006), 54–64,73 (2018)
14. Xuxin, Z., Xuesong, W., Yong, M., et al.: 2020 International research progress on driving behavior and driving risk. Chin. J. Highway Transp. 33(202)(06), 5–21 (2020)
15. Kofi, A.E., et al.: Better rested than sorry: data-driven approach to reducing drowsy driving crashes on interstates. J. Transp. Eng. Part A: Syst. 147(10), 04021067 (2021)
16. Xing, L.: Analysis of the influence of driver's age on driving safety. Energy Conserv. Environ. Protect. Transp. 12(6), 15–17 (2016)
17. Yuerong, F.: Experimental study on the differences in driving behavior characteristics of different types of drivers. Safety Environ. Eng. 27(131)(05), 208–212 (2020)

External Information Security Resource Allocation with the Non-cooperation of Multiple Cities

Jun Li[1], Dongsheng Cheng[2], Lining Xing[2], and Xu Tan[2(✉)]

[1] Academy of Hi-Tech Research, Hunan Institute of Traffic Engineering, Hengyang 421099, People's Republic of China
[2] School of Software Engineering, Shenzhen Institute of Information Technology, Shenzhen 518172, People's Republic of China
tanxu_nudt@yahoo.com

Abstract. The external information security resource allocation method is proposed considering the non-cooperation of multiple cities. In this method, the effects of different influence factors, for example, city size, probability of intrusion by illegal users and propagation probability of one-time intrusion on resource allocation is explored. Through the simulation experiment, the proposed conclusions are conveniently and clearly verified.

Keywords: Information security · External resource · Allocation method · Non-cooperation

1 Introduction

A modern smart city cannot be a closed system, and its communication will not be limited in the interior. In the actual operation, its external sharing and communication will sometimes be even more extensive than the internal communication. Therefore, it is necessary to strengthen studies on external resource allocation of the city on the premise of thorough research on internal resource allocation of the city [1–3].

With the rapid development and wide application of big data and artificial intelligence and the continuous integration and development of all walks of life [4, 5], information security has become a huge challenge for smart cities at present [6–8]. It is not an isolated and separate issue, but is ubiquitous and can easily develop into a public security problem [9–13]. The cooperation in information security and business contacts between cities make urban resources be complementary to a certain extent [14–16]. After illegal users intrude into a city, they need to intrude into another city linked to obtain the corresponding benefits.

2 Problem Description and Modelling

2.1 Problem Description

Because resources between cities are complementary, if illegal users intrude into a city, but fail to intrude into cities linked, complementary of resources guarantees all or

Z. Qian et al. (Eds.): WCNA 2021, LNEE 942, pp. 316–324, 2022.
https://doi.org/10.1007/978-981-19-2456-9_33

partial information security, so that it is difficult for illegal users to fully benefit, thus avoiding heavy loss of the cities. At present, most scholars mainly focus on the research of resource allocation to information security in cities under the condition of information sharing. In fact, cities will also consider input and output and if the disadvantages of cooperation outweigh the advantages, they tend to choose not to cooperate. Therefore, it is necessary to study the optimal resource allocation in the case of non-cooperation. This section mainly studied the problem that multiple cities with complementary external resources suffer from multiple propagation and intrusion by illegal users in the actual operation of smart cities. Firstly, the optimal resource allocation schemes were compared under non-cooperation and full cooperation situations and then government's compensation mechanisms and information sharing mechanisms were introduced. Furthermore, a numerical analysis was carried out.

2.2 Problem Modeling

Any game problem can be described as $GT = \{P, St, Ut\}$. For complementary external resources, cities are linked with each other and they may be attacked by illegal users. Even if cities are not attacked directly, they can also be attacked indirectly through propagation. Any problem of complementary external resource allocation can be transformed into a game problem through the propagation probability.

Assumption 1: When the propagation probability of one-time intrusion between cities is same and set as a, illegal users can attack another city directly linked thereto by using the probability.

Assumption 2: Illegal users do not have any prior information about the vulnerability for information security construction in cities. Therefore, the probabilities of illegal users intruding into all cities are same, and the value is β.

Assumption 3: The losses borne by cities intruded by illegal users are same, namely L.

Assumption 4: When resources are not allocated to information security in cities, the probabilities of intrusion by illegal users are same across cities and value v.

It is assumed that there are n cities forming complementary external resources and the probability of intrusion by illegal users after allocating resources to information security in the $j(j = 1, 2, \cdots n)$ th city is p_j. Moreover, the volume of resource allocation to information security is e_j, loss rescued by amount of money per unit is E and the expected loss after allocating resources to information security in cities is set as C_j. By improving the model proposed by Gordon [14], the probability p_j of intrusion by illegal users in the jth city can be obtained.

$$p_j = \beta v^{Ee_j+1} \tag{1}$$

Considering complementarity of resources between cities, that is, if illegal users intrude into one or several cities linked, but not all cities linked, it is acceptable to the whole information security system to a certain extent. Therefore, if illegal users want to maximize their profits, they have to intrude into all cities linked.

3 Resource Allocation to Information Security in Cities Under Non-cooperation

This section mainly analyses strategies for allocation of complementary external resources under non-cooperation between smart cities. Based on the assumptions in the above section and Formula (1), it is known that the probability of intrusion by illegal users in the $j(j = 1, 2, \cdots n)$ th city is $1 - (1 - p_j) \prod_{k=1,k \neq j}^{n}(1 - a^{k-1}p_k)$, so the minimum expected loss C_j of the city is taken as a loss function.

$$\text{Min} C_j = \left[1 - (1 - p_j) \prod_{k=1,k \neq j}^{n} \left(1 - a^{k-1}p_k \right) \right] L + e_j \qquad (2)$$

By substituting Formula (1) into Formula (2), the following formula can be obtained.

$$\text{Min} C_j = \left[1 - \left(1 - \beta v^{Ee_j+1} \right) \prod_{k=1,k \neq j}^{n} \left(1 - a^{k-1}\beta v^{Ee_k+1} \right) \right] L + e_j \qquad (3)$$

Because $\prod_{k=1,k \neq j}^{n}(1 - a^{k-1}\beta v^{Ee_k+1})$ in Formula (3) is independent of e_j, let $\Phi = \prod_{k=1,k \neq j}^{n}(1 - a^{k-1}\beta v^{Ee_k+1})$, the following formula can be obtained by solving the partial derivative of Formula (3):

$$\frac{\partial C_j}{\partial e_j} = \beta EL\Phi v^{Ee_j+1} \ln v + 1 \qquad (4)$$

By further solving the partial derivative of Formula (4), the second-order derivative of Formula (5) can be obtained.

$$\frac{\partial^2 C_j}{\partial e_j^2} = \beta E^2 L\Phi v^{Ee_j+1}(\ln v)^2 \qquad (5)$$

It can be seen from Formula (5) that $\frac{\partial^2 C_j}{\partial e_j^2} \geq 0$ is always established. Therefore, when $\frac{\partial C_j}{\partial e_j} = 0$, the minimum value of the loss function C_j can be obtained, thus obtaining the following Conclusion 1.

Conclusion 1: Under non-cooperation between smart cities with complementary external resources, the Nash equilibrium solution can be obtained through games when the optimal volume of resource allocation in each city is $y^* = (e_1^*, e_1^*, \cdots, e_1^*)$, in which e_1^* meets Formula (6).

$$e_1^* = \frac{-\ln(-\beta EL\Phi v \ln v)}{E \ln v} \qquad (6)$$

In accordance with Formula (6), the effects of factors, such as size of linked cities, probability of intrusion by illegal users and propagation probability of one-time intrusion on resource allocation to information security in cities can be further analysed. Based on Conclusion 1, e_1^* meets $\beta EL\Phi v_j^{Ee_1^*+1} \ln v_j + 1 = 0$. Furthermore,

$\frac{\prod_{k=1,k\neq j}^{n}\left(1-a^{k}\beta v^{Ee_{k}+1}+1\right)}{\prod_{k=1,k\neq j}^{n}(1-a^{k-1}\beta v^{Ee_{k}+1})} = 1 - a^{n}\beta v^{Ee_{k+1}+1} < 1$ is always established. For this reason, the relationship between size of linked cities and resource allocation to information security in cities is analysed by combining with characteristics of complementary resources and considering the same volume of resource allocation between smart cities under non-cooperation based on relevant assumptions in Sect. 2.2. On this basis, the following Conclusion 2 can be made.

Conclusion 2: Under non-cooperation, with the increase of size of cities linked in complementary external resources of information security, the optimal volume e_1^* of resource allocation to information security in cities reduces correspondingly, that is, e_1^* is negatively correlated with n.

The reason is that with the increase of n, $\prod_{k=1,k\neq j}^{n}\left(1 - a^{k-1}\beta v^{Ee_{k}+1}\right)$ decreases, which raises $p_j = \beta v^{Ee_j+1}$. In addition, because $v \in [0, 1]$, e_1^* is bound to decrease accordingly. This suggests that the volume of resource allocation in each city reduces correspondingly with the increase of size of cities with complementary resources. However, this can greatly increase the probability of illegal users to intrude into a single city, so that the information security level of all smart cities significantly reduces. Although more linked cities can share the risks, such a behaviour of reducing the volume of resource allocation decreases the information security level. If the size of linked cities reaches to a certain critical value, it is not necessary for smart cities to allocate resources to information security, which is unrealistic in practice. Therefore, it is necessary for the government to coordinate the relevant departments in each city and allocate resources to information security after weighing the advantages and disadvantages.

By analyzing the relationship between the probability of intrusion by illegal users and resource allocation to information security in cities, Conclusion 3 can be made as follows:

Conclusion 3: Under non-cooperation, for any probability $\beta \in [0, 1]$ of intrusion by illegal users, the optimal volume e_1^* of resource allocation to information security in cities monotonically rises, namely $\frac{\partial e_1^*}{\partial \beta} > 0$ is always established.

Conclusion 3 indicates that the volume of resource allocation to information security in cities increases with the probability of intrusion by illegal users in the model of complementary external resource allocation in smart cities, which confirms with the common sense. When the probability of intrusion by illegal users rises, cities will invest more to prevent illegal intrusion, thus raising their information security level.

By analysing the relationship between the propagation probability of one-time intrusion between cities and resource allocation to information security in cities, Conclusion 4 can be made as follows:

Conclusion 4: Under non-cooperation, for any propagation probability a ∈ [0, 1] of one-time intrusion between cities, the optimal volume of resource allocation to information security in cities monotonically reduces, that is, $\frac{\partial e_i^*}{\partial a} < 0$ is always established.

Conclusion 4 indicates that with the increase of the propagation probability of one-time intrusion between cities, the optimal volume of resource allocation to information security in cities decreases correspondingly. This verifies the conclusion proposed in the existing study [x] that network communication has a negative impact on the optimal strategy of resource allocation. This implies that the power of cities to resource allocation to information security can be reduced with the increase of the propagation probability of one-time intrusion between cities. In the case of non-cooperation, it needs to adjust the network structure between cities and try to avoid indirect intrusion by illegal users due to network connection with other cities.

Based on Conclusions 2 and 4, with the increase of city size and propagation probability of one-time intrusion be-tween cities, the probability of intrusion by illegal users in cities rises. However, through the above analysis, instead of increasing resource allocation, cities reduce investment, which leads to a vicious circle of information security in cities. The main reason is that some cities have free-riding behaviours in the construction of information security in other cities, because the resource allocation in these cities not only has an effect on information security of them-selves, but also exerts a positive influence on cities linked thereto. Due to the free-riding behaviours, marginal benefits of cities with resource allocation to information security decrease.

4   Experimental Results and Analysis

Through a simulation experiment, the above conclusions can be conveniently and clearly verified. This section mainly deeply discusses the following problems.

(1) Based on the numerical simulation, the optimal volumes of resource allocation and expected costs under non-cooperation and full cooperation of cities are compared. The influence trends of city size n, probability β of intrusion by illegal users and propagation probability a of one-time intrusion on the optimal volume of resource allocation and expected cost are numerically studied and analysed, that is, numerical analysis under different conditions.

(2) The influences of the compensation coefficient γ and sharing rate δ of information in cities on the optimal volume of resource allocation and expected cost are discussed, that is, numerical analysis of incentive mechanisms.

According to the actual conditions, there cannot be too many cities that are linked together and have complementary external resources, generally no more than four, so the city sizes are set as n = 3 and n = 4 in the numerical simulation in this section. Because it is impossible and unnecessary to consider all values of some experimental parameters in the actual numerical simulation, this section only takes several representative values into account. It is supposed that L = 400, v = 0.5 and E = 0.1.

When n $= 3$, the propagation probability a of one-time intrusion between cities and the probability β of intrusion by illegal users are set to be 0.1–0.9, with an increase amplitude of 0.1, to analyze the influences of a and β on resource allocation. The volume of resource allocation and the expected loss are listed in Tables 1 and 2. By further analysing Tables 1 and 2, when a is 0.1 and β values [0.1, 0.9] as well as β is 0.1 and a is [0.1, 0.9], the results in Figs. 1 and 2 can be obtained.

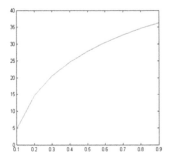

Fig. 1. Influences of β on the volume e_1^* of resource allocation

Fig. 2. Influences of a on the volume e_1^* of resource allocation

It can be obviously observed from the above figures that with the constant increase of β, the volume e_1^* of resource allocation continuously rises, which verifies the correctness of Conclusion 3; as a constantly rises, the volume e_1^* of resource allocation continuously decreases, verifying that Conclusion 4 is correct.

When n $= 4$, by setting the propagation probability a of one-time intrusion between cities as 0.1–0.9, with an increase amplitude of 0.1 and the probability β of intrusion by illegal users as 0.1, the volume of resource allocation and the expected loss are attained, as shown in Table 3.

Table 1. Influences of a and β on the volume e_1^* of resource allocation under non-cooperation

α	β								
	0.1	0.2	0.3	0.4	0.5	0.6	0.7	0.8	0.9
0.1	4.6548	14.6548	20.5044	24.6548	27.8741	30.5044	32.7283	34.6548	36.3540
0.2	4.5860	14.5860	20.4356	24.5860	27.8052	30.4356	32.6595	34.5860	36.2852
0.3	4.5055	14.5055	20.3551	24.5055	27.7248	30.3551	32.5791	34.5055	36.2048
0.4	4.4130	14.4130	20.2626	24.4130	27.6323	30.2626	32.4866	34.4130	36.1123
0.5	4.3078	14.3078	20.1575	24.3078	27.5271	30.1575	32.3814	34.3078	36.0071
0.6	4.1894	14.1894	20.0390	24.1894	27.4086	30.0390	32.2629	34.1894	35.8886
0.7	4.0567	14.0567	19.9063	24.0567	27.2760	29.9063	32.1303	34.0567	35.7560
0.8	3.9089	13.9089	19.7585	23.9089	27.1282	29.7585	31.9825	33.9089	35.6082
0.9	3.7447	13.7447	19.5944	23.7447	26.9640	29.5944	31.8183	33.7447	35.4440

Table 2. Effects of a and β on the expected loss under non-cooperation

α	β								
	0.1	0.2	0.3	0.4	0.5	0.6	0.7	0.8	0.9
0.1	20.6745	30.6745	36.5241	40.6745	43.8938	46.5241	48.7481	50.6745	52.3738
0.2	22.5016	32.5016	38.3512	42.5016	45.7209	48.3512	50.5752	52.5016	54.2009
0.3	24.6258	34.6258	40.4754	44.6258	47.8450	50.4754	52.6993	54.6258	56.3250
0.4	27.0537	37.0537	42.9033	47.0537	50.2730	52.9033	55.1273	57.0537	58.7530
0.5	29.7939	39.7939	45.6435	49.7939	53.0132	55.6435	57.8674	59.7939	61.4931
0.6	32.8566	42.8566	48.7062	52.8566	56.0759	58.7062	60.9302	62.8566	64.5559
0.7	36.2545	46.2545	52.1041	56.2545	59.4737	62.1041	64.3280	66.2545	67.9537
0.8	40.0026	50.0026	55.8522	60.0026	63.2219	65.8522	68.0762	70.0026	71.7019
0.9	44.1193	54.1193	59.9690	64.1193	67.3386	69.9690	72.1929	74.1193	75.8186

By comparing results in Table 3 with Tables 1 and 2, it can be seen that with the increase of n, the volume e_1^* of resource allocation reduces, while the expected loss increases, verifying that Conclusion 2 is correct. By comparing results in Table 3 with Tables 1 and 2, with the increase of n, the volume e_1^* of resource allocation decreases, while the expected loss rises, proving that Conclusion 2 is correct.

Table 3. Partial results of the volume of resource allocation and expected loss when n = 4 under non-cooperation

α	Resource allocation e_1^*	Expected loss
0.1	4.6542	20.6885
0.2	4.5817	22.6138
0.3	4.4910	25.0069
0.4	4.3782	27.9642
0.5	4.2385	31.5893
0.6	4.0668	35.9963
0.7	3.8568	41.3140
0.8	3.6006	47.6927
0.9	3.2882	55.3142

5 Conclusions

This research mainly discussed the methods for resource allocation in the cases of non-cooperation of multiple cities. In addition, the effects of different influence factors, such as city size, propagation probability of one-time intrusion and probability of intrusion by illegal users on resource allocation was also explored.

Acknowledgements. This research work is supported by the National Social Science Fund of China (18BTQ055), the Youth Fund of Hu-nan Natural Science Foundation (2020JJ5149, 2020JJ5150) and the Innovation Team of Guangdong Provincial Department of Education (2018KCXTD031). It is also supported by the Program of Guangdong Innovative Research Team (2020KCXTD040), the Pengcheng Scholar Funded Scheme, and the Basic Research Project of Science and Technology Plan of Shenzhen (SZIITWDZC2021A02, JCYJ20200109141218676).

Conflicts of Interest. The authors declare that they have no conflict of interest.

References

1. Nazareth, D.L., Choi, J.: A system dynamics model for information security management. Inf. Manage. **52**(1), 123–134 (2015)
2. Houmb, S.H., Franqueira, V.N.L., Engum, E.A.: Quantifying security risk level from CVSS estimates of frequency and impact. J. Syst. Softw. **83**(9), 1622–1634 (2010)
3. Feng, N., Li, M.: An information systems security risk assessment model under uncertain environment. Appl. Soft Comput. J. **11**(7), 4332–4340 (2011)
4. Kong, H.K., Kim, T.S., Kim, J.: An analysis on effects of information security investments: a BSC perspective. J. Intell. Manuf. **23**(4), 941–953 (2012)
5. Li, S., Bi, F., Chen, W., et al.: An improved information security risk assessments method for cyber-physical-social computing and networking. IEEE Access **6**(99), 10311–10319 (2018)
6. Basallo, Y.A., Senti, V.E., Sanchez, N.M.: Artificial intelligence techniques for information security risk assessment. IEEE Lat. Am. Trans. **16**(3), 897–901 (2018)
7. Grunske, L., Joyce, D.: Quantitative risk-based security prediction for component-based systems with explicitly modelled attack profiles. J. Syst. Softw. **81**(8), 1327–1345 (2008)
8. Gusm, O.A., Silval, C.E., Silva, M.M., et al.: Information security risk analysis model using fuzzy decision theory. Int. J. Inf. Manage. **36**(1), 25–34 (2016)
9. Baskerville, R.: Integration of information systems and cybersecurity countermeasures: an exposure to risk perspective. Data Base Adv. Inf. Syst. **49**(1), 69–87 (2017)
10. Huang, C.D., Hu, Q., Behara, R.S.: An economic analysis of the optimal information security investment in the case of a risk-averse firm. Int. J. Prod. Econ. **114**(2), 793–804 (2008)
11. Yong, J.L., Kauffman, R.J., Sougstad, R.: Profit-maximizing firm investments in customer information security. Decis. Support Syst. **51**(4), 904–920 (2011)
12. Li, J., Li, M., Wu, D., et al.: An integrated risk measurement and optimization model for trustworthy software process management. Inf. Sci. **191**(9), 47–60 (2012)
13. Benaroch, M.: Real options models for proactive uncertainty-reducing mitigations and applications in cybersecurity investment decision-making. Soc. Sci. Electron. Publ. **4**, 11–30 (2017)

14. Gao, X., Zhong, W., Mei, S.: Security investment and information sharing under an alternative security breach probability function. Inf. Syst. Front. **17**(2), 423–438 (2015)
15. Liu, D., Ji, Y., Mookerjee, V.: Knowledge sharing and investment decisions in information security. Decis. Support Syst. **52**(1), 95–107 (2012)
16. Gao, X., Zhong, W., Mei, S.: A game-theoretic analysis of information sharing and security investment for complementary firms. J. Oper. Res. Soc. **65**(11), 1682–1691 (2014)

Leveraging Modern Big Data Stack for Swift Development of Insights into Social Developments

He Huang[1,4(✉)], Yixin He[3,4], Longpeng Zhang[3], Zhicheng Zeng[4], Tu Ouyang[2], and Zhimin Zeng[4]

[1] University of Melbourne, Parkville, VIC 3010, Australia
`hhhu@student.unimelb.edu.au`
[2] Computer and Data Science department, Case Western Reserve University, 10900 Euclid Ave., Cleveland, OH 44106, USA
`tu.ouyang@case.edu`
[3] University of Electronic Science and Technology of China, ChengDu, China
`yixinhe09@std.uestc.edu.cn, zlp1988@uestc.edu.cn`
[4] Zilian Tech Inc., ShenZhen, China
`zengzc@ziliantech.net, zengzm@tsingzhi.cn`

Abstract. Insights of social development, presented in various forms, such as metrics, figures, text summaries, whose purpose is to summarize, explain, and predict the situations and trends of society, is extremely useful to guide organizations and individuals to better realize their own objectives in accordance with the whole society. Deriving these insights accurately and swiftly has become an interest for a range of organizations, including agencies governing districts, city even the whole country, they use these insights to inform policy-makings. Business investors who peak into statistical numbers for estimating current economical situations and future trends. Even for individuals, they could look at some of these insights to better align themselves with macroscopical social trends. There are many challenges to develop these insights in a data-driven approach. First, required data come from a large number of heterogeneous sources in a variety of formats. One single source's data could be in the size of hundreds of Gigabytes to several TeraBytes, ingesting and governing such huge amount of data is not a small challenge. Second, many complex insights are derived by domain human experts in a trail-and-error fashion, while interacting with data with the aid of computer algorithms. To quickly experiment various algorithms, it asks for software capabilities for infusing human experts and machine intelligence together, this is challenging but critical for success.

By designing and implementing a flexible big data stack that could bring in a variety of data components. We address some of the challenges to infuse data, computer algorithm and human together in Zilian Tech company [20]. In this paper we present the architecture of our data stack and articulate some of the important technical choices when building such stack. The stack is designed to be equipped with scalable storage that could scale up to PetaBytes, as well as elastic

---

H. Huang and Y. He—Contribute equally, their work were done when authors interned in Zilian Tech.

© The Author(s) 2022
Z. Qian et al. (Eds.): WCNA 2021, LNEE, pp. 325–333, 2022.
https://doi.org/10.1007/978-981-19-2456-9_34

distributed compute engine with parallel computing algorithms. With these features the data stack enables *a)* swift data analysis, by human analysts interacting with data and machine algorithms via software support, with on-demand question answering time reduced from days to minutes; *b)* agile building of data products for end users to interact with, in weeks if not days from months.

Keywords: Cloud · Data stack · Social development

1   Introduction

The potential benefits are immense by drawing on large-scale online and commercial data to construct insights of social development, for example, trends in economic and business development, emerging patterns of people's daily life choices, comparative technology advances of competing regions, population sentiment to social events and so on. These insights are valuable, sometimes critical, in scenarios like helping government agencies for more objective policy making, aiding decision-making of investors before pulling money into certain business in certain regions, even helping individuals who might just want to check cities and companies' outlooks before settling among several job offers.

Recent years have seen many articles to investigate various aspects of social activities and developments based on data and models. Bonaventura et al. [23] construct a worldwide professional network of start-ups. The time-varying network connects start-ups which share one or more individuals who have played a professional role. Authors suggest such network has predictive power to assess potential of early stage companies. [26] investigates foreign interference found on twitter, during the 2020 US presidential election. Natural language processing models are used to classify troll accounts, network flow statistics are leveraged to reveal super-connectors. Drawn on top of analysis results drawn from these models, this report is able to quantify prevalence of troll and super-connector accounts in various politics-inclined communities and these accounts' influence among these communities. Jia et al. [24] devise a risk model of covid-19 based on aggregate population flow data, the model is to forecast the distribution of confirmed cases, identify high risk regions threatened by virus transmission, one such model is built and verified using major carrier data of mobile phone geolocations from individuals leaving or transiting through Wuhan between 1 January and 24 January 2020. Authors suggests the methodology can be used by policy-makers in any nations to build similar models for risk assessment.

To realize many of aforementioned applications, a large amount of data need to be acquired, stored and processed, a scalable and efficient big data processing platform is the key. In our company, we have built such a data platform. We argue that the data stack of our platform provides enough flexibility to incorporate a variety of modern data component implementations and products from different vendors and bring them together to enable data applications to solve our use cases. Mainly two categories of applications are enabled by the design of the data stack: analytics-oriented applications and real-time transactional applications (usually customer-facing). These two application categories suite different use cases when developing data applications for extracting insights of social development. We showcase two concrete applications: one is a notebook-like analytics tool for analysts to examine research publications of a country with the world's biggest population. The other is a customer-facing search application one of whose function is to retrieve and summarize companies' patent statistics in past 20 years of the same big country.

This paper's main contributions are not on advancing techniques of individual data components, but more of a practical study on how to incorporate appropriate data techniques under a flexible stack framework we propose, to enable real-world data-oriented user cases with minimum time-to-market. We document technical trade-offs we made for choosing the right set of components and technologies, from many existing ones, we use these components to compose a cohesive platform that suits our use cases.

In the following of this paper, Sect. 2 presents the architecture of the big data stack, then dive into the technical reasoning to choose concrete techniques for several key components. Section 3 shows two example applications and explain how the big data stack enable swift development, followed by the conclusion in Sect. 4.

2   The Big Data Stack

Figure 1 depicts a high-level view of what are in the big data stack, the stack is composed of five key components. In the past decade, we have seen a blossom of technologies that could possibly be used to implement the components of proposed stack. Too many techniques sometimes bring no help, but on the contrary quite a lot challenges for a system architect, who need to carefully compare and make trade-offs between several technologies and eventually decide on the right one to have it incorporated into one single cohesive stack.

The applications that we want the techniques to enable are mainly two categories: analytics-oriented and real-time customer facing. To enable these two categories, we set out with a number of goals for choosing the techniques to implement the data stack. First major goal is *flexibility*, we strive to the keep our options open to be able to switch to a different technique in the future in needed and avoid being locked into certain set of techniques. *Scalability* and *agility* are two goals for analytics-oriented applications. *Responsiveness* is one goal for real-time applications, "real-time" means the processing time is within the order of sub-second.

Below we dive into technical reasoning in each component of our stack, about the choices of concrete techniques. Note that, the index numbers of the list items correspond to the labels of components in Fig. 1.

1. **Data Governance**

   The social development data could come as structured, e.g., files with clearly defined schema, e.g., CSV, parquet [28] files; or semi-structured, like XML and JSON; or unstructured, e.g., pictures, audio files, videos. The existing and emerging storage technologies to choose include: structured-data-only traditional database, that aims to store key operational data only; data warehouse that are designed to stores all your data but mainly structured data, snowflake [17] and Oracle [10] are examples of such warehouse providers. Recent data lake technologies [25], that promise to be able to store huge amount of structured and unstructured data. Data lakehouse [21]

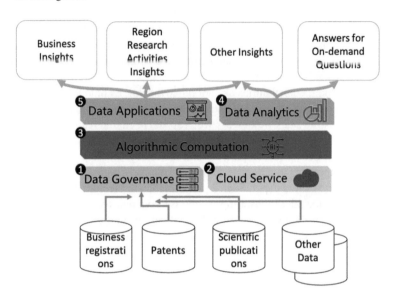

Fig. 1. The conceptual view of the big data stack of our data intelligence platform Each component in this stack figure has a corresponding text paragraphs of the same label for more detail.

is another recent data storage paradigm attempting to unify both data warehouse and data lake. We keep an open mind in choosing storage technologies since we believe at this time not a single existing technology mature enough to solve all the cases. When picking technologies for our stack, we decide on data lake storage techniques for raw data storage for analytics-oriented applications that meet the goal of *scalability*. While for real-time applications, we integrate traditional relational databases for its optimized transaction handling for *responsiveness*.

2. **Cloud Service**

   Fifteen years after the launch of AWS, we now enjoy a competitive cloud service provider market. There are global leading providers like AWS [2] and Azure [8], as well as region challengers like AliCloud [1], OVHCloud [11], the cloud services offered by different providers are more or less overlapped and converged gradually. The choice of providers sometimes more rely on business factors, like the availability of that providers in the region of target markets. We build internal software tools to abstract away the native cloud services from our applications as much as possible, we invest on Kubernetes technologies [27] as the application runtime environment so that we keep the option open to later evolve the stack for hybrid or multi-clouds if needed. Using cloud service enables *scalability* both in storage and computation.

3. **Algorithmic Computation**

   Distributed data computation engine that provides parallel-processing capabilities is key to analytics-oriented applications processing massive datasets. Spark [4] and Flink [3] are two leading techniques. Flink is from the beginning a streaming-oriented data processing engine while Spark is more popular engines for batch processing and is catching up in streaming. We choose Spark as the our stack's compute engine, because we consider Spark is better positioned in the whole data processing ecosystem. Many technologies come with existing solutions to integrate with Spark, with that we could enjoy more flexibility on choosing other techniques and know they will integrate well with the compute engine. This computation component is related to, and interleaved with the data analytics component described below.

4. **Data Analytics**

   Many open source tools to choose from for data analysis, tools used in single machine include Pandas [12], Scipy, sklearn [15]. We prioritize to support tools in the stack that are able to run on multiple machines in order to harvest distributed computing power provided by the cloud, to support *agility* for analytics-oriented applications. Spark is our chosen technique that provides the desirable distributed computation capability, additionally Spark provides APIs in SQL semantic that is familiar to many data-analysis specialists already.

   Tensorflow [18] and PyTorch [13] are two machine learning tools that we aim to integrate into our platform.

   The design principle in this data analytics component is not to lose the flexibility and being able to integrate more tools in the future if necessary. We try to best to avoid locking into a handful of tools pre-maturely. Tools that have low learning curves are preferred, because *agility* is one main goal. We try to reduce as much as possible the unnecessary effort of an analyst to wrestle with unfamiliar tooling concepts or APIs.

5. **Data Applications**

   We leverage open-source frontend Jupyter [7] to build analytics-oriented applications. We also use data visualization tools directly from some cloud vendors, e.g., PowerBI from Azure. When choosing such a specific data visualization tool from one vendor, we usually examine whether it supports many data input/output techniques rather than only those from the same vendor. We decide on frontend frameworks such as Vue and ReactJS [14, 19], and backend frameworks such as NodeJS and Django [6, 9], to build customer-facing real-time applications. These techniques have matured, they have been integrated and tested in cloud environments for many years. In addition there are existing open source data connectors for the frameworks we choose, for connecting them to different data storage techniques so that we keep the *flexibility* and not being locked into certain techniques. Another principle we have is to bias the choices on those that we could quickly prototype with, and then iterate on the prototype with fast turn-around time, this helps to achieve our *agility* goal.

3   Two Example Applications Enabled by the Stack

In this section, we showcase two example applications built on top of our data stack.

One analytics-oriented application shown in Fig. 2a is to investigate academic paper publication trends in each major city of China, for assessing cities' research activity levels. The research publication data we collected contains around 8 millions entries, organized as JSON files in ~80 GB. We use one cluster that consists of 30 nodes, each node of which has 14GB memory and 8 CPU cores, for data processing. Spark, the compute engine running on this cluster, orchestrates distributed computation tasks of analysis code. We choose a browser-based Jupyter [7] notebook environment for analyst to program analysis code, analysis results returned by the compute engine are also shown on the same UI. The programing API is a combination of SQL and DataFrame [22], both are familiar to experienced analysts, in fact our analysts put these new tools in use in a matter of a few hours' learning. Figure 2a shows the UI of this browser-based programming tool for analyst's use, backed by a powerful distributed cluster underneath. After loading the data into the cluster memory within minutes, analyst could use family APIs to program and then execute analytics tasks on the cluster. One example task is to group the publications by individual cities, then sort the cities by the publication numbers this particular analysis task takes less than one minute on the whole 80 GB dataset. With swift turn-around time of many such analytics tasks, analysts feel enabled and motivated to explore more analysis questions and experiment more analysis approaches to solve same questions.

Another application depicted in Fig. 2b is a customer-facing information-search web application. This application provides a search function for companies with their patent statistics in different regions of China. We leverage a cluster of 60 nodes, each has 14GB memory and 4 CPU cores, for running routine batch jobs to calculate key metrics that power the search. One of most expensive task in these routine batch jobs is to calculate region-specific company patent metrics, which needs to perform an expensive *SQL JOIN* of two large datasets: one is company registration data of past 20 years, consisting of ~36MM entries the other dataset is patent data of past 20 years that includes ~28MM patent entries. First a *SQL JOIN* of these two large dataset and then a *SQL GROUPBY* to group companies with patents by different regions. In total this task takes around 12 min by Spark engine on this cluster. The resulting metrics are then inserted into a PostgrepSQL relational database, which in turns powers the web search application. The search portal responses to users' search with results in few seconds. Figure 2b shows on such search result page, a geographical map of all regions is on the left side, where each region is colored according to its magnitude of numbers of company that have patents, on the right side is the top 10 regions. We are able to build this data application, from ingesting raw data, to setting up batch jobs for analysis, then eventually having web search application powered by a relational database, in weeks. The cohesive data stack connects a number of data storage and compute technologies together, enabling this swift development.

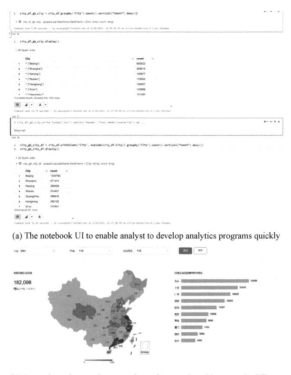

(a) The notebook UI to enable analyst to develop analytics programs quickly

(b) A search result page shows numbers of companies with patents in different regions

Fig. 2. Two applications built on top of the big data stack

4   Conclusion

We present a design of big data stack that collectively function as data intelligence platform, for swiftly deriving social development insights from huge amount of data. We present the concrete techniques to implement this stack, as well as the underlying reasonings on why choosing them among many other choices. The two showcases exemplify two categories of applications this data stack enables: analytics-oriented applications and real-time applications.

We hope to spur discussions on related topics in the community that would also benefit future development of our stack. The better the stack, the better it serves the purpose of providing insights and intelligence to aid informed decision-making of the society.

For future developments, one direction we are looking at is data-mesh like architectural paradigm [5,16], the purpose is to unlock access to a growing number of domain-specific datasets located within different organizations. Another direction is to ingest and process streaming data in near real-time. For example, extracting information real-time news feed. We consider this a great technical challenge to our data stack and we need to bring in new techniques carefully. Should it be implemented in our data stack, many interesting applications became feasible. We believe the impact, particularly to present decision-makers with near real-time insights from data, would be huge.

Acknowledgments. This work is partially supported by National Social Science Foundation of China (Grant No. 20CJY009).

References

1. Alibaba cloud services. https://www.aliyun.com
2. Amazon web services (aws) - cloud computing services. https://aws.amazon.com
3. Apache flink: Stateful computations over data streams. https://flink.apache.org/
4. Apache spark - unified analytics engine for big data. https://spark.apache.org/
5. Data mesh principles and logical architecture. https://martinfowler.com/articles/data-mesh-principles.html
6. Django: The web framework for perfectionists with deadlines. https://www.djangoproject.com/
7. Jupyter notebook. https://jupyter.org/
8. Microsoft azure: Cloud computing services. https://azure.microsoft.com
9. Node.js. https://nodejs.org
10. Oracle data warehouse. https://www.oracle.com/database/technologies/datawarehouse-bigdata.html/
11. Ovhcloud. www.ovh.com
12. pandas - python data analysis library. https://pandas.pydata.org/
13. Pytorch - an open source machine learning framework. https://pytorch.org/
14. React - a javascript library for building user interfaces. https://reactjs.org/
15. scikit-learn. https://scikit-learn.org
16. Service mesh. https://www.redhat.com/en/topics/microservices/what-is-a-service-mesh
17. Snowfalke, data cloud. https://www.snowflake.com/
18. Tensoflow - an end-to-end open source machine learning platform. https://www.tensorflow.org/
19. Vue js framework. https://vuejs.org
20. Zilian tech, Shenzhen, China. http://tsingzhi.cn/About-Us/
21. Armbrust, M., Ghodsi, A., Xin, R., Zaharia, M.: Lakehouse: a new generation of open platforms that unify data warehousing and advanced analytics. CIDR (2021)
22. Armbrust, M., et al.: Spark SQL: relational data processing in spark. In: Proceedings of the 2015 ACM SIGMOD International Conference on Management of Data, pp. 1383–1394 (2015)
23. Bonaventura, M., Ciotti, V., Panzarasa, P., Liverani, S., Lacasa, L., Latora, V.: Predicting success in the worldwide start-up network. Sci. Rep. **10**(1), 1–6 (2020)
24. Jia, J.S., Lu, X., Yuan, Y., Xu, G., Jia, J., Christakis, N.A.: Population flow drives spatio-temporal distribution of COVID-19 in china. Nature **582**(7812), 389–394 (2020)
25. Khine, P.P., Wang, Z.S.: Data lake: a new ideology in big data era. In: ITM Web of Conferences, vol. 17, p. 03025. EDP Sciences (2018)
26. Marcellino, W., Johnson, C., Posard, M.N., Helmus, T.C.: Foreign interference in the 2020 election: Tools for detecting online election interference. Technical report, RAND CORP SANTA MONICA CA SANTA MONICA United States (2020)
27. Sayfan, G.: Mastering kubernetes. Packt Publishing Ltd (2017)
28. Vohra, D.: Apache parquet. In: Practical Hadoop Ecosystem, pp. 325–335. Springer (2016). https://doi.org/10.1007/978-1-4842-2199-0_8

Design of the Electric Power Spot Market Operation Detection System

Min Zeng[✉] and Qichun Mu

Chengdu Polytechnic, Chengdu, Sichuan, China
748601807@qq.com

Abstract. With the continuous deepening of the construction of the electric power pot market, it is necessary to optimize the operation mechanism of the spot market according to the operation of the spot market, study the information interaction and data integration technology of the spot market to support the coordinated operation of multiple markets, and design the overall architecture, application architecture, functional architecture of the information interaction and data integration platform of the spot market for the coordinated operation of multiple markets Hardware architecture and security protection system provide technical support for information interaction and data integration of multiple market coordinated operation of the power spot market. Through data visualization technology, this paper realizes the data visualization and background management of the provincial power spot market operation detection system, which is convenient for decision-makers to carry out data analysis and management.

Keywords: PDO · Transaction mechanism · PhpSpreadsheet

1 Introduction

With a large number of new energy connected to the grid and the rapid growth of electricity demand in some areas, China's power supply structure and supply and demand situation has changed, which puts forward a greater demand to solve the problem of system peak regulation and trans-provincial surplus and deficiency regulation. Therefore, it is urgent to further deepen inter-provincial spot transactions, optimize the allocation of resources in a wider range, discover the time and space value of electric energy, and realize the sharing of peak regulation resources and inter-provincial surplus and deficiency adjustment by market means.

At the same time, with the continuous deepening of the construction of the spot market, it is necessary to optimize the operation mechanism of the spot market according to the operation of the spot market, study the information interaction and data integration technology of the spot market to support the coordinated operation of multiple markets, and design the overall architecture, application architecture, functional architecture, and data integration platform of the spot market for the coordinated operation of multiple markets Hardware architecture and security protection system provide technical support

© The Author(s) 2022
Z. Qian et al. (Eds.): WCNA 2021, LNEE 942, pp. 334–341, 2022.
https://doi.org/10.1007/978-981-19-2456-9_35

for information interaction and data integration of multi market coordinated operation of the power spot market.

In order to support the construction of inter provincial electricity spot market, it is necessary to develop a visual system with perfect function and friendly interface on the basis of technology research and development.

The minimum configuration of front-end display hardware recommended by this system is CPU Intel i7-7700k, memory 8GB DDR4, disk 300 gb, graphics card GTX 1060, display standard resolution 1920 × 1080. The minimum configuration of database server is CPU Intel Xeon e5–4650, memory above 16 GB DDR4 and disk 1 TB.

The required software environment includes: Microsoft operating system, HTML5 standard browser, PHP development environment, Apache server environment, relational database.

2 System Design

The whole system is divided into front-end visualization system and background data management system, as shown in Fig. 1.

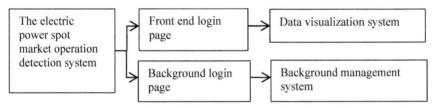

Fig. 1. System structure

The front end of the system has two major functional modules, data overview and operation data. All modules are developed with HTML5 technology such as webgl and canvas, and the mainstream web framework is used. The data interface is provided by the background of PHP to obtain the data of MySQL database for visualization.

The system is divided into 11 pages: login page, transaction statistics, channel path, declaration status, declaration statistics, declaration details, channel available capacity, node transaction result, channel transaction result, path transaction result and personal center. The front end structure is shown in Fig. 2.

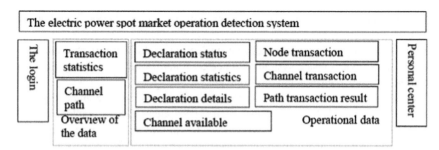

Fig. 2. Schematic diagram of front end structure

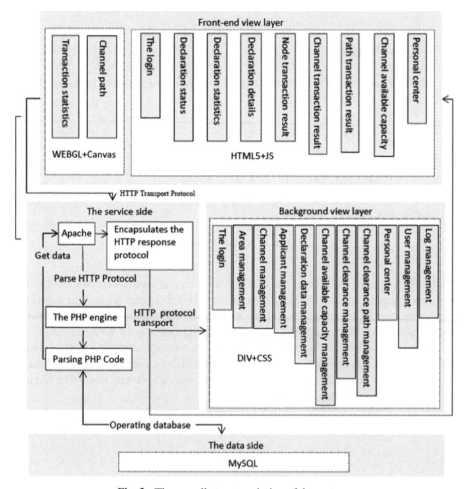

Fig. 3. The overall structure design of the system

The background of this system uses the mainstream Web back-end framework. The data interface is provided by the background of PHP to obtain the data of MySQL database for visual presentation.

The system is divided into 11 pages: login page, area management, channel management, applicant management, declaration data management, channel available capacity management, channel clearance management, channel clearance path management, personal center, log management.

Because of the huge and complex data of the power system, the data source of this system is provided by the management personnel in the background through the way of importing Excel files. The Excel file is uploaded and submitted by the provinces, and then imported by the backstage management personnel according to the unit of day.

The detailed structural design of this system is shown in Fig. 3.

3 The Specific Implementation

This system content is more, limited to the length, this paper after the Taiwan management system as an example, detailed introduction of the implementation of specific functions.

3.1 Import of Excel File

The background management part of this system is developed by PHP7.3. In PHP7, the best way to import Excel is to use third-party plug-ins. PHPSpreadsheet is one of the most powerful and easy to use plug-ins and is recommended for use.

PHPSpreadsheet is a library written in pure PHP that provides a set of classes that allow you to read and write different spreadsheet file formats. PHPSpreadsheet provides a rich API that allows you to set up many cell and document properties, including styles, images, dates, functions, etc. You can use it in any Excel spreadsheet you want. The document formats supported by PHPSpreadsheet are shown in Table 1.

Table 1. Formatting sections, subsections and subsubsections.

Format	Reading	Writing
Open document format(.ods)	✓	✓
Excel 2007 and above	✓	✓
Excel 97 and above	✓	✓
Excel 95 and above	✓	
Excel 2003	✓	
HTML	✓	✓
CSV	✓	✓
PDF		✓

To use phpSpreadsheet, your system requires a PHP version greater than 7.2. In your project, you can use Composer to install PHPSpreadsheet with the following command:

```
composer require phpoffice/phpspreadsheet
```

To install PHPSpreadsheet, if you need to use documents and examples, use the following command:

```
composer require phpoffice/phpspreadsheet --prefer-source
```

The basic use of phpSpreadsheet is very simple. When the plug-in is downloaded and installed, you just need to introduce the autoload.php file into your project. The following code is a simple example that generates an Excel file and populates the cells with the specified content.

```php
<?php
require 'vendor/autoload.php';
use PhpOffice\PhpSpreadsheet\Spreadsheet;
use PhpOffice\PhpSpreadsheet\Writer\Xlsx;
$spreadsheet = new Spreadsheet();
$sheet = $spreadsheet->getActiveSheet();
$sheet->setCellValue('A1', 'Hello World !');
$writer = new Xlsx($spreadsheet);
$writer->save('hello world.xlsx');
```

In this project, we need to make an auxiliary page for uploading Excel files. In order to simplify the operation of the manager, the system supports the import of multiple Excel files at one time, and it only needs to add multiple attribute in the File field.

```
<input type="file" name="file[]" multiple="">
```

After uploading the file, create a corresponding PHPSpreadsheet reader based on the extension of the Excel file, set up read-only operations, and read the contents of the file into an array.

```
/** Create a reader **/
if ($ext == 'xls') {
    $reader = new \PhpOffice\PhpSpreadsheet\Reader\Xls();
} else {
    $reader = new \PhpOffice\PhpSpreadsheet\Reader\Xlsx();
}
$reader->setReadDataOnly(true);  //Just the data, not the format
$spreadsheet = $reader->load($inputFileName);
$data = $spreadsheet->getActiveSheet(0)->toArray();
```

After reading the contents of the file, the next step is to verify that the table header, row, and column data are correct according to the template requirements. After all the data is correct, it can be written to the appropriate database.

In the operation of the database, due to the complex structure of the Excel file, there are a lot of data to be verified. There will be several operations on the database, and there will be correlation between each other. In order to maintain the consistency of data, we use the transaction mechanism of PDO to deal with this part of content.

The transaction mechanism of PDO supports four characteristics: atomicity, consistency, isolation, and persistence. In general terms, any operation performed within a transaction, even if performed in stages, is guaranteed to be applied to the database safely and without interference from other connections at commit time. Transactional operations can also be undone automatically on request (assuming they haven't been committed), which makes it easier to handle errors in the script.

We can use Begin Transaction to enable transactions, Commit to commit changes, and roll Back to and from operations. Here's the relevant demo code:

```php
<?Php
try{
$dbh=new
PDO('odbc:demo,'mysql','mysql',array(PDO::ATTR_PERSISTENT=>true));
    Echo "Connected\n";
    }
    Catch (Exception $e){
    die("Unable to connect:".$e->getMessage());
    }
try{
    $dbh->setAttribute(PDO::ATTR_ERRMODE,PDO::ERRMODE_EXCEPTION);
    $dbh->beginTransaction();
    $dbh->exec("insert into table1 (id,first,last) values
(23,'mike','Bloggs')");
    $dbh->exec("insert into tabel2 (id,amount,date) values
(23,50000,time())");
    $dbh->commit();
    }
    catch(Exception $e){
    $dbh->rollBack();
    Echo "Failed:".$e->getMessage();
    }
```

3.2   Editing of Imported Data

After the Excel data is imported into the data, it should be possible to edit and modify the data according to the user's needs. Due to the large quantity, in order to facilitate editing and modification, we use the DataGrid in the EasyUI framework for processing.

EasyUI is a set of user interface plug-ins based on jQuery. Using easyUI can greatly simplify our code and save the time and scale of master web development. While EasyUI is simple, it is powerful.

The EasyUI front-end framework contains many commonly used front-end components, among which the DataGrid is distinctive. The EASYUI Data Grid (DataGrid) displays data in a tabular format and provides rich support for selecting, sorting, grouping, and editing data. Data grids are designed to reduce development time and do not require specific knowledge of the developer. It's lightweight, but feature-rich. Its features include cell merging, multi-column headers, frozen columns and footers, and more. For back-end data editing on our system, the DataGrid is best suited.

We use JS to generate the static content of the data table, and then request the data interface through Ajax, and then render the data table after getting the data, so as to get the results we want.

This system focuses on the use of data table editor, you can achieve online editing table data. To edit the data, when initializing the DagGrid, you need to add an edit button in the last column using the formatting function, as follows:

```
formatter: function (value, row, index) {
            if (row.editing) {
            var s = '<span style="cursor: pointer; float: left;
background: #5c641b;color: #ffffff;padding: 1px 35px 1px
35px;margin:5px;display: inline-block;height: 40px;line-height:
40px;" onclick="saveRow(this)">save</span> ';
            var c = '<span style="cursor: pointer; float: left;
background: #349564;color: #ffffff;padding: 1px 35px 1px
35px;margin:5px;display: inline-block;height: 40px;line-height:
40px;" onclick="cancelRow(this)">cancel</span>';
            return s + c;
            } else {
            var e = '<span style="display:inline-block;cursor:
pointer; background: #3d70a2;color: #ffffff;padding: 1px 35px 1px
35px;height: 40px;line-height: 40px;margin:5px;"
onclick="editRow(this)">edit</span> ';
            return e;
    }
}
```

After editing is complete, you can change the database through the event OnAfterEdit.

4   Conclusion

By connecting MySQL database with PHP and cooperating with DataGrid of Easy UI, we completed the design and implementation of the background management system of the operation and detection system of the electric spot market. The key content of this system is to use PHP to import Excel files, and verify the validity of data format and content, and then use the transaction mechanism of PDO to complete the data writing. Data is displayed through the data network function of Easy UI, and the editor is used to complete the data editing.

With the data, in the front end can be through the API interface, access to background data, and display in the front end.

References

1. Tatroe, K., MacIntyre, P.: PHP Programming. Electronic Industry Press (2021)
2. Zandstra, M.: An In-Depth Look at PHP Object Orientation, Patterns, and Practices. Posts and Telecommunications Press (2019)
3. Yu, G.: PHP Programming from Entry to Practice. Posts and Telecommunications Press (2021)

4. Tang, Q.: Practical Application of PHP Web Security Development. Tsinghua University Press (2018)
5. Lei, C.: PHP 7 Low-Level Design and Source Code Implementation. China Machine Press (2018)

FSTOR: A Distributed Storage System that Supports Chinese Software and Hardware

Yuheng Lin[2], Zhiqiang Wang[1(✉)], Jinyang Zhao[3], Ying Chen[2], and Yaping Chi[1]

[1] Cyberspace Security Department, Beijing Electronic Science and Technology Institute, Beijing, China
wangzq@besti.edu.cn
[2] Department of Cryptography and Technology, Beijing Electronic Science and Technology Institute, Beijing, China
[3] Beijing Baidu T2Cloud Technology Co. Ltd., 15A#-2nd Floor, En ji xi yuan, Haidian District, Beijing, China

Abstract. In order to develop a distributed storage system that adapts to Chinese software and hardware, build a cloud computing platform that is independently usable, safe and reliable, data utilization is more concentrated and intelligent, and service integration is more unified and efficient. This paper designed and implemented a distributed storage system that supports Chinese software and hardware, which is compatible with Chinese mainstream CPU, operating system, database, middleware and other software and hardware environments. After a lot of experiments and tests, it is confirmed that the system has high availability and high reliability.

Keywords: Cloud computing platform · Distributed storage system · Localization

1 Introduction

The distributed storage system is a data storage technology that distributes data on multiple independent devices, and provides storage services as a whole externally1,2. It has the characteristics of scalability, high reliability, availability, high performance, high resource utilization, fault tolerance and low energy consumption3. Its development process can be roughly divided into three stages. One is the traditional network file system, which is typically represented by Network File System (NFS), etc., the second is the general cluster file system, such as Galley, Shared File System (GPFS), etc., and the third is the object-oriented transit distributed file system, such as Google File System (GFS), Hadoop Distributed File System (HDFS), etc. NFS4,5 is a UNIX presentation layer protocol developed by SUN; GPFS6,7 is IBM's first shared file system. GFS8 is a dedicated file system designed by Google to store massive search data. The above-mentioned typical distributed storage systems are all developed by foreign companies, and all have incompatibility with Chinese software and hardware.

Z. Qian et al. (Eds.): WCNA 2021, LNEE 942, pp. 342–350, 2022.
https://doi.org/10.1007/978-981-19-2456-9_36

In response to the above problems, this paper designed and implemented a localized distributed software-defined storage system named FSTOR, which is based on B/S architecture, has standard interfaces and supports various localized operating systems and virtualization systems, and both servers and databases are localized facility. The system implements distributed cloud storage block storage services, snapshot management, full-user mode intelligent cache engine, cluster dynamic expansion, pooled storage function, fault self-check and self-healing functions.

The organization structure of this article is as follows: The first part introduces the relevant research background of the system; the second part introduces the system architecture; the third part describes the functional architecture of the system; the fourth part tests the system and analyzes the test results; the fifth part summarizes full text.

2   System Structure

The detailed system architecture is shown in Fig. 1. The overall technology and software system can run normally on the Chinese CPU. The Chinese x86 architecture Zhaoxin, the ARM architecture Feiteng and the Alpha Shenwei can be used, and the operating system Kylin or CentOS can be used. The system can use automated operation and maintenance technology to ensure daily operation and maintenance management, including but not limited to data recovery, network replacement, disk replacement, host name replacement, capacity expansion, inspection, failure warning, capacity warning, etc.

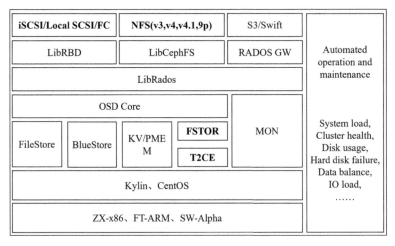

Fig. 1.   System architecture diagram

(1) **LibRBD**

A module that supports localized block storage, abstracts the underlying storage, and provides external interfaces in the form of block storage. LibRBD supports the localized virtualization technology to be mounted to the localized operating system through the RBD protocol, and is provided to some localized databases.

(2) **Libcephfs**
A module that supports localized Posix file storage, supports the Kylin and the CentOS operating system to mount the file system locally to the Chinese operating system through the mount command and provide it for use.

(3) **RADOS GW**
In order to support a gateway module for localized object storage, two different object storage access protocols, S3 and Swift, are provided. Localized software can use these two protocols to access the object storage services provided by the system.

(4) **Librados**
A module supporting blocks, files, and object protocols is responsible for interacting with the core layer of the Chinese storage system. It is a technical module of the interface layer.

(5) **MON**
The brain of the system. The management of the storage system cluster is handed over to MON.

(6) **OSD Core**
Responsible for taking over the management of a physical storage medium.

(7) **FileStore**
An abstract module that manipulates the file system. The system accesses business data through the Poxis standard vfs interface. The space management of the physical disk is handed over to the open source xfs file system to manage.

(8) **BlueStore**
A small Chinese file system. It can replace the xfs file system to manage the physical disk space, reducing some performance problems caused by the xfs file system being too heavy.

(9) **T2CE**

A Chinese smart cache module. The system can make full use of physical hardware resources to improve storage performance. Its intelligent caching engine can perceive data characteristics and frequency, and store data that meets a predetermined strategy on high-speed devices, and store data that does not meet the predetermined strategy on slow devices. Under the premise of not significantly increasing hardware costs, use high-speed equipment to drive low-speed equipment to ensure business performance requirements.

The intelligent cache engine revolves around the close cooperation between multiple core modules such as IO feature perception, intelligent aggregation, disk space allocation and defragmentation, and maximizes the combination of high-speed and low-speed devices between performance and capacity to achieve a perfect balance. The smart cache uses a large number of efficient programming models and algorithms to maximize the performance of high-speed devices.

3 Function Architecture

The system function framework is shown as in Fig. 2. The system includes a hardware abstraction layer, a unified storage layer, a storage service layer, an interface protocol layer and an application layer. The unified storage layer includes multiple copies, pooling, tiered storage, linear expansion, fault medical examination, data recovery QoS, erasure coding, strong data consistency, intelligent caching, dynamic capacity expansion, fault domain and fault self-healing. The storage service layer includes snapshot cloning, data link HA, data stream QoS, encryption compression, quota control, thin provisioning, multipart upload, permission control, version control, multi-tenancy, data tiering, and write protection. The interface protocol layer includes block storage interface, object interface and file storage interface. The application layer includes virtualization, unstructured data and structured data.

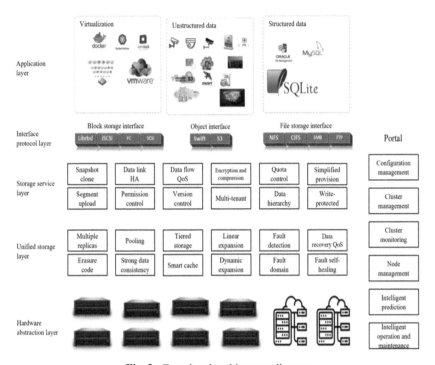

Fig. 2. Functional architecture diagram

(1) **Object Storage Segmented Upload**

Segmented upload is the core technology of breakpoint continuingly functions. When the fault is restored, avoid re-uploading the content of the uploaded file and cause unnecessary waste of resources. Users can also implement user-side QoS functions based on the multipart upload function. The multipart upload function will verify the content of the uploaded file, and the parts that fail the verification will be re-uploaded.

(2) **Dynamic Capacity Expansion and Reduction Without Perception**

The system supports dynamic capacity expansion and contraction without perception, and can respond to changes in application requirements in a timely manner without perception of the application, ensuring the continuous operation of the business. In addition, the performance also increases linearly with the increase of the number of nodes, giving full play to the performance of all hardware

(3) **Data Redundancy Protection Mechanism**

The system provides two different pool data redundancy protection mechanisms: replica and erasure code to ensure data reliability.

Replica mode is a data redundancy realized by data mirroring, with space for reality. Each replica keeps complete data, and users can pool 1–3 replicas according to specific business requirements to maintain strong consistency. The greater the number of replicas, the higher the fault tolerance allowed, and the consumed capacity increases proportionally.

Erasure code mode is an economical redundancy scheme, which can provide higher disk utilization. Users can choose K + m combination according to the specific business requirements. K represents to store the original data in K blocks, and M represents to generate M pieces of coded data. The size of each piece of coded data is the same as that of the block. The K pieces of block data and M pieces of coded data are stored separately to achieve data redundancy. According to any k pieces of data in K + m, the original data can be reconstructed.

(4) **Troubleshooting**

The system supports a variety of different levels of fault domain design, the smallest fault is the tiered disk, and the largest fault tier can be the data center. It is common to use the cabinet as the fault level, and the user can divide it according to the actual situation. The fault domain can ensure the failure level of data redundancy. Whether it is a failure of a disk, a rack, or a data center, the reliability of the data can be guaranteed. At the same time, the system also supports intelligent fault detection and fault self-healing and alarms to avoid manual intervention, and supports intelligent data consistency verification to avoid data loss due to silent errors.

4 System Test

4.1 Test Environment

The test environment topology is shown in Fig. 3. Four node servers and a notebook are used. The server and notebook are connected to the switch. FIO 2.2.10 (cstc10184742) is used as the test tool.

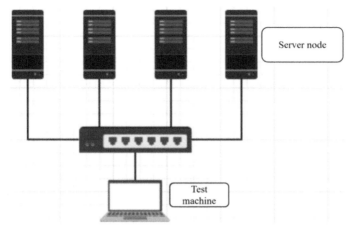

Fig. 3. System test network topology

The model and configuration of server and client are shown in Table 1. In the test, the model and configuration of the four node servers are the same, all of them are Kylin system, and the CPU is FT1500a@16c CPU.The notebook is the ultimate version of Windows 7 system, the model is ThinkPad T420, and the notebook is equipped with Fio.

Table 1. Environment configuration

Equipment name	Model and configuration	Operating system	Software configuration
Node server (4)	CPU: FT1500a@16c CPU 1.5 GHz RAM: 64 GB hard disk: 1.8TB	Kylin V4.0	FSTOR distributed storage system MariaDB V10.3 RabbitMQ V3.6.5
notebook (1)(CSTC10124326)	model: Thinkpad T420 CPU: Intel Core i5-2450M 2.50 GHz RAM: 4 GB hard disk: 500GB	Windows 7 Ultimate	Google Chrome 52.0.2743.116 Fio 2.2.10

4.2 Test Content

The content of system test is shown in Table 2. IOPs (input/output operations per second) is the input/output volume (or read/write times) per second, used for computer storage device performance test. The test results show that the system realizes the functions designed in all functional architectures.

Table 2. Test Content

Technical index	Test results
Block storage service	The block storage volume can be successfully created and the storage volume can be mapped to the virtual machine File system can be created for storage volume
Snapshot management	Supports the snapshot function of storage volumes, and clones new storage volumes through snapshots You can perform a rollback operation on the storage volume that has been snapshotted
Smart cache engine	The smart cache engine storage pool can be successfully created
Cluster dynamic expansion	A new storage server or hard disk can be added to the storage cluster
Pool storage function	Can create storage pools with different performance
Fault self-checking and self-healing	Delete an object storage device and kick it out of the cluster, and cluster business will not be interrupted
Web storage mount	Web storage can be mounted via NFS protocol
4k random write	4k random write without cache IOPS: 1694 4k random write IOPS with cache: 5149
4k random read	4k random read without cache IOPS: 2474 4k random read IOPS with cache: 6507
4k mixed random read and write	4k mixed random read without cache IOPS: 1944 4k mixed random read with cache IOPS: 4863 4k mixed random write without cache IOPS: 648 4k mixed random write buffered IOPS: 1621

4.3 Test Results

(1) **System Structure**

The system is based on B/S architecture, the server adopts Kylin v4.0 operating system, the database adopts MariaDB V10.3, the middleware adopts RabbitMQ v3.6.5, and the bandwidth is 1000Mbps. The client operating system is the ultimate version of Windows 7, and the browser adopts Google Chrome 52.0.2743.116.

(2) **Performance Efficiency**

The system performance is as follows: 4K random write without cache IOPs: 1694; 4K random write buffer IOPs: 5149; No IOPs: 4K random read cache; 4K random read buffer IOPs: 6507; 4K mixed random read without cache IOPs: 1944; 4K mixed random read buffer IOPs: 4863; 4K mixed random write without cache IOPs: 648.

5   Conclusions

Aiming at the problem that the distributed storage system needs localization and supports Chinese software and hardware, this paper designed and implemented a distributed storage system namedFSTOR, which runs on the Chinese operating system and CPU, and each module supports localization. The system ensures the daily operation and maintenance management by realizing automatic operation and maintenance, and ensures the reliability of data through two pool data redundancy protection mechanisms and fault or division methods: copy and erasure code. After a large number of tests, the system runs stably, realizes complete functions, and achieves high reliability and high availability.

Acknowledgments. This research was financially supported by National Key R&D Program of China (2018YFB1004100), China Postdoctoral Science Foundation funded project (2019M650606) and First-class Discipline Construction Project of Beijing Electronic Science and Technology Institute (3201012).

References

1. Zhu, Y., Fan, Y., Yubin, W., et al.: An architecture design integrating distributed storage. Henan Sci. Technol. **40**(36), 22–24 (2021)
2. Lin, C.: Research and Implementation of Replica Management in Large-scale Distributed Storage System. University of Electronic Science and Technology of China (2011)
3. Li, G., Yang, S.: The analysis of the research and application of distributed storage system. Network Secur. Technol. Appl. **2014**(09), 73+75 (2014)
4. Sandberg, R.: The sun network filesystem: design, implementation and experience. In: Proceedings of USENIX Summer Conference, pp. 300–313. University of California Press (1987)
5. Huang, Y.: Docker data persistence and cross host sharing based on NFS. North University of China, pp. 22–24 (2021)
6. Schmuck, F., Haskin, R.: GPFS: A shared-disk file system for large computing clusters. In: Proceedings of the Conference Oil File and Storage Technologies (FAST 2002), 28–30 January 2002, Monterey, CA, pp. 231–244 (2002)
7. Zhang, X.-N., Wang, B.: Installation configuration and maintenance of GPFS. Comput. Technol. Dev. **28**(05), 174–178 (2018)
8. Ghemawat, S., Gobioff, H., Leung, S.T.: The Google file system. ACM SIGOPS Operat. Syst. Rev. **37**(5), 29–43 (2003)

Wireless Sensor Networks

Joint Calibration Based on Information Fusion of Lidar and Monocular Camera

Li Zheng[1](✉), Haolun Peng[2], and Yi Liu[2]

[1] Automation College, Chengdu Technological University, Chengdu, China
zhengli@mail.cdtu.edu.cn
[2] Control Engineering College, Chengdu University of Information Technology, Chengdu, China

Abstract. To solve the problem of joint calibration in multi-sensor information fusion, a joint calibration technique based on three-dimensional lidar point cloud data and two-dimensional gray image data is proposed. Firstly, by extracting the corner information of the gray image data, the two-dimensional coordinates of the corner were obtained, and the calibration of the monocular camera was completed by using the corner information, and its internal and external parameters were obtained. Then, by extracting the corner information of the point cloud data obtained by lidar, the corresponding corner points are matched. Finally, the rotation and translation matrix from lidar coordinate system to image coordinate system is generated to realize the joint calibration of lidar and camera.

Keywords: Multisensor · Joint calibration · Corner · Feature point matching

1 Introduction

Multi-sensor data fusion is a novel technology for collecting and processing information. With the development and application of unmanned system technology, intelligent equipment needs to realize information perception of the surrounding environment based on external sensors [1], in order to realize unmanned operation. Lidar can obtain the distance of the target and provide precise and accurate three-dimensional point cloud data, but it can not get rich other environmental information;

Monocular camera can collect various environmental information, but it can not obtain accurate distance information. Considering the characteristics of both, the fusion of lidar and monocular camera sensing information can well obtain various environmental information around intelligent equipment and provide necessary information feedback for unmanned operation of intelligent equipment. To complete information fusion, the first thing to do is to conduct joint calibration among multiple sensors [2]. This is in order to obtain the relative position between the respective sensors, and find out the conversion relationship between the coordinates of each sensor [3]. In this paper, a joint calibration method based on LIDAR point cloud data and two-dimensional data of gray image is proposed. A rectangular standard plate is used as the calibration plate to verify the effectiveness of the method.

© The Author(s) 2022
Z. Qian et al. (Eds.): WCNA 2021, LNEE 942, pp. 353–366, 2022.
https://doi.org/10.1007/978-981-19-2456-9_37

2   Monocular Camera Calibration

The purpose of monocular camera calibration is to realize the rapid conversion between monocular sensor coordinate system and world coordinate system, obtain the relative position relationship between them, and obtain the internal and external parameters of monocular sensor.

2.1   Pinhole Camera Model

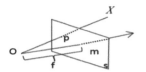

Fig. 1. Linear camera model

As shown in Fig. 1, a point O in space is the projection center of the pinhole camera, F, that is OP represents the distance from point O to point P on the plane. Project point X in space onto planes can obtain projection point P.

The image plane of the camera is plane s, where the optical center of the camera is point O and the focal length of the camera is OM, which can be expressed by f, the optical axis of the camera is a ray emitted outward with the optical center of the camera as the starting position, also known as the main axis. The optical axis of the camera is perpendicular to plane s, and the optical axis has an intersection with image plane s, which is called the main point of the camera.

$$\lambda \begin{bmatrix} \bar{x} \\ \bar{y} \\ 1 \end{bmatrix} = K \begin{bmatrix} x_c \\ y_c \\ x_c \end{bmatrix} = \begin{bmatrix} f_u & s & u_0 \\ 0 & f_v & v_0 \\ 0 & 0 & 1 \end{bmatrix} \begin{bmatrix} x_c \\ y_c \\ z_c \end{bmatrix} \tag{1}$$

In Formula 1, matrix K is the internal parameter matrix of the camera. We can do a very fast transformation from the camera coordinate system to the image coordinate system through the internal reference matrix. (f_u, f_v) is the focal length parameter of the camera. The focal length is the distance between the world and the image plane. Under the pinhole camera model, the two values are the same. (u_0, v_0) is the offset of the main point from the image plane. When the U-axis of the image coordinate system is not completely perpendicular to the v-axis, the s generated is called distortion factor.

2.2   Camera Calibration Principle

Camera Calibration Principle [4]:

If ranging is carried out through gray image, In order to obtain the three-dimensional coordinates of a point on an object in space and its corresponding point in the camera image more quickly and accurately, and get the change and conversion between them, we need to establish a geometric model based on gray image, and the parameters of the camera constitute a basic parameter of the geometric model. Through a lot of calculation and practice, these parameters can be solved and given accurately. This process is called the camera calibration process.

2.3 Coordinate System Under Camera

Coordinate System:

Four coordinate systems in the camera imaging model:

a. World coordinate system: a coordinate system established with a reference point outside, the coordinate points are (XW, YW, ZW)
b. camera coordinate system: a coordinate system established with the optical center of monocular camera as the reference point, and the coordinate points are (x, y, z)
c. Image coordinate system: the optical center is projected on the imaging plane, and the obtained projection point is used as the reference point to establish a rectangular coordinate system. The coordinate point is (x, y)
d. pixel coordinate system: the coordinate system that can be seen by the end user. The origin of the coordinate system is in the upper left corner of the image, and the coordinate point is (u, v)

Various transformation relations from the world coordinate system to the pixel coordinate system are shown in Fig. 2:

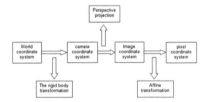

Fig. 2. Conversion from world coordinate system to pixel coordinate system

The conversion relationship between coordinates is shown in Fig. 3:

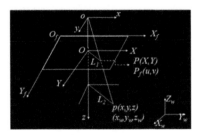

Fig. 3. Schematic diagram of coordinate system relationship

a) The transformation formula between the world coordinate system and the camera coordinate system is shown in Eq. 2:

$$
\begin{bmatrix} x \\ y \\ z \\ 1 \end{bmatrix} = \begin{bmatrix} R & T \\ 0^T & 1 \end{bmatrix} \cdot \begin{bmatrix} X_w \\ Y_w \\ Z_w \\ 1 \end{bmatrix} = \begin{bmatrix} r_{11} & r_{12} & r_{13} & t_x \\ r_{21} & r_{22} & r_{23} & t_y \\ r_{31} & r_{32} & r_{33} & t_z \\ 0 & 0 & 0 & 1 \end{bmatrix} \cdot \begin{bmatrix} X_w \\ Y_w \\ Z_w \\ 1 \end{bmatrix}
\tag{2}
$$

where matrix R is the rotation matrix. And R meets the following conditions:

$$
\begin{cases} r_{11}^2 + r_{12}^2 + r_{13}^2 = 1 \\ r_{21}^2 + r_{22}^2 + r_{23}^2 = 1 \\ r_{31}^2 + r_{32}^2 + r_{33}^2 = 1 \end{cases}
\tag{3}
$$

The R matrix contains three variables, R_x, R_y, R_z, t_x, t_y, t_z which together are called the external parameters of camera.

b) The transformation relationship between the image coordinate system and the camera coordinate system is as follows:

$$
z \begin{bmatrix} X \\ Y \\ 1 \end{bmatrix} = \begin{bmatrix} f & 0 & 0 & 0 \\ 0 & f & 0 & 0 \\ 0 & 0 & 1 & 0 \\ 0 & 0 & 1 & 0 \end{bmatrix} \begin{bmatrix} x \\ y \\ z \\ 1 \end{bmatrix}
\tag{4}
$$

This conversion relationship is from 3D to 2D, which belongs to the relationship of perspective projection. After this conversion, the monocular of the projection point is not converted to pixels, so the next conversion is carried out.

c) The actual relationship between image coordinate system and pixel coordinate system is as follows:

$$
\begin{cases} u = \frac{X}{dx} + u_0 \\ v = \frac{Y}{dy} + v_0 \end{cases}
\tag{5}
$$

$$
\begin{cases} u - u_0 = \frac{X}{dx} = s_x \cdot X \\ v - v_0 = \frac{Y}{dy} = s_y \cdot Y \end{cases}
\tag{6}
$$

Because both the image coordinate system and the pixel coordinate system are located on the image plane, they are only different in scale. Except for the origin and their respective units, they are the same.

d) Transformation between camera coordinate system and pixel coordinate system.

$$
\begin{cases} u - u_0 = \frac{fs_x x}{z} = f_x x/z \\ v - v_0 = \frac{fs_y y}{z} = f_y y/z \end{cases}
\tag{7}
$$

f_x is the focal length in the axial direction and f_y is the focal length in the axial direction, f_x, f_y, u_0, v_0. It are called the internal parameters of the camera, because these four elements are related to the structure of the camera itself.

e) Transformation relationship between pixel coordinate system and world coordinate system:

$$z \begin{bmatrix} u \\ v \\ 1 \end{bmatrix} = \begin{bmatrix} f_x & 0 & u_0 & 0 \\ 0 & f_y & v_0 & 0 \\ 0 & 0 & 1 & 0 \end{bmatrix} \begin{bmatrix} R & T \\ 0^T & 1 \end{bmatrix} \begin{bmatrix} X_w \\ Y_w \\ Z_w \\ 1 \end{bmatrix} = M_1 \cdot M_2 \cdot X = M \cdot X \qquad (8)$$

Using the above mathematical expression, we can uniquely determine the internal parameters of the camera, correspond the collected corner coordinates with their image point coordinates one by one, and calculate the internal and external parameters of the camera to complete the calibration of the camera.

Specific implementation steps:

1. Preprocessing the image
2. Edge detection
3. Extracting the contour of the calibration plate
4. Corner detection
5. Calibration

The corner point, internal parameter and external parameter matrix of the camera are shown in the Figs. 4 and 5 below:

image_width: 640
image_height: 480
camera_name: narrow_stereo
camara_matric:
  rows: 3
  cols: 3
  data: [887.7844629183168 , 0.000000 , 319.9060924025313 ; 0.000000 , 887.9671237624945 , 235.2051452903424 ;
  0.000000 , 0.000000 , 1.000000]
distortion_model: plumb_bob
distortion_coefficients:
  rows: 1
  cols: 5
  data: [-0.369649369605705 , -0.436141758075861 , -9.7593070017402e-05 , 0.0002778575622676858 , 4.38382300069067]
rectification_matric
  rows: 3
  cols: 3
  data: [1.000000 , 0.000000 , 0.000000 , 0.000000 , 1.000000 , 0.000000 , 0.000000 , 0.000000 , 1.000000]

Fig. 4. Camera calibration corner diagram

Fig. 5. Camera calibration parameters

3   Lidar Calibration

Line scan lidar is selected in this scheme, and 16 line specifications are selected. The operation principle of the lidar is as follows: the target distance is measured through the transceiver of the laser signal. The lidar controls the scanning of the lidar by controlling the rotation of the internal motor - scanning the linear array to the external environment, the distance from the lidar to the target object is calculated according to the TOF flight principle. There is a laser transmitter and a laser receiver inside the lidar. During operation, the lidar emits the laser. At the same time, the internal timer starts timing. When the laser hits the target, the reflection occurs, and the laser returns to the laser receiver. The timer records the arrival time of the laser, The actual movement time is obtained by subtracting the start time from the return time. Because of the principle of constant speed of light (TOF), the actual distance can be obtained through calculation.

The lidar coordinate system depicts the relative position of the object relative to the lidar, as shown in Fig. 6:

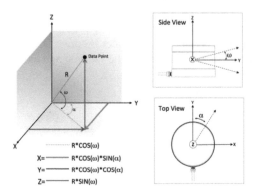

Fig. 6. Schematic diagram of lidar coordinate system

When collecting data, the laser line ID can be used through Table 1. Because the laser point has its own specific ID, the unique laser line inclination can be obtained. The query table is shown in Table 1. According to the distance value r actually measured by the lidar, the coordinate x_0 of the laser point in the scanning plane coordinate system can be obtained through formula 9 [5].

$$X_0 = \begin{bmatrix} x_0 \\ y_0 \\ 0 \end{bmatrix} = \begin{bmatrix} r\,sin\,\omega \\ r\,cos\,\omega \\ 0 \end{bmatrix} \qquad (9)$$

Table 1. Vertical angles (ω) by laser ID and model

Laser ID	Vertical angel VLP-16	Vertical angel puck LITE	Vertical correction (mm)	Vertical angel puck Hi-Res	Vertical correction (mm)
0	$-15°$	$-15°$	11.2	$-10.00°$	7.4
1	$1°$	$1°$	-0.7	$0.67°$	-0.9
2	$-13°$	$-13°$	9.7	$-8.67°$	6.5

(continued)

Table 1. (*continued*)

Laser ID	Vertical angel VLP-16	Vertical angel puck LITE	Vertical correction (mm)	Vertical angel puck Hi-Res	Vertical correction (mm)
3	3°	3°	−2.2	2.00°	−1.8
4	−11°	−11°	8.1	−7.33°	5.5
5	5°	5°	−3.7	3.33°	−2.7
6	−9°	−9°	6.6	−6.00°	4.6
7	7°	7°	−5.1	4.67°	−3.7
8	−7°	−7°	5.1	−4.67°	3.7
9	9°	9°	−6.6	6.00°	−4.6
10	−5°	−5°	3.7	−3.33°	2.7
11	11°	11°	−8.1	7.33°	−5.5
12	−3°	−3°	2.2	−2.00°	1.8
13	13°	13°	−9.7	8.67°	−6.5
14	−1°	−1°	0.7	−0.67°	0.9
15	15°	15°	−11.2	10.00°	−7.4

When the lidar is scanning, a scanning angle can be obtained α, This is the angle between the scanning plane and the lidar coordinate plane. The scanning plane coordinates are transformed into lidar coordinates, and the rotation matrix is

$$R_x = \begin{bmatrix} 1 \\ 0 \\ 0 \\ \sin a \\ -\cos a \\ \cos a \\ \sin a \end{bmatrix} \tag{10}$$

Obtain the coordinates of the target corner in the lidar coordinate system:

$$X_C = \begin{bmatrix} x \\ y \\ z \end{bmatrix} = R_x \times X_0 \tag{11}$$

4 Joint Calibration of Lidar and Camera:

The camera coordinate system and lidar coordinate system are established to obtain the target corner coordinates in their respective field of view. In the lidar coordinate system, it is a 3D corner coordinate, while in the camera coordinate system, it is a 2D corner.

Lidar coordinate system to camera coordinate system:

$$z \begin{bmatrix} u \\ v \\ 1 \end{bmatrix} = \begin{bmatrix} f_x & 0 & u_0 & 0 \\ 0 & f_y & v_0 & 0 \\ 0 & 0 & 1 & 0 \end{bmatrix} \cdot \begin{bmatrix} R & T \\ 0^T & 1 \end{bmatrix} \cdot \begin{bmatrix} X \\ Y \\ Z \\ 1 \end{bmatrix} \tag{12}$$

Joint calibration can be realized by the following methods:

1. Correspondence between 3D points and 2D planes [6]
2. Calibration based on multi-sensor motion estimation [7]
3. Calibration is completed by maximizing mutual information between lidar and camera [8]
4. Volume and intensity data registration based on geometry and image [9]

To complete the transformation from 3D points to 2D points, I choose to use PNP algorithm [10] (complete the matching of 3D points to 2D points) to calculate the rotation and translation vectors between the two coordinate systems. The final conversion relationship is as follows:

$$X_c = MX + H \tag{13}$$

In the above formula, M is the rotation matrix, which records the transformation relationship between the lidar coordinate system and the camera coordinate system, and H is the translation vector, which records the transformation relationship between the origin of the lidar coordinate system and the camera coordinate system. Finally, the joint calibration between lidar and camera can be completed by unifying the obtained 3D points and 2D points.

PNP algorithm: Taking the lidar coordinate system as the world coordinate system, select the three-dimensional feature points in the lidar coordinate system and the coordinate points of the feature points projected into the image coordinate system through perspective, so as to obtain the pose relationship between the camera coordinate system and the lidar coordinate system, including R matrix and t matrix, and complete the matching of 3D points to 2D points.

Requirements for feature points: it is necessary to know not only the coordinates in the three-dimensional scene, but also the coordinates in the two-dimensional image, so that a certain solution can be obtained for perspective projection. We select four corners

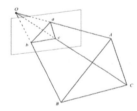

Fig. 7. Pose diagram of camera coordinate system relative to lidar coordinate system

of the rectangular board as feature points, 3D points are A, B, C, D, and 2D points are a, b, c, d. Triangles have the following similar relationships (Fig. 7):

where: $\triangle Oab - \triangle OAB$, $\triangle Oac - \triangle OAC$, $\triangle Obc - \triangle OBC$.

(1) According to the cosine theorem:

$$OA^2 + OB^2 - 2 \cdot OA \cdot OB \cdot \cos <a, b> = AB^2 \tag{14}$$

$$OA^2 + OC^2 - 2 \cdot OA \cdot OC \cdot \cos <a, c> = AC^2 \tag{15}$$

$$OB^2 + OC^2 - 2 \cdot OB \cdot OC \cdot \cos <b, c> = BC^2 \tag{16}$$

(2) Eliminate the above formula, that is, divide by OC2 at the same time, and $x = \frac{OA}{OC}$, $y = \frac{OB}{OC}$. You can get:

$$x^2 + y^2 - 2 \cdot x \cdot y \cdot \cos <a, b> = AB^2/OC^2 \tag{17}$$

$$x^2 + 1 - 2 \cdot x \cdot y \cdot \cos <a, c> = AC^2/OC^2 \tag{18}$$

$$y^2 + 1 - 2 \cdot x \cdot y \cdot \cos <b, c> = BC^2/OC^2 \tag{19}$$

(3) Let $u = (AB^2)/(OC^2)$, $v = (BC^2)/(AB^2)$, $w = (AC^2)/(AB^2)$ then:

$$x^2 + y^2 - 2 \cdot x \cdot y \cdot \cos <a, b> = u \tag{20}$$

$$x^2 + 1 - 2 \cdot x \cdot y \cdot \cos <a, c> = wu \tag{21}$$

$$y^2 + 1 - 2 \cdot x \cdot y \cdot \cos <b, c> = vu \tag{22}$$

(4) Simplified:

$$(1 - w)x^2 - w \cdot y^2 - 2 \cdot x \cdot \cos <a, c> + 2 \cdot w \cdot x \cdot y \cdot \cos <a, b> + 1 = 0 \tag{23}$$

$$(1 - v)x^2 - v \cdot y^2 - 2 \cdot y \cdot \cos <b, c> + 2 \cdot v \cdot x \cdot y \cdot \cos <a, b> + 1 = 0 \tag{24}$$

What we need to do is to solve the coordinates of A, B and C in the camera coordinate system through the above formula, in which the image position of 2D points and $\cos <a, b>$, $\cos <a, c>$, $\cos <b, c>$ are known, and u and w can also be obtained. Therefore, it is transformed into the solution of the above binary quadratic equation.

The specific solution process of the above binary quadratic equations is as follows:

1. The two binary quadratic equations are equivalent to a set of characteristic columns, and the equivalent equations are as follows:

$$a_4x^4 + a_3x^3 + a_2x^2 + a_1x^1 + a_o = 0 \tag{25}$$

$$b_1y - b_0 = 0 \tag{26}$$

2. According to Wu's elimination method, we can get that a1-a4 are all known and obtain the values of x and y.
3. Calculate the values of *OA, OB* and *OC*

$$x^2 + y^2 - 2 \cdot x \cdot y \cdot \cos <a,b> = AB^2/OC^2 \tag{27}$$

where: $x = OA/OC$, $y = OB/OC$.

4. Obtain the coordinates of *A, B* and *C* in the camera coordinate system:

$$A = \overrightarrow{a} \cdot \|PA\| \tag{28}$$

Using PNP algorithm, because I use three groups of corresponding points and can get four groups of solutions, I use point d to verify the results and judge which group of solutions is the most appropriate.

The joint calibration results are shown as follows (Fig. 8):

rotation:
[0.2419071067896962, -0.9698935918727406, -0.02806015197450512;
0.03233768293969008, 0.0369617945989984, -0.9987933219651168;
0.9697603961529522, 0.2407078024996598, 0.04030543225241662]
translation:
[-0.5664374195245468; -0.01315429680654068; -0.2699806670236895]

Fig. 8. Joint calibration parameters

5 Experiments

Verify the algorithm through the following experiments.

5.1 Experimental Equipment

This experiment selects velodyne 16 line lidar, narrow_sterto monocular camera. in the experiment, we fixed the relative position of the lidar and the camera. The fixing diagram of the calibration plate is shown in the figure, and the calibration plate is located in front of the lidar (Fig. 9).

Fig. 9. Schematic diagram of placing lidar, camera and calibration plate

The selected experimental equipment is shown in Table 2:

Table 2. Experimental equipment

Equipment name	Model	Main technical indicators
Lidar	Velodyne-VLP16	16 wire, point frequency 320 kHz
Monocular camera	narrow_stereo	640×480 pixel
Computer	PC	Intel-i5

5.2 Experimental Results

According to the algorithms in the previous sections, we completed the following experiments:

(1) The lidar and camera are fixed at corresponding positions respectively. The height of the camera is 1.3 m and the height of the lidar is 1.2 m
(2) we used a fixed 12 * 9 chessboard grid calibration board which the distance of each grid is 30 mm. It is placed about 4 or 5 m away from the front of the lidar. The lidar and the camera collect images at the same time. In addition, a rectangular wooden board is used to complete the image acquisition.
(3) Move the position of the calibration plate and board, and then re collect the image.
(4) We can obtained the two-dimensional corner coordinates of the four corners of the board in the camera image and the three-dimensional coordinates of the lidar image.
(5) We used the 11 * 8 corners of the chessboard calibration board to complete the separate calibration of the camera, and then the coordinate values of the four corresponding corners of the rectangular board are used to complete the joint calibration of the two.

The individual calibration results of the camera are shown in Table 3 below. Because there are too many chessboard corners, which are 11 * 8, 10 of them are selected:

The results obtained after joint calibration of lidar and camera are shown in Table 4 below:

Table 3. Camera calibration results

Corner coordinate measurement (x, y)	Calculated value (x', y')
(429.91098, 400.1738)	(429.971, 400.128)
(400.2941, 402.55194)	(400.223, 402.733)
(370.27206, 405.14648)	(370.26, 405.195)
(340.36008, 407.48935)	(340.162, 407.506)
(310.02762, 409.57397)	(310.015, 409.659)
(279.65219, 411.5592)	(279.901, 411.648)
(249.73608, 413.40631)	(249.906, 413.466)
(220.27628, 414.97083)	(220.111, 415.112)
(190.50243, 416.59354)	(190.594, 416.587)
(161.46346, 417.70218)	(161.419, 417.9)
(132.61649, 418.76422)	(132.63, 419.075)

Table 3 shows the results of camera calibration separately. After obtaining the measured values of image corner coordinates, the calculated values of specific image corners are obtained by re projection, using the three-dimensional coordinates of corners under the camera and the internal and external parameter matrix of the camera. Compared with the measured values, the average error of camera calibration is 0.0146333 pixels.

Table 4. Joint calibration results

Lidar measurements (x, y, z)	Camera measurements (x, y)	Calculated value (x', y')
(4.16499, 1.07492, 0.53206)	(194,145)	(192.676,145.234)
(4.07381,0.0667174,-0.503505)	(410,375)	(415.134,375.327)
(3.71897, -0.38916, 0.459004)	(516,130)	(513.415,128.142)
(3.69492, 1.07548, -0.468488)	(156,380)	(154.752,376.406)

Table 4 shows the conversion results after joint calibration. This result is that the rapid conversion from the coordinate system of lidar to the pixel coordinate system corresponding to the camera can be completed by using the R, T matrix between lidar and camera and the internal parameter matrix of camera. Compared with the measured values of camera, it is concluded that the average error of joint calibration is 1.81792 pixels.

It is obvious from the above two tables that the accuracy of the camera itself is still quite accurate, with an average error of 0.0146333 pixels, which meets the required accuracy requirements. However, because the lidar itself is not very accurate and its quantization accuracy is decimeter level, the joint accuracy obtained after joint calibration is compared with the calibration accuracy of the camera, The accuracy of joint calibration is slightly poor.

6 Conclusion

In order to realize the multi-sensor fusion of lidar and camera, a joint calibration method between lidar and camera sensors based on rectangular board is proposed in this paper. The experimental results show that this method has certain practical significance.

References

1. Zhu, H., Yuen, K.V., Mihaylova, L., et al.: Overview of environment perception for intelligent vehicles. IEEE Trans. Intell. Transp. Syst. **18**(10), 2584–2601 (2017)
2. Veľas, M., Španěl, M., Materna, Z., et al.: Calibration of RGB camera with velodyne lidar (2014)
3. Jianfeng, L., Tang, Z., Yang, J., et al.: Joint calibration method of multi-sensor. Robot **19**(5), 365–371 (1997)
4. Shu, N.: Research on Camera Calibration Method. Nanjing University of Science and Technology, Nanjing (2014)
5. Huang, X., Ying, Q.: Obstacle identification based on LiDAR and camera information fusion. Comput. Meas. Control **28**(01), 184–188+194 (2020)
6. Verma, S., Berrio, J.S., Worrall, S., et al.: Automatic extrinsic calibration between a camera and a 3D Lidar using 3D point and plane correspondences. In: 2019 IEEE Intelligent Transportation Systems Conference (ITSC), pp. 3906–3912. IEEE (2019)
7. Ishikawa, R., Oishi, T., Ikeuchi, K.: Lidar and camera calibration using motions estimated by sensor fusion odometry. In: 2018 IEEE/RSJ International Conference on Intelligent Robots and Systems (IROS), pp. 7342–7349. IEEE (2018)
8. Pandey, G., McBride, J., Savarese, S., et al.: Automatic targetless extrinsic calibration of a 3D lidar and camera by maximizing mutual information. In: Proceedings of the AAAI Conference on Artificial Intelligence, vol. 26, no. 1 (2012)
9. Cobzas, D., Zhang, H., Jagersand, M.: A comparative analysis of geometric and image-based volumetric and intensity data registration algorithms. In: Proceedings 2002 IEEE International Conference on Robotics and Automation (Cat. No. 02CH37292), vol. 3, pp. 2506–2511. IEEE (2002)
10. Moreno-Noguer, F., Lepetit, V., Fua, P.: Accurate non-iterative O(n) solution to the PNP problem. In: 2007 IEEE 11th International Conference on Computer Vision, pp. 1–8. IEEE (2007)

On Differential Protection Principle Compatible Electronic Transducer

Guangling Gao[1]([✉]), Keqing Pan[2], Zheng Xu[3], Xianghua Pan[3], Xiuhua Li[1], and Qiang Luo[1]

[1] State Grid of China Technology College, Jinan 250000, China
gao_gl@163.com
[2] Shandong University of Finance and Economics, Jinan 250000, China
[3] Shandong Electric Power Corporation, Jinan 250000, China

Abstract. The technology of digital interface compatible electronic transducer is studied. The measuring and protective equipment is explored to make a new application of current and voltage signals from the electronic transducer so that electronic transducer differential signal is directly used as a protection input. The new differential protection principle based on the differential input signal is put forward. And the theoretical analysis and simulation shows that the protection principles proposed are feasible.

Keywords: Differential protection · Digital interface · Electronic transducer

1 Introduction

Transducers are used to monitor the primary device and provide reliable electric quanti-ties to secondary equipment. The traditional transient electromagnetic transducers have the issues of saturation and low accuracy. The electronic transducer has low output, suf-ficient bandwidth, good linearity, simple structure, and other advantages. At the same time the electronic transducer does not require direct contact with the measured current circuit. The output of the electronic transducer is a digital signal, which is essentially different from the analog signal output of the traditional transducers and will have a profound impact on secondary equipment.

In the paper, the characteristics of two different interfaces are analyzed and the differential protection principle based on the differential input signals are proposed. Then the simulation tests are made by using PSCAD. The simulation results show that the differential protection principle based on the differential input signals can correctly identify the internal fault and external fault. The electronic transducer differential signals are applied to protection algorithm directly without an integral circuit, which can give full play to the advantages of electronic transducer and improve the reliability and accuracy of protection.

© The Author(s) 2022
Z. Qian et al. (Eds.): WCNA 2021, LNEE 942, pp. 367–374, 2022.
https://doi.org/10.1007/978-981-19-2456-9_38

2   Overview of Digital Interface

2.1   Structure of Electronic Transducer

According to IEC60044-8 "Electronic Current Transducer" standards, the electronic transducer includes one or many current sensors and voltage sensors which connect the transmission system to the secondary converter. The measured current and voltage is exported with analog or digital signals and is transmitted proportionally to the protection systems and other secondary measurement and control instruments. As what is shown in Fig. 1, the analog signal of the transducer is supplied directly to the secondary devices, and the digital signal is combined by a merging unit and exported to the secondary devices. Electronic transducers can be divided into two types: active electronic transducers and passive optical transducers, depending on if the transducer require a power supply.

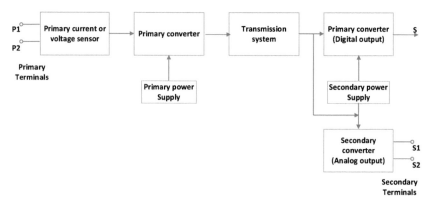

Fig. 1.   Structure of electronic transducer

2.2   Two Modes of Electronic Transducer Interface

Interface with Integral Circuit. The first interface is shown in Fig. 2. Firstly, the output optical digital signals of the transducer are transported to the low voltage side through optical fibers. Then the signals are carried to the relay protection system after being further processed in the merging unit. Because the outputs of the electronic current transducer based on a Rogowski coil and the resistive-capacitive divider voltage transducer are differential signals. In order to reflect the voltage and current, the outside integral circuit is increased in the sensor system of electronic transducer.

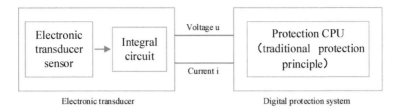

Fig. 2. Interface with integral circuit

The interface model with integral circuit has many advantages: the interface is simple; the protection system hardware requires few changes; the cost of the protection system change is low; and the protection system software algorithm can be used without adjustment. This interface model has disadvantages too. A digital integrator is achieved entirely by software, so it requires high operation speed and greater hardware cost. In addition, the integral circuit limits the measurement band of the electronic transducer.

Interface Without Integral Circuit. The second interface is shown in Fig. 3. In this approach, the differential signals from electronic transducer are used in the protection algorithm directly. The transducer integral part is omitted and the traditional protection algorithm is modified.

The interface model without integral circuit has many advantages: the system reliability is increased and takes advantage of the electronic transducer to improve the reliability and accuracy of protection system. On the other hand, the software algorithm of traditional protection system must be adjusted with this interface mode.

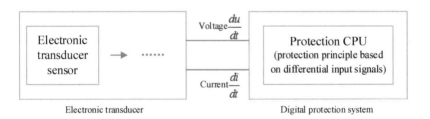

Fig. 3. Interface without integral circuit

3 Differential Protection Basing on Differential Input Signals

Transmission line current differential protection determines whether there is a short circuit fault protection on the protected line by comparing current phase at both ends of the line. Differential protection can cut the fault quickly and is not affected by the power operating mode of single side, mutual inductance in parallel lines, system oscillations, line series capacitor compensation, TV disconnection, etc. Differential protection has become the primary choice for EHV transmission line main protection because of its ability to choose phase. The conventional differential has a big problem. The secondary side current of traditional electromagnetic transducer (CT) is used to make protection to work. At the condition of external short circuit fault, the core may be saturated, which causes the traditional transducer transient current to be distorted and results in a large imbalance current and differential protection malfunction. Electronic current transducer (ECT) has non-magnetic saturation, simple and reliable insulation, wide measuring range, etc.

In electronic current transducer based on a Rogowski coil, after removing the integral link, input signal sent to computer protection system is a current differential signal $\frac{di(t)}{dt}$, on the basis of which the differential protection principle and criterion is analyzed.

Assuming line current at both sides are following.

$i_m(t) = \sqrt{2}I_m \sin(\omega t + \varphi_m)$, $i_n(t) = \sqrt{2}I_n \sin(\omega t + \varphi_n)$.

Then the corresponding current are $\dot{I}_m = I_m \angle \varphi_m$, $\dot{I}_n = I_n \angle \varphi_n$.

If i_{mj} is represented as $i_{mj} = \frac{di_m(t)}{dt} = \sqrt{2}\omega I_m \cos(\omega t + \varphi_m) = \sqrt{2}\omega I_m \sin(\omega t + \frac{\pi}{2} + \varphi_m)$ then the corresponding current are as following: $\dot{I}_{mj} = \omega I_m \angle \frac{\pi}{2} + \varphi_m$.

And if i_{nj} is can be represented as $i_{nj} = \frac{di_n(t)}{dt} = \sqrt{2}\omega I_n \cos(\omega t + \varphi_n) = \sqrt{2}\omega I_n \sin(\omega t + \frac{\pi}{2} + \varphi_m)$ then $\dot{I}_{nj} = \omega I_n \angle \frac{\pi}{2} + \varphi_n$.

Thus we can produce Eq. (1)

$$\left|\dot{I}_{mj} + \dot{I}_{nj}\right| = \omega\left|\dot{I}_m + \dot{I}_n\right| \tag{1}$$

Compared with conventional phase current differential protection, input signal amplitude at both sides of differential protection based on differential input expands ω times, phase shifts $\frac{\pi}{2}$, and the current relative relationship on both sides do not change. When line is normal, external fault, internal short circuit fault, current waveform, and phase diagram at both sides are shown in Fig. 4:

It can be concluded that compared with conventional phase current differential protection, protection differential signal as input signal, because the current in line ends has a phase shift at the same time and the relative phase relationship of both sides of the current do not change, the current differential protection principle based on the differential input signals is same as conventional one. At any moment, current phasor summation is zero at both ends of the normal or external fault line. The mathematical formula is expressed as follows: $\sum \dot{I} = 0$. When an internal line fault occurs, there is a short circuit current flowing. If current positive direction is from bus to line, current phasor summation at both ends is equal to the current flowing into the fault point without considering the impact of distributed capacitance, namely $\sum \dot{I} = \dot{I}_{dj}$.

Using electromagnetic transient simulation software PSCAD to build a double-ended single line power supply system, the paper has simulated the single-phase grounding, two-phase grounding, the two-phase short-circuit, and three-phase short-circuit failures; F1 is set up at the N-terminus of the line as the external fault, F2 serves as the internal fault, and the fault type and fault time can be set flexibly. The simulation system model is shown in Fig. 5.

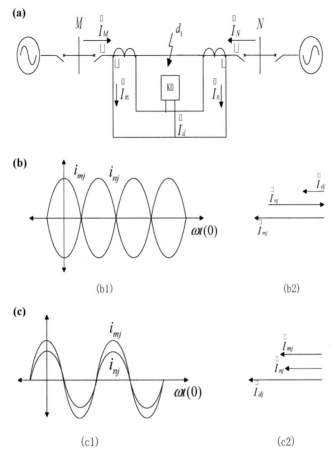

Fig. 4. (a) Principle diagram (b) current waveform and phase of normal operation and external fault (c) current waveforms and phase of internal short fault

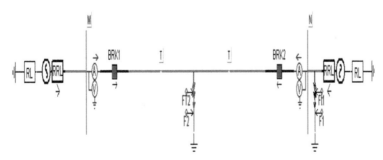

Fig. 5. Differential input current differential protection fault simulation model

Typical fault simulation examples are given as follows. i_{ma}, i_{mb}, i_{mc} express the three phase currents of M side;

i_{na}, i_{nb}, i_{nc} express the three phase currents of N side;

dma, dmb, dmc express the three phase currents differential of M side, namely: $\frac{di_{ma}}{dt}$, $\frac{di_{mb}}{dt}$, $\frac{di_{mc}}{dt}$;

$$\text{Restraint current} \begin{cases} S_{ja} = \left| \frac{Dma \angle Pma - Dna \angle Pna}{2} \right| \\ S_{jb} = \left| \frac{Dmb \angle Pmb - Dnb \angle Pnb}{2} \right| \\ S_{jc} = \left| \frac{Dmc \angle Pmc - Dnc \angle Pnc}{2} \right| \end{cases}$$

Examples: A phase ground short internal fault (F2/AN).

As shown in Fig. 6, three-phase current, differential current, and braking current waveforms are simulated respectively when point A phase ground short circuit fault in the F2 region occurs. Figure 6 shows that when the internal single-phase ground fault occurs, the differential current of the fault phase (A phase) is more than the braking current; the differential current and the braking current of non-fault phase (B, C phase) are small.

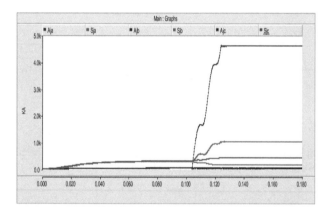

Fig. 6. Three-phase operating current and restraint current waveforms

The fundamental phase is calculated according to the current sample value after fault, and then the differential current and the breaking current are obtained whose trajectory curve operating point is shown as Fig. 7. It can be seen that the operating point of faulty phase (A phase) is in action area and protection work reliably.

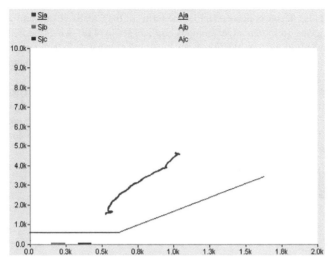

Fig. 7. A phase ground short internal fault operating characteristic curves diagram

4   Conclusions

In this paper, a new differential protection principle is proposed based on the differential input signal of an electronic transducer. The differential signal of the transducer is applied directly to the protection algorithm, which allows the integral part of the transducer to be omitted so that the full potential of an electronic transducer can be realized. It is proved through theoretical analysis and simulation that the protection principles proposed are correct and feasible.

References

1. Gao, G.L., Pan, X.H., et al.: Study on adaptive protection principle based on electronic transducer. In: Proceeding of 2015 4th International Conference on Energy and Environmental Protection, pp. 2465–2471 (2015)
2. Gao, G., Pan, X., et al.: Study on cluster measurement and control device of intelligent substation. In: The IEEE Conference on Energy Internet and Energy System Integration, pp. 2938–2942 (2018)
3. Gu, H., Zhang, P.: Influence of optical current transducer on line differential protection. Electric Power Autom. Equip. **27**(5), 61–64 (2007)
4. Han, X., Li, W.: Applying electronic current transformer to transformer differential protection. Proc. CSEE **27**(4), 47–53 (2007)
5. Brunner, C.: The impact of IEC 61850 on protection. developments in power system protection (DPSP), pp. 14–19 (2008)

Technological Intervention of Sleep Apnea Based on Semantic Interoperability

Ying Liang[✉], Weidong Gao, Gang Chuai, and Dikun Hu

Information and Communication Engineering, Beijing University of Posts and
Telecommunications, Beijing, China
liangying@bupt.edu.cn

Abstract. Sleep apnea is an important factor that could affect sleep quality.
A great number of existing monitoring and intervention devices, such as the
polysomnography, mature heart rate respiratory monitoring bracelets and ventilator headgear can improve breathing in sleep, but are all functioning separately,
with their data being disconnected, which fails to achieve multi-parameter fusion
or a greater variety of applications. With the development of the Internet of Things
(IoT), information interaction between IoT devices to facilitate integration of IoT
devices has become a hot research topic. This paper focuses on the interoperability
information model and technology for establishing interoperability information
model among sleep and health devices for sleep apnea syndrome. This paper analyzes the heterogeneity of the knowledge organization system in sleep health data
information through the abstract representation of data information, establishes
the mapping relationship between data, information, and devices, and realizes
the semantic heterogeneity elimination. It also defines inference rules about sleep
apnea scenarios, achieves semantic interoperability between monitoring devices
and other health devices, and finally realizes an unmonitored closed-loop control
system for sleep apnea intervention. According to the test results, the system can
react quickly in sleep apnea scenarios.

Keywords: Sleep apnea syndrome · Intervention · Semantic interoperability

1 Introduction

Sleep is a complex process that plays an important and irreplaceable role in people's life
and particularly in their physiological activities. Multiple organs perform detoxification
during sleep, such as the liver and the kidney, which helps people recover their physical
strength and energy. Additionally, high-quality sleep can effectively enhance the people's
immune system. However, studies have shown that the quality of people's sleep has been
declining in recent years, with sleep disorders being an important cause for the increasing
severity of sleep quality problems, among which sleep apnea is particularly prominent.
Sleep apnea syndrome is a medical condition in which the airflow between the nose and
mouth disappears or is weakened for more than ten seconds during sleep, and includes
Obstructive Sleep apnea (OSA), Central Sleep Apnea (CSA), and Mixed Sleep Apnea

© The Author(s) 2022
Z. Qian et al. (Eds.): WCNA 2021, LNEE 942, pp. 375–386, 2022.
https://doi.org/10.1007/978-981-19-2456-9_39

(MSA) [1] Patients suffering from sleep apnea snore during sleep and are likely to experience a brief respiratory arrest during sleep, which leads to insufficient oxygen supply in the blood, reduced sleep quality, daytime drowsiness, memory loss, and in severe cases, psychological and intellectual abnormalities, and may even cause other diseases, such as arrhythmias, cerebrovascular accidents, and coronary heart disease. To address these problems, research in scientific and timely monitoring of sleep apnea and the possibility of providing timely intervention to patients is of extreme value [2].

Polysomnography (PSG) is considered the "gold standard" for diagnosing apnea events and some other sleep disorders. However, PSG devices are costly and require electrodes to be attached to the patient and tension sensors to be worn, which may lead to First Night Effect of the users and dislodgement of devices in the middle of the night. In addition, in the market, there are already mature heart rate respiratory monitoring bracelets or head-mounted respirators that can improve breathing problems during sleep, but because all these devices can interfere with human activity to varying degrees, thus having an impact on sleep quality on the other hand [3]. There is thus an urgent need for a contactless, effective, and more accessible assistive device for monitor and intervention. A very important medical indicator to detect the occurrence of apnea events is called the arterial oxygen saturation (SaO2). Given that the accurate measurement of SaO2 requires the facilitation from an oximeter, the interconnection of sleep monitoring devices with an oximeter is a subject worth investigating. Additionally, existing sleep health devices can detect the occurrence of disease but cannot timely conduct any relief or rescue treatment. Therefore, if the monitoring equipment and rescue equipment can be interconnected, the disease will be relived in a timely manner. For example, homecare devices can alleviate certain reaction caused by acute symptoms and provides help for the subsequent hospital treatment [4]. However, the health devices are currently developed separately by different companies, which means that different conceptual expression models and languages, and different degrees of formalization with the overlapping of knowledge in different domains will lead to multiple inconsistencies and disconnection [5]. As a result, a multi-parameter fusion among the devices to provide richer applications become impossible. Interoperability can solve the problems of multiple device network heterogeneity, data format conflicts, and incompatible interfaces, eventually realizing data sharing and collaborative work among information systems. It is thus extremely important to carry out study on the interoperability between heterogeneous devices [6].

2   Related Work

As of now, related departments and research institutions have presented various evaluation models to evaluate interoperability, among which Levels of Conceptual Interoperability Model (LCIM) is highly representative. It has six levels, namely no interoperability, technical interoperability, syntactic interoperability, semantic interoperability, pragmatic interoperability, and conceptual interoperability [7]. Semantic technology targets integration and collaboration of heterogeneous systems by providing unified descriptions, and it is now very popular in recent years to study how to attach semantics to IoT systems. In 2006, Brock proposed the concept of SWOT (Semantic Web of Things, SWOT), advocating that IoT should be called the Semantic Internet of Things.

He believes that the internet, as a bridge between the physical world and the information world, should have an underlying sensing device of its own system that can provide information being aware of context and capable of reasoning, rather than focus on the changes of the objects themselves. They should also be able to "communicate" and "understand" as human beings do, and to communicate collaboratively between devices through registration, addressing, auto-discovery and search [8].

Saman Iftikhar [9] studied the feasibility of semantic interoperability among various semantic languages and realizes interoperability between semantic information exchange and resultant information systems across services. Shusaku Egami [10] investigates an ontology-based approach to semantic interoperability data integration for air traffic management. A domain ontology that is based on the flight, aviation and weather information exchange model is built, while an approach is proposed to integrate heterogeneous domain ontologies. As a result, interoperability of exchanging information about aircraft operations between different systems and operators in global air traffic management is solved, while the interoperability and coordination of all kinds of information in global operations is enhanced. Soulakshmee Devi Nagowah [11] put forward an approach based on new paradigms such as the Internet of Things and pedagogical concepts such as Learner Analysis, which is to build an ontology of IoT smart classrooms for university campuses to improve semantic interoperability in smart campus environments.

Wanmei Li [12] from China University of Mining and Technology put forward a semantic interoperability system for mining equipment based on distributed query, using semantic technology to propose a somaticized description model for IoT in mines, and a task matching scheme based on compound reasoning, which enables mutual understanding and interaction between equipment and production systems. It has combined semantic technology, distributed system and edge computing framework and applied the integration in which is applied in mine production activities with an aim to reduce humanized mine production and improve automatic production efficiency of coal mines.

In health, Bozhi Shi [13] studied the interoperability characteristics of heart monitors and researched their data information exchange capability. To summarize, the existing interoperability studies are in the process of development, and there is not a complete standard applicable to the health field in terms of the depth of related research. In addition, there are even fewer studies about the interoperability system of health equipment, so the research of interoperability needs more attention (Fig. 1).

3   Overview of Design Model

This paper focuses on the interoperability information model and technology of devices that monitor and intervene with sleep apnea. Through analysis of the requirements of interoperability of sleep apnea monitoring and intervention devices, an information model is constructed to design a specific method to achieve the semantic interoperability. The specific research content is as follows:

An ontology-based semantic description model of sleep monitoring devices is proposed from four aspects, namely the basic information, status, function, and operation control, so that device information can be represented by a semantic document in a unified syntax format.

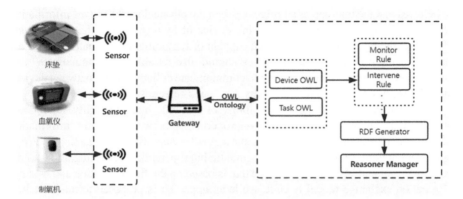

Fig. 1. Overall flow chart of model.

In terms of the need of monitoring and intervention tasks, a semantic description model of monitoring and intervention tasks is proposed to semantically describe the task information. Meanwhile, a task matching scheme based on compound reasoning is proposed to strengthen the autonomy of the sleep device interoperability system. The study integrates the relevant theories and technologies of ontology, extracts the information of the device or task ontology, and then inputs it into the reasoning ontology, and guides the output device according to the designed reasoning rules.

By interoperating the non-contact mattress and the oximeter, the heart rate and respiration rate calculated from the mattress and the initial judgment of whether an apnea event has occurred are combined with the results of the real-time oxygen saturation from the oximeter, which are then input into the intervention task ontology and the inference rule. If the apnea symptoms are serious, the oxygen production can be increased to help the human body keep the normal functioning; when the oxygen production is detected to have reached a normal degree or no apnea event occurs for a long time, the oxygen production can be reduced or turned off. As a result, it provides a higher discriminant accuracy than single mattress-based signal processing or single oximeter measurement results, offering higher medical reference value.

4   Implementation

4.1   Creating an Ontology

In 1998, Tim Berners-Lee, the founder of the World Wide Web, first proposed the concept of Sematic Web, and then the World Wide Web Consortium (W3C) developed a series of technological specifications related to the Semantic Web, including Web Ontology Language (OWL), Resource Description Framework (RDF). With the development of the Semantic Web, "ontology" has been introduced into computer science and given a completely different meaning in recent years. An ontology is a systematic explanation of things in the objective world through a formal language, while the OWL provides a way for users to write formal descriptions of concepts [14]. OWL consists of three elements,

Class: a collection of individuals with certain properties; Property: a binary relationship between a class and another class; Individual: an instance of a class, which inherits the properties of the class and facilitates the definition of data for reasoning. The OWL is used in this paper as the preferred language for ontology, while Protégé, an open-source ontology editor designed by Stanford University is chosen to facilitate the research and development of ontologies.

4.2 The Process of Creating an Ontology

To support autonomous and coordinated interactions among devices in an interoperable system, this section applies the powerful expressive power of semantic technologies to modeling in health. From the aspect of practical application of apnea intervention, the devices, the discrimination and intervention tasks, and the execution progress of the tasks in the sleep environment are semantically described, which results in a sleep health environment ontology system consisting of two domain ontologies, a sleep health device ontology, and a task ontology. This study combines the seven-step approach of ontology creation and METHONTOLOGY [15] as follows:

Identification of the domain and scope of the ontology. The sleep health system description ontology constructed in this study aims to provide the semantic support for intelligent collaboration between multiple devices in apnea discrimination and intervention tasks. The model mainly consists of two parts: device description model and task description model.

Reuse of existing ontologies. The ontology model related to sleep health system is extracted from the existing related ontologies, while the category attributes of related concepts and their inter-concept binary relations are integrated. In the process of creating ontology, the scalability of the ontology model can be enhanced by the mapping between related concepts.

Normalization of concepts. Firstly, class concepts are defined, and divided into classes of a hierarchy, i.e., important concepts are extracted from the corpus knowledge to form a glossary dedicated to the sleep environment, and a hierarchy is assigned to the concepts in the glossary. Secondly, the attributes of classes and their related constraints are defined according to the hierarchy. Finally, cases are built on the basis of the glossary to complete the creation of ontology.

Validation and evaluation of ontology. The ontology editor is used to build the relevant glossaries and their related ontologies, while the ontologies are validated according to the indexes of practicality, cohesion, and accuracy, continuously improving the ontology model.

Device Description Model. SSN (Semantic Sensor Network Ontology, SSN) is an ontology model issued by W3C. It is to describe sensors and provides a unified high-level semantic description of sensors in terms of deployment environment, functional role, and observed properties. The modeling for sleep health discriminative interventions in this study refers to the SSN ontology model and adds to it some control functions and other concepts. Based on the SSN ontology model and the analysis of the role of the device in the sleep health IoT system, the device is described semantically in four aspects:

basic information, device function, status, and control, forming a unified representation model, and providing semantic level support for the sleep health interoperability system.

The basic information refers to the description of some information that the device has since it was made by the manufacturer, such as the name, parameters, model and parts of the health device (oximeter, oxygen generator, mattress).

The device status describes the real-time situation of devices. The main consideration in modeling the concept of device status is the relationship between the device and the task, such as which operational state the oxygen generator is in and whether it is conditioned to perform the intervention task. In response to these questions, this paper provides description in terms of operational state and perceived state.

The device function refers to the specific tasks that the device can perform. This study describes the functions in control, measurement, input, and output of the three devices, namely oximeter, mattress, and oxygen generator, and the discrimination and intervention tasks.

The control describes the interaction between the devices and the control of the devices. The control operation in this study refers to the control of the ventilator based on the physiological parameters generated by the oximeter and the mattress. Therefore, the control operation is conducted through the on and off state of the oxygen generator (Fig. 2).

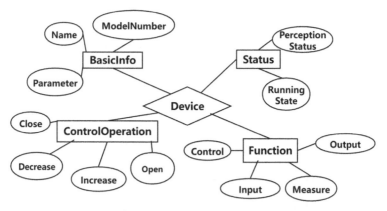

Fig. 2. The entity-relationship diagram of device model.

Equipment Model Evaluation. The quality of current ontology model can be evaluated in terms of its structure, operability, and maintainability, while its structure can be further divided into cohesiveness, redundancy, and coupling [16] Cohesiveness is the most frequently measured feature and can be quantified by the degree of independence of each module in the model and the correlation between internal concepts. The higher the cohesiveness, the better the cohesiveness of the system and the higher the degree of closeness between concepts. The cohesiveness of an ontology model is mainly influenced by the inheritance relationship between concepts within the ontology.

In this study, M is used to simplify the conceptual model of the device ontology, so $M1$, $M2$, $M3$, and $M4$ represent the conceptual model of its basic information, the conceptual model of its state, the conceptual model of its function, and the conceptual model of its control, respectively. The cohesiveness of the conceptual model of the device ontology is represented by $C(M)$, which is calculated as:

$$C(M)x = \begin{cases} \frac{2\sum_{i=1}^{i=n}\sum_{j>i}^{j=n} r(c_i,c_j)}{n(n-1)} & n > 1 \\ 1 & n = 1 \end{cases} \tag{1}$$

where n represents the number of nodes in the ontology model, r represents the relationship strength between two concepts in an ontology, c represents a class in the concept model ontology. If the two classes are directly inherited or indirectly inherited, then r equals to 1. If the number of concepts in the ontology model is 0, then the cohesiveness is 0. If there is only one concept in the model, the cohesiveness is 1 because the concept itself is the most compact structure in the model and does not depend on any other concept.

$$AVG = \frac{\sum_{i=1}^{m} C(M_i)}{m} \tag{2}$$

In this study, the device ontology is divided into four conceptual models, and the average cohesion AVG formula of the device ontology is calculated, and the cohesion of each conceptual model can be calculated according to the above formula, $C(M1) = 0.82$, $C(M2) = 0.71$, $C(M3) = 0.63$, and $C(M4) = 0.62$, and the average cohesion of the four models is obtained as 0.7, from which it can be considered that the concepts are more closely related to the topic of sleep health devices.

Task Description Model.   This study creates a model of task first, and then describes the discriminative and intervention task concepts in terms of basic information, conditional constraints, and inter-task relatedness. The semantic description of discriminative intervention tasks and execution progress information enables the device to directly understand the process of the current working task, so that it can determine whether to participate in the execution of the task and the prerequisites needed for execution. Among them, the basic information is the most basic description of the task, including task name, ID, and attributes, with name and ID being used to identify the task, and task attributes being used to describe the execution environment of the task. Task constraints include state constraints and timing constraints, and only devices that satisfy these constraints are qualified to claim the task. Task correlation is a concept used to judge the relationship between tasks, including temporal sequence and dependency. The tasks that come later in the temporal sequence can only be executed after the previous task is completed. The mutual dependency is mainly reflected in the data dependency between two tasks. For example, the execution of the intervention task requires the results of the monitoring task. The ontology and entity settings for the discrimination and intervention tasks in the sleep health system ontology are shown in the following figure (Fig. 3):

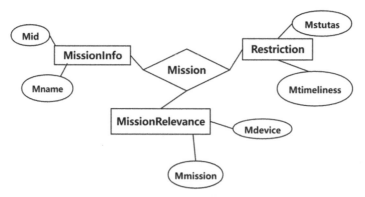

Fig. 3. The entity-relationship diagram of task model.

4.3   Reasoning

Contradictory knowledge may appear in the process of model creating, which leads to inconsistency of the ontology and affects the subsequent knowledge inference. The consistency of ontology is represented in three aspects: structural consistency, logical consistency, and user-defined consistency, referring to the ontology's syntactic structure, syntactic logic, and a series of constraints specified by the user to comply with the constraints of the language syntax model respectively. To uphold the ontology consistency, it is important to ensure that classes, attributes, and case individuals that have been created in the ontology are logically and structurally consistent. This step can further perform the rule reasoning. This study chooses HermiT and Pellet, two reasoners of Protégé to perform consistency testing of the ontology, imports the completed device ontology model and monitoring intervention task ontology into Protégé, and then performs the testing in HermiT and Pellet. No error message is suggested in the testing results, which proves that the term set and cases of the completed ontology system information are consistent.

The rules of reasoning need to be clarified before reasoning. Apnea is medically defined as the absence of or significant reduction of nasal or oral airflow for more than 10 s during sleep, accompanied by a sustained respiratory effort and a decrease in oxygen saturation. As the mattress can collect human physiological signals to obtain real-time heart rate and respiratory values, the signal processing can initially assess whether the user has apnea or not. Even if the user doesn't have apnea, it proves that the user's heart rate and respiratory shift is slightly abnormal. Thus, semantic interconnection with the oxygen machine can automatically turn on the oxygen generator and release a small amount of oxygen to avoid an acute anoxia. In addition, the oxygen saturation results measured by the oximeter are also considered to determine whether an apnea has occurred, and if so, to increase the oxygen concentration. When the values of the user's heart rate, respiration and blood oxygen saturation recover to the normal range, it means that the physiological parameters are more normal during this time, and the increase in oxygen in the air will lead to the opposite effect. Therefore, the oxygen generator should automatically be adjusted to the non-operating state, finally forming a closed-loop system (Fig. 4).

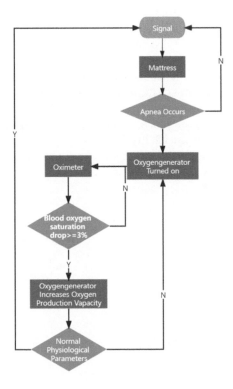

Fig. 4. The overall reasoning process.

5 Experiments

5.1 Experiment Settings

This study chooses local inputs instead of sensors, and preset values instead of mattress and oximeter operating performance and status. Considering only the prediction and discrimination of obstructive apnea syndrome, SWRL inference rules are set up in Protégé based on the above-mentioned reasoning. According to the reasoning of Pellet, 20 rules of the rule base are applied. When the output of the mattress ontology shows the occurrence of apnea, or when the decrease of blood oxygenation on the oximeter ontology reaches or exceeds 3%, the oximeter ontology will increase the generation of oxygen. When the value of the mattress ontology and oximeter ontology normalizes, the oximeter sill stop performing the task.

5.2 Performance

Assume the patient is in a bedroom of 15 m², where the oxygen generator is placed at about 3 m from the human body during sleep. The attendant will turn the oxygen generator on when there are signs of apnea and turn it off when the respiratory and heart rate recover to the normal level through the observation of the instruments. In the test, each instrument works separately, so the attendant must observe and judge the physiological parameters before deciding on the status of the oxygen generator. The whole process can be divided into three steps: observation, judgment and action, and the time spent in each step is different, with the most time spent in action, which greatly increases the length of time spent on the intervention. This study has conducted multiple sets of tests, assuming that the attendant can switch on the oxygen generator in the fastest speed, then the average time consumed, minimum time consumed, and maximum time consumed were 1.883 s, 1.49 s and 2.26 s respectively. In Protégé, the average response time, minimum response time and maximum response time were 15.385 ms, 15.063 ms and 15.612 ms respectively. The system performance would be better if the tasks were performed in binary (Fig. 5).

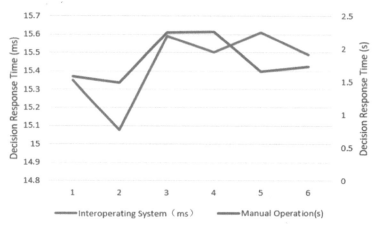

Fig. 5. Comparison of 6 sets of data on the decision response time of the two operations.

6 Conclusion

Semantic interoperability is a very challenging research issue. This paper aims to address the collaborative interaction between sleep health devices to achieve semantic-level interoperability between monitoring devices and other health devices, ultimately building an unmonitored closed-loop system for sleep apnea intervention. The discrimination and intervention has been simply implemented in the platform of Protégé, and the ontology design and rule base need to be enriched specifically in the future research to support more complex scenarios. The testing of the system is also realized by simulation in an experimental environment, which is inevitably too ideal, while real sleep environment can be highly unpredictable. Thus, further validation of the system in actual scenarios is needed in the future.

Acknowledgements. This work is supported by National Key R&D Program of China under grant number 2020YFC203303.

References

1. Gislason, T., Benediktsdóttir, B.: Snoring, apneic episodes, and nocturnal hypoxemia among children 6 months to 6 years old. An epidemiologic study of lower limit of prevalence. Chest **107**(4), 963–966 (1995)
2. Sharma, S.K., Kumpawat, S., Banga, A., Goel, A.: Prevalence and risk factors of obstructive sleep Apnea syndrome in a population of Delhi, India. Chest **130**(1), 149–156 (2006)
3. Peppard, P.E., Young, T., Palta, M., Skatrud, J.: Prospective study of the association between sleep-disordered breathing and hypertension. N. Engl. J. Med. **342**, 1378–1384 (2000)
4. Magalang, U.J., Chen, N.H., Cistulli, P.A., et al.: Agreement in the scoring of respiratory events and sleep among international sleep centers. Sleep **36**(4), 591–596 (2016)
5. "W3C Semantic Web Activity": World Wide Web Consortium (W3C), November 7, 2011, Retrieved 26 November 2011)
6. Jambhulkar, S.V., Karale, S.J.: Semantic web application generation using Proté´ge´ tool. In: 2016 Online International Conference on Green Engineering and Technologies (IC-GET), Coimbatore, pp. 1–5 (2016)
7. Canellas, M.C., Feigh, K.M., Chua, Z.K.: Accuracy and effort of decision-making strategies with incomplete information: implications for decision support system design. IEEE Trans. Hum. Mach. Syst. **45**(6), 686–701 (2015)
8. Lakka, E., Nikolaos, E.: End-to-End Semantic Interoperability Mechanisms for IoT. Foundation for Research and Technology. Hellas (FORTH). IEEE (2019)
9. Iftikhar, S.: Agent based semantic interoperability between agents and semantic web languages. In: 22nd International Conference on Advanced Information Networking and Applications. Workshops. IEEE (2008)
10. Egami, S.: Ontology-based data integration for semantic interoperability in air traffic management. In: 2020 IEEE 14th International Conference on Semantic Computing (ICSC). IEEE (2020)
11. Nagowah, S.D.: An ontology for an IoT-enabled smart classroom in a university campus. In: 2019 International Conference on Computational Intelligence and Knowledge Economy (ICCIKE). IEEE (2019)
12. Li, W.: Research on Semantic Interoperability System of Mine Equipment Based on Distributed Query (2020)
13. Shi, B.: Research on Interoperability Framework of Heart Ability Monitor for Personal Health Field (2017)
14. Ornelas, T., Braga, R., David, J.M.N., et al.: Provenance data discovery through semantic web resources. Concurr. Comput. Pract. Exper. **30**(1), e4366 (2017)
15. Corcho, Ó., Fernández-López, M., Gómez-Pérez, A., et al.: Building legal ontologies with METHONTOLOGY and WebODE. In: International Seminar on Law & the Semantic Web: Legal Ontologies, Methodologies, Legal Information Retrieval, & Applications (2003)
16. Gangemi, A., Catenacci, C., Ciaramita, M., et al.: Modelling ontology evaluation and validation. In: Semantic Web: Research & Applications, European Semantic Web Conference, Eswc, Budva, Montenegro, June 2016. Springer-Verlag (2006). https://doi.org/10.1007/117 62256_13

A Novel Home Safety IoT Monitoring Method Based on ZigBee Networking

Ning An[1,4], Peng Li[1,2,3,4(✉)], Xiaoming Wang[1,2,3,4], Xiaojun Wu[1,2,3,4], and Yuntong Dang[2,5]

[1] School of Computer Science, Shaanxi Normal University, Xi'an 710119, China
lipeng@snnu.edu.cn
[2] Key Laboratory of Intelligent Computing and Service Technology for Folk Song, Ministry of Culture and Tourism, Xi'an 710119, China
[3] Key Laboratory of Modern Teaching Technology, Ministry of Education, Xi'an 710062, China
[4] Engineering Laboratory of Teaching Information Technology of Shaanxi Province, Xi'an 710119, China
[5] School of Music, Shaanxi Normal University, Xi'an 710119, China

Abstract. This paper realizes the design of home safety early warning system by studying the wireless communication networking technology of ZigBee and WiFi, as well as sensor communication technology, which is based on taking home safety monitoring as the application background. In this study, CC2530 chip was used as ZigBee wireless communication module. A novel home security IoT monitoring method was proposed through sensor triggering, human activity trajectory perception algorithm design, and wireless networking and communication optimization. Meanwhile, the safety early warning and remote monitoring of home staff can be realized, and home safety can be guaranteed. The system can achieve the purpose of home monitoring and early warning with low software and hardware cost through the experimental design and result analysis. It can not only provide reference for the design of sensor communication system, but also provide technical reference for aging society and response.

Keywords: Wireless sensor networks · ZigBee · OneNet cloud platform · Communication network · WiFi

1 Introduction

With the rapid development of society, science and technology, people have higher and higher requirements for their quality of life. In particular, people pay great attention to home safety. Therefore, designing a home safety IoT monitoring system, which uses ZigBee and WiFi technology to collect and transmit data between nodes and between nodes and platforms. The sensor nodes form a wireless sensor network which distribute in every corner of the home. The system can not only realize the real-time monitoring of

© The Author(s) 2022
Z. Qian et al. (Eds.): WCNA 2021, LNEE 942, pp. 387–398, 2022.
https://doi.org/10.1007/978-981-19-2456-9_40

the home environment, but also ensure the safety of the elderly living alone preliminarily and reduce their need for care at home which provides great convenience for their children [1, 2].

In this system, CC2530 is used as the core of wireless transceiver and processing module [3]. CC2530 is an integrated chip, which uses the 8051 core and encapsulates the Z-stack protocol stack [4–6]. It can be used to transmit data in wireless sensor networks. The system uses CC2530 module to establish a small ZigBee network [7–9], which is composed of three node types: coordinator node, router node and sensor node.

With the changing needs of people, wireless access technology is more and more in line with the development trend of society. Therefore, people's demand for wireless sensor networks is increasing exponentially. Wireless sensor networks (WSN) adopts a distributed sensor network, which fully combines various advanced technologies such as distributed information processing technology, modern network and wireless communication technology [10, 11]. It can cooperate with each other to detect and collect all monitored area data in real time, and process the collected data. Then the data is transmitted wirelessly and transmitted to users in the form of wireless Ad Hoc network and multi hop network [12–15].

2   System Architecture Design

In the home security IoT monitoring system, it uses the low-cost and low-power ZigBee low-speed and short-distance wireless network protocol to detect the security parameters of the detected location. The system is mainly composed of coordinator, router, terminal, gateway, server, client and other components. The coordinator is in charge of creating Zigbee network at the mobile terminal, initializing the network, assigning an address to the mobile terminal node that initially needs to join the network and controlling the joining of the mobile terminal node. It can upload the collected data and realize the automation function of remote control of the terminal at the mobile terminal. The terminal equipment includes temperature and humidity sensor, MQ2 smoke sensor and human infrared sensor, which can realize indoor data acquisition, storage and transmission. The router is responsible for forwarding messages from other nodes.

In the system architecture design, the terminal collects the required data, and the coordinator receives the data through ZigBee sensor node networking. The coordinator uploads the data to the gateway through the serial port, and then the gateway sends its data to the computer. The WiFi module can also be driven through the protocol stack. The WiFi module can communicate with mobile phones, computers and routers, and load the collected data into HTTP format and send it to the cloud service OneNet cloud platform. The sensing layer of the system sends the data which collected by the sensor to the application layer through the network layer. The application layer analyzes and processes the data, and monitors it in real time. When the monitoring data is abnormal, it will send out alarm prompt information in time, so as to realize the management and monitoring of home safety. The systematic software flow chart is described in Fig. 1.

The architecture of the whole IoT system consists of three parts: IoT device end, device cloud platform and web background server [16]. The Internet of things device cloud platform is based on OneNet device cloud. The main steps of OneNet cloud platform accessing the development process are as follows [17]:

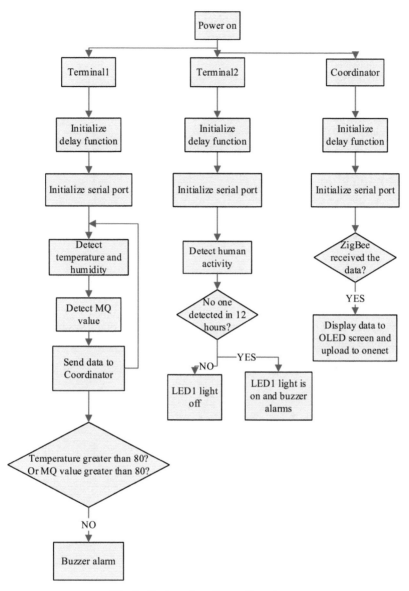

Fig. 1. Systematic software flow chart.

1) Registered product information;
2) Create equipment list;
3) Establish TCP connection and upload data;
4) View the data flow.

The device access flow chart of OneNet cloud platform is shown in Fig. 2.

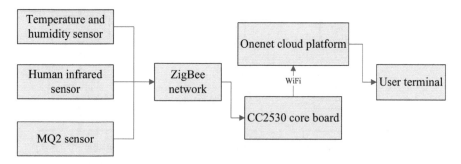

Fig. 2. Onenet cloud platform device access process.

3 Hardware Platform

The design of home IoT monitoring system is mainly composed of sensor, ZigBee gateway design and OneNet cloud platform [18]. The design of the systematic hardware architecture is shown in Fig. 3.

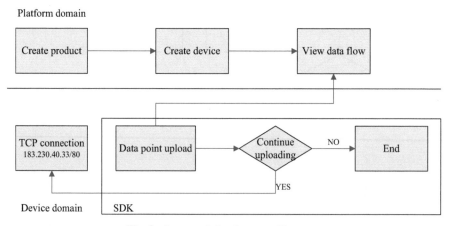

Fig. 3. Systematic hardware architecture.

In the design of nodes, we mainly refer to several commonly used sensors in home security to meet the requirements of the system. The human infrared sensor adopts HC-SR501 [19] model, and its sensing range is less than 7m. We usually add a Fresnel lens to the sensor module to improve the sensitivity of human detection. DHT11 contains a temperature and humidity sensor with calibrated digital signal output [20, 21]. The module realizes the collection of temperature and humidity data by controlling the timing. It is necessary to wait 1 s after the sensor is powered on to ensure the accuracy of the measured data. MQ2 sensor is mainly used to detect gas leakage [22]. It has the advantages of high sensitivity, good anti-interference and long service life. In the setting of the system, if the concentration of natural gas leakage is higher, the voltage output from AO pin will be higher. Thus, the value after ADC conversion will be larger. The ESP8266 WiFi module has low power consumption, supports transparent transmission and does not have serious packet loss. It can not only realize data transmission, but also connect to a designated router as a WiFi client [23]. The buzzer of the active module is selected. The active module is driven by triode, which is triggered at low level, that is, when the I/O port inputs low level, the buzzer makes a sound.

4   Algorithm Design and Implementation

The system uses IAR Embedded Workbench platform to realize ZigBee data communication through the design of ZigBee connection algorithm. In this system, the terminal enters the SampleApp_ProcessEvent() event firstly, and then the terminal calls SampleApp_SendTheMessage() function collects data. In this function, it sends the data by calling AF_DataRequest() function. If the data sent by the terminal is received through the ZigBee coordinator, it will enter SampleApp_ProcessEvent() event, which triggers SampleApp_MessageMSGCB() function in turn, receives the data sent by the terminal, and then its data is displayed on the OLED screen.

In SampleApp.c, configuring the product apikey, device ID, router account and password of OneNet cloud platform to realize the data interaction between WiFi module and OneNet cloud platform. The configuration code is as follows:

```
#define devkey "Ea=PgE0QU=fpzA44Zn88zyD6XKY=" //Onenet platform product apikey
#define devid  "699539810"                //Onenet platform device ID
#define LYSSID "3314"                      //SSID of router
#define LYPASSWD "computer3314"            //Router password
```

MCU can use ESP8266 WiFi module to send AT command to realize the configuration of WiFi transmission module. The configuration command is shown in Table 1.

Table 1. WiFi transmission module configuration.

Function	Instruction format
Set to STA+AP mode	AT+CWMODE = 3
Connect to the server	AT+CIPSTART = \"TCP\",\"183.230.40.33\",80
Transparent transmission mode	AT+CIPMODE = 1
Instruction to send data	AT+CIPSEND

Since the data packet of DHT11 sensor is composed of 5 bytes [24] and its data output is uncoded binary data, the temperature and humidity data need to be processed separately. The calculation formulas of temperature and humidity values are shown in (1) and (2), where byte4 is the integer of humidity, byte3 is the decimal of humidity, byte2 is the integer of temperature, and byte1 is the decimal of temperature.

$$humi = byte4.byte3 \tag{1}$$

$$temp = byte2.byte1 \tag{2}$$

The resistance calculation of MQ2 smoke sensor is shown in formula (3), where Rs is the resistance of the sensor, Vc is the loop voltage, Vrl is the output voltage of the sensor, and Rl is the load resistance. The calculation of resistance Rs and the concentration C of the measured gas in the air is shown in formula (4), where m and n are constants. The constant n is related to the sensitivity of gas detection. It will change with the sensor material, gas type, measurement temperature and activator [25]. For combustible gases, most values of the constant m are between 1/2 and 1/3 [26]. According to the above formula, the output voltage will increase with the increase of gas concentration.

$$Rs = (\frac{Vc}{Vrl} - 1) \cdot Rl \tag{3}$$

$$logRs = mlogC + n \tag{4}$$

The human infrared sensor uses the algorithm of timer T1 query mode, and its safety alarm logic judgment steps are as follows. The function realization process of the alarm program is shown in Fig. 4.

1) The InitT1() function initializes the timer.
2) To configure the three registers T1CTL, T1STAT and IRCON of timer T1, that is, set T1CTL = 0x0d (the working clock is 128 frequency division, and the automatic reload is 0x0000-0xFFFF), T1STAT = 0x21(the status is channel 0, the interrupt is valid), and IRCON = 1 (you can judge whether the storage space is full by querying).
3) To judge whether a person is detected and set DATA_PIN = 1 is detected.
4) If no one is detected, judge whether the storage space is full.
5) If the storage space is full, IRCON > 0, clear it, set IRCON = 0, and judge whether the unattended time count is within 12 h, so as to know whether there is any abnormality.
6) If count > = 12 h, it is considered that the elderly living alone have an abnormal state, the buzzer gives an alarm and LED1 is off.

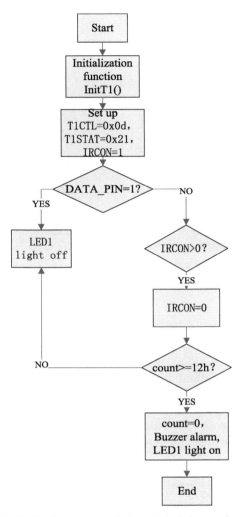

Fig. 4. Realization process of alarm logic judgment function.

5 Experimental Analysis

5.1 Sensor Data

After the software and hardware of the system are designed, data acquisition is carried out in the laboratory. The temperature, humidity and MQ data measured by terminal 1 are shown in Fig. 5. If humidity or MQ value is detected excessively, the buzzer will sound an alarm. The information detected by terminal 2 is shown in Fig. 6. If no person detected is displayed in the detection results for a long time, LED1 light will be on and the buzzer will alarm.

Fig. 5. Temperature, humidity and MQ values.

Fig. 6. Human body detection.

5.2 OneNet Cloud Platform Data

Selecting the baud rate of 115200 on the serial port debugging tool after the configuration of OneNet cloud platform is completed. The configuration results are shown in Fig. 7. The WiFi module uses STA+AP mode. The WiFi serial port module establishes a TCP connection, configures a server with IP 183.230.40.33 and port number 80. In the transparent transmission mode, the data is transmitted, and the module is connected to the network through the router, so as to realize the remote control of the equipment by the computer.

```
CoordinatorZB
120AT

OK
ZIGBEE-WIFI OK
AT+CWMODE=3

OK
WIFI CONNECTED
AT+CWJAP="3314", "computer3314"
WIFI DISCONNECT
WIFI CONNECTED
WIFI GOT IP

OK
AT+CIPSTART="TCP", "183.230.40.33", 80
CONNECT

OK
AT+CIPMODE=1

OK
AT+CIPSEND

OK

>Send data to server [..]
```

Fig. 7. OneNet configuration results.

After the system is docked through WiFi module and OneNet cloud platform, the temperature and humidity sensor uploads the collected data to the cloud platform successfully, as shown in Fig. 8. I take 10 groups of data as an example through the long-term collection of temperature and humidity data in the laboratory, as shown in Fig. 9.

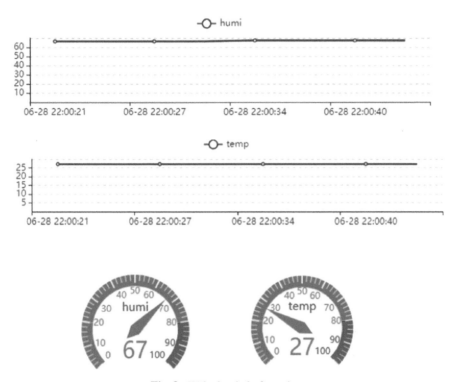

Fig. 8. Web cloud platform data.

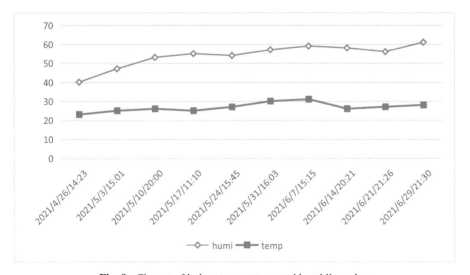

Fig. 9. Change of indoor temperature and humidity value.

6   Conclusion

This paper takes ZigBee technology as the core through the combination of ZigBee wireless Ad Hoc network and WiFi communication technology. The home IoT monitoring system is studied and designed, which integrates the Internet, intelligent alarm, communication network and other scientific and technological means effectively. The system adopts temperature and humidity sensor, human infrared sensor and MQ2 smoke sensor to realize the data acquisition of the home environment. For this data, if there is any abnormality, the buzzer will give an alarm. The system adopts ZigBee technology with low cost, low power consumption and strong networking ability, which not only increases the practicability of the system, but also can monitor home safety in real time for a long time, so as to avoid safety accidents and reduce losses.

Acknowledgements. This work is partly supported by the National Key R&D Program of China under grant No. 2020YFC1523305; the National Natural Science Foundation of China under Grant No. 61877037, 61872228, 61977044, 62077035; the Key R & D Program of Shaanxi Province under grant No. 2020GY-221, 2019ZDLSF07–01, 2020ZDLGY10–05; the Natural Science Basis Research Plan in Shaanxi Province of China under Grant No. 2020JM-302, 2020JM-303, 2017JM6060; the S&T Plan of Xi'an City of China under Grant No. 2019216914GXRC005CG006-GXYD5.1; the Fundamental Research Funds for the Central Universities of China under Grant No. GK201903090, GK201801004; the Shaanxi Normal University Foundational Education Course Research Center of Ministry of Education of China under Grant No. 2019-JCJY009; the second batch of new engineering research and practice projects of the Ministry of Education of China under Grant No. E-RGZN20201045.

References

1. Mei, M., Shen, S.: A data processing method in ZigBee life assistance system. Comput. Technol. Dev. (030): 005 (2020)
2. Das, R., Bera, J.N.: ZigBee based small-world home area networking for decentralized monitoring and control of smart appliances. In: 2021 5th International Conference on Smart Grid and Smart Cities (ICSGSC), pp. 66–71. IEEE, Tokyo (2021)
3. Bernatin, T., Nisha, S.A., Revathy, Chitra, P.: Implementation of communication aid using zigbee technology. In: 2021 5th International Conference on Intelligent Computing and Control Systems (ICICCS), pp. 29–32. IEEE, Madurai (2021)
4. Jia, N., Li, Y.: Construction of personalized health monitoring platform based on intelligent wearable device. Comput. Sci. **46**(6A), 566–570 (2019)
5. Cen, R., Jiang, Q., Hu, J., Sun, M.: ZigBee WiFi gateway for smart home applications, 26 (1), 232–235 (2017)
6. Mamadou, A.M., Chalhoub, G.: Enhancing the CSMA/CA of IEEE 802.15.4 for better coexistence with IEEE 802.11. Wireless Netw. **27**(6), 3903(2021)
7. Wang, D., Jiang, S.: A novel intelligent curtain control system based on ZigBee. In: 2020 5th International Conference on Mechanical, Control and Computer Engineering (ICMCCE), pp. 1010–1013. IEEE, Harbin (2020)
8. Abdalgader, K., Al Ajmi, R., Saini, D.K.: IoT-based system to measure thermal insulation efficiency. J. Ambient Intell. Hum. Comput. (2010)

9. Han, N., Chen, S., Zhang, X., Zhou, Y., Zhang, K., Feng, J.: Open architecture design of smart home integrated sensing device. Power Inf. Commun. Technol. **18**(04), 104–108 (2020)

10. Li, P., Liu, H., Guo, L., Zhang, L., Wang, X., Wu, X.: High-quality learning resource dissemination based on opportunistic networks in campus collaborative learning context. In: Guo, S., Liu, K., Chen, C., Huang, H. (eds.) CWSN 2019. CCIS, vol. 1101, pp. 236–248. Springer, Singapore (2019). https://doi.org/10.1007/978-981-15-1785-3_18

11. Yan, X., Ruan, Y., Wen, Z.: Design of smoke automatic alarm system based on wireless infrared communication. Modern Electron. Technol. **44**(8), 24–28 (2021)

12. Samijayani, O.N., Darwis, R., Rahmatia, S., Mujadin, A., Astharini, D.: Hybrid ZigBee and WiFi wireless sensor networks for hydroponic monitoring. In: 2020 International Conference on Electrical, Communication, and Computer Engineering (ICECCE), pp. 1–4. IEEE, Istanbul (2020)

13. Klobas, J.E., McGill, T., Wang, X.: How perceived security risk affects intention to use smart home devices: a reasoned action explanation. Comput. Secur. **87** (2019)

14. Impedovo, D., Pirlo, G.: Artificial intelligence applications to smart city and smart enterprise. Appl. Sci. **10**(8), 2944 (2020)

15. Zhan, Q., He, N., Chen, Z., Huang, Z.: Research on ZigBee-based remote water temperature monitoring and control system. In: 2021 IEEE 2nd International Conference on Big Data, Artificial Intelligence and Internet of Things Engineering (ICBAIE), pp. 1074–1077. IEEE, Nanchang (2021)

16. Li, J., Zhang, Y., Man, J., Zhou, Y., Wu, X.: SISL and SIRL: two knowledge dissemination models with leader nodes on cooperative learning networks. Physica A **468**, 740–749 (2017)

17. Parida, D., Behera, A., Naik, J.K., Pattanaik, S., Nanda, R.S.: Real-time Environment Monitoring System using ESP8266 and ThingSpeak on Internet of Things Platform. 2019 International Conference on Intelligent Computing and Control Systems (ICCS), pp. 225–229. IEEE, Madurai (2019)

18. Wang, D., Yuan, W., Wu, D., Liu, S.: Library environment monitoring system based on WiFi internet of things. Comput. Sci. **45**(11), 532–5349 (2018)

19. Yongyong, Y., Chenghao, H.: Design of data acquisition system of electric meter based on ZigBee Wireless Technology. In: 2020 IEEE International Conference on Advances in Electrical Engineering and Computer Applications (AEECA), pp. 109–112. IEEE, Dalian (2020)

20. Qin, Z., Sun, Y., Hu, J., Zhou, W., Liu, J.: Enhancing efficient link performance in ZigBee under cross-technology interference. Mob. Networks Appl. **25**(1), 68–81 (2019). https://doi.org/10.1007/s11036-018-1190-0

21. Kinoshita, K., Nishikori, S., Tanigawa, Y., Tode, H., Watanabe, T.: A ZigBee/Wi-Fi cooperative channel control method and its prototyping. Web Sci. **103**(3), 181–189 (2020)

22. Shao, C., Hoorin, P., Roh, H., Wonjun, L.: DOTA: physical-layer decomposing and threading for ZigBee/Wi-Fi co-transmission. Web Sci. **8**(1), 133–136 (2019)

23. Yasmine, B.A., Balaji, M., Vishnuvardhan, G., Harshavardhan, G., Lazer, M.T.: Development of animal collar for state of health determination of livestock. J. Inf. Optim. Sci. **41**(2), 489–497 (2020)

24. Vallabh, B., Khan, A., Nandan, D., Choubisa, M.: Data acquisition technique for temperature measurement through DHT11 sensor. In: Goyal, D., Chaturvedi, P., Nagar, A.K., Purohit, S.D. (eds.) Proceedings of Second International Conference on Smart Energy and Communication. AIS, pp. 547–555. Springer, Singapore (2021). https://doi.org/10.1007/978-981-15-6707-0_53

25. Hernández, C., Villagrán, S., Gaona, P.: Predictive model for detecting MQ2 gases using fuzzy logic on IoT devices. In: Jayne, C., Iliadis, L. (eds.) EANN 2016. CCIS, vol. 629, pp. 176–185. Springer, Cham (2016). https://doi.org/10.1007/978-3-319-44188-7_13
26. Gautam, A., Verma, G., Qamar, S., Shekhar, S.: Vehicle pollution monitoring, control and challan system using MQ2 sensor based on internet of things. Wireless Personal Communications **116**, 1071–1085 (2021)

PCCP: A Private Container Cloud Platform Supporting Domestic Hardware and Software

Zhuoyue Wang[1], Zhiqiang Wang[1(✉)], Jinyang Zhao[2], and Yaping Chi[1]

[1] Beijing Electronic Science and Technology Institute, Beijing, China
wangzq@besti.edu.cn

[2] Beijing Baidu T2Cloud Technology Co., Ltd., 15A#-2nd Floor, En ji xi yuan, Haidian district, Beijing, China

Abstract. With the widespread use of container cloud, the security issue is becoming more and more critical. While dealing with common security threats in cloud platforms and traditional data centres, there are some new security issues and challenges in the container cloud platform. For example, there are significant challenges in network isolation and resource management. This paper proposes a private container cloud platform PCCP based on Docker supporting domestic software and hardware to solve these security problems. This paper introduces the system architecture and functional architecture of the platform. The system has been tested and confirmed to have high availability and high reliability. The platform gives full play to the value of domestic software and hardware and is better able to serve the information construction of our country.

Keywords: Cloud computing · Container · Virtual network · Localization

1 Introduction

Cloud computing is an Internet-based computing approach. In this way, the hardware and software resources shared can be provided to various computer terminals and other on-demand devices [1]. The cloud computing architecture covers three-tier services, and they are IaaS, PaaS, and SaaS [2]. IaaS has low resource utilization, and the scenario needs to be considered. PaaS uses container technology, does not rely on virtual machines, and is highly scalable [3]. Docker was proposed as an open-source tool in October 2014. It can package applications and their dependencies into containers, and it solves the compatibility problem. However, Docker also faces many problems. For example, the application iteration is slow, the operation and maintenance management are more and more complex [4]. Under this background, container cloud technology is proposed. The container cloud is divided into containers for resources and encapsulates the entire software run-time environment. And it provides the developers and system administrators with a platform for creating, publishing, and running distributed applications [5]. When the container cloud focuses on resource sharing and isolation, container orchestration, and deployment, it is closer to the concept of IaaS. When the container cloud penetrates the application support and run-time environment, it is closer to the idea of PaaS.

© The Author(s) 2022
Z. Qian et al. (Eds.): WCNA 2021, LNEE 942, pp. 399–407, 2022.
https://doi.org/10.1007/978-981-19-2456-9_41

To solve the problems such as the slow application iteration and the more complex operation and maintenance management, a private container cloud platform PCCP supporting domestic hardware and software based on Docker is designed and implemented. The system is based on B/S architecture. The server and database are all made in China. And the functions of cluster management, mirror management, and so on are realized. This paper first introduces the research background of the PCCP container cloud platform, then introduces the system testing of the PCCP container cloud platform, and finally summarizes this paper.

2   System Architecture Design

2.1   Functional Architecture

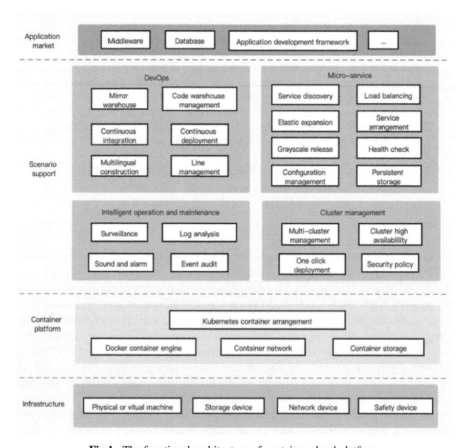

Fig.1. The functional architecture of container cloud platform

A container is a change from an existing application that is run by a physical or virtual machine to the application that deploy with the containers. And the container runs in the container runtime environment of the cloud operating system. Combined with other DevOps tools such as continuous integration, cloud-based rapid deployment, elastic scaling, and increased resource utilization can be achieved [6]. The functional architecture of the PCCP container cloud platform designed according to the system requirements is shown in Fig. 1.

2.2   Scenario Support

(1) DevOps: Help companies achieve the process of DevOps
(2) Micro-service: Support for a micro-service framework to meet the enterprise from a single architecture to the transformation of micro-service architecture.
(3) Intelligent operation and maintenance: It mainly includes multi-index and multi-dimension monitoring alarm, logs analysis, and event audit.
(4) Cluster management: Visual cluster management support multi-cluster management and container security policy development.
(5) Application market: Provide out-of-the-box application market. Users can easily use a variety of middleware, database, and application development framework.

Core Function. PCCP container cloud platform has several functions, including multi-tenant authority management, cluster management, application management, mirror management, storage management, resource management, pipeline management, load balancing, service discovery, application market, monitoring alarm, log management [7]. The functions and implementations are shown in Table 1.

Table 1. The core functions of the PCCP container cloud platform.

Functions	Implementations
Multi-tenant rights management	Independent quota and application resources Isolated network, logbook, and surveillance
Cluster management	Graphically deploy K8S clusters, manage nodes and view cluster resource usage
Application management	One-click deployment, upgrade rollback, elastic scaling, health checks, resource constraints, and so on
Mirror management	Mirror warehouse management, mirror upload, and download
Storage management	File storage, object storage, and other storage resources management to provide application persistence support
Resource management	Centralized management of application resources such as configuration, cipher-text, certificate

(*continued*)

Table 1. (*continued*)

Functions	Implementations
Line management	Achieve the automation process of source acquisition, compilation, build, and deployment
Load balancing	Apply traffic forwarding to the cluster to improve the high availability of services
Service discovery	Add DNS to enable callers of micro-services to find instances of micro-services dynamically
Application market	A large number of out-of-the-box application templates that support adding a private Helm repository
Surveillance alert	Multilevel and multidimensional monitoring alarm, support email, SMS, and other notification methods
Log management	Automatically collect application logs and retrieve, analyze, and display the record

2.3 Technical Architecture

The container cloud platform uses a container scheduling engine to pool resources such as computing, network, storage, and so on to provide application management capabilities at the distributed data center level. And it is no longer limited to the single mode for the application to give the required types of resources. The resource utilization can be greatly improved, and the IT cost can be reduced based on the lightweight container technology and the scheduling algorithm [8]. Depending on the features such as self-healing, health check, and elastic scaling, the stability and availability of the applications deployed on it can be significantly improved. Relying on the characteristics of orchestration, configuration management, service discovery, and load balancing can dramatically reduce the complexity of application deployment and operation, especially when the application scale is enormous. With these essential applications, you can focus more on business logic and deliver business value more quickly. The hierarchical design and hierarchical structure of the overall architecture are as follows:

(1) The first layer is the application system for business services deployed on the platform.
(2) The second layer is the platform service layer, which provides the platform level service support for the upper layer application to consider more business logic. And turn the deployment, extension, high availability, monitoring, and maintenance work of the application to the platform layer. The platform service layer provides an application development framework and middle-ware, application and service directory, software custom network, performance monitoring, and log management, automated cluster deployment and management, container scheduling, application cluster elastic scaling, abnormal self-healing, persistent volume, service discovery, configuration management, and other functions. The functions provided by the container platform service layer can guarantee the high availability, high scalability,

and stability of the applications running on it. And it can send a warning before service failure, which can help IT staff quickly locate and solve problems [9].

(3) The primary component layer contains the underlying core components of the container cloud platform and the components that run with a container. It provides uniform packaging standards for applications and isolation between applications. The network component is used to implement the inter-node container network communication and network isolation policy, and the storage component is used to provide storage support for stateful service.

(4) The infrastructure layer is primarily a physical or virtual machine cluster. It provides the computing, networking, and storage resources needed by the container cloud platform. The platform is compatible with domestic hardware and operating system.

The technical architecture diagram of the container cloud platform is shown in Fig. 2.

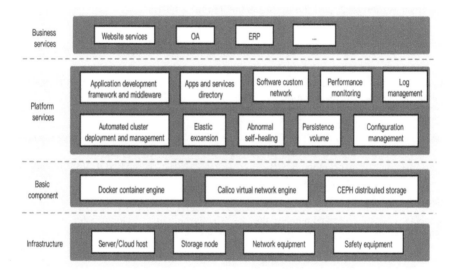

Fig. 2. Technical architecture diagram of PCCP container cloud platform

3   System Testing

3.1   Test Environment

The test environment topology is shown in Fig. 3. The test uses a node server and a laptop. They are both connected to the switchboard.

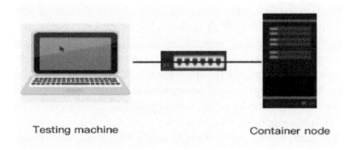

Testing machine   Container node

Fig. 3. PCCP container cloud platform test network topology

The model and configuration of the server and client are shown in Table 2. In the test the node server is Kylin system. The CPU is FT1500a@16c CPU. The laptop is the flagship of Windows 7, and the model is the ThinkPad T420.

Table 2. The test environment configuration table.

The name of the equipment	Model and configuration	Operating system	Software configuration
Server			
Node server(1)	CPU: FT1500a@16c CPU 1.5GHz Memory: 32GB Hard disk: 140GB	Kylin V4.0	PCCP container cloud platform MySQL V5.7.14etcd V3.2.24
Client			
Laptop(1) (CSTC10124326)	Model number: the ThinkPad T420 CPU: Intel Core i5-2450M 2.50GHz Memory: 4GB Hard disk: 500GB	The flagship of Windows 7	Google Chrome 52.0.2743.116

3.2 Test Content

The contents of the system test are shown in Table 3. In the test results, "·"is the coincidence term, and it conforms to the requirements of the system requirements specification. "*" is the nonconformity. "#" is the coincidence term after modifying. As can be seen from the table, all the test results in this test meet the requirements of the system requirements specification.

Table 3. Text content.

Technical specification	Test results
Container application management	You can create, edit, pause/resume, and delete containers Supports editing configurations for mirroring, environment variables, storage volume mounts, port mappings, and container commands
Console management interface	Support the management platform graphical interface directly bring up the container console The container can be manipulated through the container console
Configuration version management	Support for application configuration state rollback
Customized scheduling mechanism	It can set up independent scheduling rules for application and can select all, partial or priority, to meet three scheduling conditions
Log management	The log output of the service application can be tracked in real- time
Start a single application container	It takes an average of 1.8 s to start a single application container
Create 20 copies of the application container	It takes an average of 8.5 s to create 20 copies of an application container at the same time

3.3   Test Results

In this paper, we test the "PCCP container cloud platform" from the functional performance efficiency. The test results are as follows:

1. System architecture. The system is based on B/S architecture. The server adopts Kylin V4.0 operating system, the database adopts MySQL V5.7.14, the middleware adopts etcd V3.2.24, and the bandwidth is 1000Mbps. The client operating system is the flagship of Windows 7, and the browser uses Google Chrome 52.0.2743.116.
2. System function. The system realizes the container application management, console management interface, configuration version management, customized scheduling mechanism, and log management.
3. Performance efficiency. Starting a single application container took an average of 1.8

Seconds, creating 20 application container copies at the same time took an average of 8.5 s.

4   Conclusion

This paper takes the container cloud platform as the research object. A private container cloud platform PCCP based on Docker is proposed by analyzing the current problems and challenges. PCCP supports domestic software and hardware. The platform uses a container scheduling engine to pool resources such as computing, network, storage, and so on to provide application management capabilities at the distributed data center level. And the platform is no longer limited to the single mode for the application to give the required types of resources. After testing, the system runs stably and has a complete function.

Acknowledgements. This research was financially supported by National Key R&D Program of China (2018YFB1004100), China Postdoctoral Science Foundation funded project (2019M650606) and First-class Discipline Construction Project of Beijing Electronic Science and Technology Institute (3201012).

References

1. Katal, A., Dahiya, S., Choudhury, T.: Energy efficiency in cloud computing data center: a survey on hardware technologies. Clust. Comput. **25**(1), 675–705 (2021). https://doi.org/10.1007/s10586-021-03431-z
2. Meng, Z.Y.: Research on cloud computing technology of computer network in the new era. Comput. Program. Skills Maint. **417**(03), 93–94+107 (2020)
3. Chen, X.Y.: Design and implementation of network resource management and configuration system based on container cloud platform. Zhejiang University (2016)
4. Parast, F.K., Sindhav, C., Nikam, S., Yekta, H.I., Kent, K.B., Hakak, S.: Cloud computing security: A survey of service-based models. Comput. Secur. **114**, 102580 (2022)
5. Alouffi, B., Hasnain, M., Alharbi, A., Alosaimi, W., Alyami, H., Ayaz, M.: A systematic literature review on cloud computing security: threats and mitigation strategies. IEEE Access **9**, 57792–57807 (2021). https://doi.org/10.1109/ACCESS.2021.3073203
6. Feng, W.C.: Design of network resource configuration management system for container cloud platform. Industrial Instrumentation and Automation (2018)
7. Cai, L., Lu, J.N., Cai, Z.G., et al.: Resource quota prediction method for container cloud platform based on historical data analysis, CN110990159A[P] (2020)
8. Zheng, B.: Design of enterprise container cloud platform based on Kubernetes. Digital Technology and Application, **37**(348(06)), 148+151 (2019)
9. Li, J.Z., Zhao, Q.C., Yang, W.: A one-click deployment of big data and deep learning container cloud platform and its construction method, CN111274223A[P] (2020)

From Data Literacy to Co-design Environmental Monitoring Innovations and Civic Action

Ari Happonen[1](✉) , Annika Wolff[1] , and Victoria Palacin[2]

[1] Software Engineering, School of Engineering Science, LUT University, 53850 Lappeenranta, Finland
{ari.happonen,annika.wolff}@lut.fi
[2] Social Computing Research Group, Faculty of Social Sciences, University of Helsinki, 00014 Helsinki, Finland
victoria.palacin@helsinki.fi

Abstract. SENSEI is an environmental monitoring initiative run by Lappeenranta University of Technology (LUT University) and the municipality of Lappeenranta in south-east Finland. The aim was to collaboratively innovate and co-design, develop and deploy civic technologies with local civics to monitor positive and negative issues. These are planned to improve local's participation to social governance issues in hand. These issues can be e.g. waste related matters like illegal dumping of waste, small vandalism into city properties, alien plant species, but on the other hand nice places to visits too. This publication presents initiatives data literacy facet overview, which is aimed at creating equitable access to information from open data, which in turn is hoped for to increase participants motivation and entrepreneurship like attitude to work with the municipals and the system. This is done by curating environmental datasets to allow participatory sensemaking via exploration, games and reflection, allowing citizens to combine their collective knowledge about the town with the often-complex data. The ultimate aim of this data literacy process is to enhance collective civic actions for the good of the environment, to reduce the resource burden in the municipality level and help citizens to be part of sustainability and environmental monitoring innovation activities. For further research, we suggest follow up studies to consider on similar activities e.g. in specific age groups and to do comparisons on working with different stage holders to pin point most appropriate methods for any specific focus group towards collaborative innovation and co-design of civic technologies deployment.

Keywords: Environmental monitoring · Collaboratively innovate · Co-design innovation · Data literacy · Civic technologies · Open data

1 Introduction

In the last decade, civic technologies such as citizen sensing (also known as ICT enabled citizen science or crowdsensing) have been a popular means for empowering citizen participation and citizen engagement [1]. Specially the civic technologies have popular

Z. Qian et al. (Eds.): WCNA 2021, LNEE 942, pp. 408–418, 2022.
https://doi.org/10.1007/978-981-19-2456-9_42

in context of management and governance of cities, by augmenting both formal and informal aspects of civic life, government and public services [2]. The up shift in popularity has definitely drawn part of it suggest from global digitalization and sustainability trends [3, 4], the new level of awareness in general population against unnecessary waste and improvement in waste processing capabilities of municipalities [5], growth in public – private sector collaboration [6], and miniaturization and quality improvement in IT and sensor technologies [7].

This article summarizes an environmental monitoring initiative named as SENSEI [8]. Core of the summary is the role of data literacy within the project for mobilizing people to take civic action. SENSEI aimed to co-design, develop and deploy environmental sensing technologies in collaboration with citizens. Sensei shows how hardware, software and participatory practices can be combined to create civic technologies for local communities to monitor their environment, make sense of datasets and solve problems collectively. SENSEI technologies are being designed to monitor relevant positive and negative environmental issues (e.g. alien plant species, abandoned items and places citizens appreciate) for both citizens and decision makers. Lot of other examples are available from different cultural, social and physical environments [9–13]. We selected those monitoring areas, which are natural for our experiments local living environment as the goal was for the local community to collect, share and act upon available data [14]. Also, citizens will be able to monitor issue of their own interest as private monitoring targets they control and share when considered relevant. The aim of SENSEI is to prompt civic actions to enhance public participation and the environmental management of the town and try to generate long term effects [15] from the citizen sensing project.

This initiative followed the "a city in common" framework by [14]. We started with a collective identification of potential issues in town, using a series of ideation and co-design workshops with local citizens. Goal was to deploy an environmental monitoring of issues of common and individual interest during June-September 2018. Next, citizens were supported to enhance their ability to understand, make sense and solve collective issues with resources created during the initiative such as data, prototypes and social networks. Also, a data exhibition in a public space was organized. The exhibition supports participatory sensemaking by curating the data collected during the monitoring, allowing local citizens (including the ones who were not actively monitoring) to explore and make sense of the data, which was collected to enhance civic actions. This paper describes our approach, addressing the challenges attached to the design and orchestration of activities to support people to informally acquire or use existing data literacy skills. In case one would be arranging similar activities for data collection, and assuming possible data quality issues, we suggest on referring "data quality issue to solution mechanism table", by Vaddepalli et al. [16].

2 The SENSEI Data Exhibition

To get the participants in speed with the formerly unknown data, SENSEI data exhibition was used to welcome visitors with different data literacy skills and ability to interpret the data. During the exhibition, visitors were invited to frame questions related to relevant issues and opportunities in the town, from their own point of view. This was done through exploration and ideation around curated datasets. People who did not collect data them-selves or have not had previous data collection experiences, could face challenges during this stage [17]. Therefore, the exhibition goal was to create an enjoyable and equitable sense-making event in terms of access to information and ability to participate. In gen-eral, it is critical that the event design supports informal learning of data literacy skills for whoever needs them. Finally, the event design should naturally support collaboration and participatory sense-making to enhance civic action and to reduce ending up having non-wanted challenges and to be able to focus on solutions and new opportunities [18].

Whilst several definitions of data literacy can be found (e.g. [19, 20]), in this article data literacy is defined as follows: "the ability to ask and answer real-world questions from large and small data sets through an inquiry process, with consideration of ethical use of data. It is based on core practical and creative skills, with the ability to extend knowledge of specialist data handling skills according to goals. These include the abilities to select, clean, analyze, visualize, critique and interpret data, as well as to communicate stories from data and to use data as part of a design process." [20]. See Fig. 1.

Fig. 1. Data literacy pool (taken from [20])

The research questions related to the design and development of this data literacy process are:

1. Are participants who have actively monitored issues more likely to be engaged with the data? Does this participation lead to better sensemaking?
2. Can urban data games help visitors, especially non-data collectors, get up to speed and become engaged with the data?
3. How does the design of the space and activities support participatory sensemaking?
4. Can an initiative such as Sensei, including both the participatory sensing and sensemaking, lead to mobilization of citizens around important topics?

As participation is based on semi structured activities, evaluation cannot happen in a controlled experiment as controlling might generate unwanted behavior such as the Hawthorne effect [21]. Instead we provide an experience which is both playful to explore and informative in relation to issues that citizens are truly interested in. Attending and all engagement actions are entirely voluntary. Since intervening with questions or questionnaires could distract the attention from participation, the data capturing was designed to be unobtrusive and integrated to the event themes.

2.1 Capturing the Visitor Experience

Behavior data collection starts with a visitor number linked to a badge, onto which visitor can add self-selected ribbons. These ribbons were visitor descriptors / participant classificators as data-expert, data-collector, volunteer or citizen. Badge number and the ribbon choices will be noted with information whether they participated in data collected or not. Visitors can also pick up ribbons as they leave, which will be noted. Visitors receive an event related activity game (linked to badge number) which encourages them to visit each activity station and use a stamp there and a pen to mark some additional data to the card. Stamping captures the participation order in the stations. When visitors write questions, or create artefacts, they will also use their visitor ID (and name, if they choose). This will help with additional data capturing. Visitors handing the card are rewarded with a small prize related to number of stamps and a lottery participation with the chance to win a bigger prize. If possible, other metrics are collected too, to identify visitor hotspots/participation time details, either with facilitators help or with technology solutions. In addition, interacting with data exhibits leaves traces of participants actions, which can be captured. For example, time spent exploring data, quantity and quality of questions asked and stories told from data. The data collected should help to answer to the set questions.

3 Designing the SENSEI Data Exhibition Experience

The event is curated as an interactive exhibition, with a number of activities related to the Lappeenranta environmental monitoring designed to encourage and support visitors to engage and collaborate in data sensemaking actions. Additionally, general information related to monitoring themes and some additional craft activities aimed mainly at younger visitors are also included. These are e.g. arts table to draw pictures inspired by displayed material. Results were photographed and uploaded to a Sensei online exhibition (with approvals from the participants).

Free exploration is allowed, but knowledge of museum curation strategies will be used in designing the space to prompt visitors to follow a path that takes them through several distinct phases of interaction with data, with increasingly less constrained data exploration. We hope that this will also help us in follow up stages with the collected data and digital curation of it [22]. Stages are shown in Fig. 2.

Designing the space, where it is easy for people to collaborate, is important for participatory sense-making support. This leads to the communal property of civic intelligence, as defined by Schuler et al. [23]. Each stage builds on work conducted within a

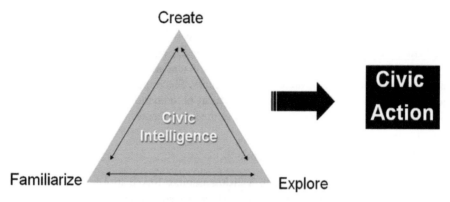

Fig. 2. Staged data exploration to build civic intelligence and enhance civic action.

UK data literacy initiative, that developed a number of Urban Data Games [24, 25] and founded a set of principles to support building data literacy from complex data sets in formal (e.g. classrooms) and informal (e.g. museum) settings. The principles were:

- Guide a data inquiry,
- Expand out from a representative part of dataset,
- Work collaboratively (STEAM approach) on creative activities and
- Balance screen activities with tangible ones [26].

3.1 Familiarize

The familiarization stage can consists of a number of interactive games; speed data-ing (Fig. 3), shark-bytes (Fig. 4) and top data-trumps (Fig. 5), for visitors to play. These would help visitors to know what types of data they can explore and what they might find. This is specially designed for non-data collecting visitors.

Speed data-ing is designed to help visitors get to know the different collected datasets. Visitors have only 30 s getting to know the open data types from the environmental dataset (decided by the city or by the citizen's, during the monitoring period). A short time period is used, as positive time-based stress helps people to focus on most important aspects and as such helps productivity too [27]. Key information will be a) the name and icon used to consistently identify the dataset in SENSEI platform and in the exhibition b) the types of places to look for instances of the data c) the most likely time periods containing data.

Shark-bytes is a play on the US television show Card Sharks (Play your cards right in the UK). The play starts with a random playing card. Contestant must guess if the following subsequent card (facing downwards) would be higher or lower. In this case, key datasets are the line of cards, in timeline order. Players predict whether the value for that datatype went up, or down (in total) in each following week. A player 'wins' by getting to the end of the line of cards without error. It is anticipated that players in general will discuss how they base their prediction, using their knowledge both of the town and also knowledge of human behavior e.g. by knowing popular holidays, player

Fig. 3. Speed data-in.

might predict lower values when those monitoring may not collect data. The aim is to support visitors in thinking about the importance of finding and analyzing data trends and to cause reflection on how data is collected, what sort of cultural, societal, human behavior and so on matters can affect the results and may also lead to 'errors' in data.

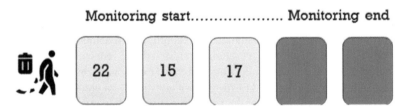

Fig. 4. Shark-bytes. 3 cards shown, the visitor predicting the next 2 values.

Top data-trumps is based on the original Top Trumps card game. Data-trump cards relate to places in Lappeenranta. Values relate to the data types and the total value for that data type in each place within the monitoring period. This game teaches data comparison skills. In general, utilization of different activation and idea generations support means and methods are all designed to make exploration of the complete datasets easier and more meaningful / understandable task.

Fig. 5. Top data-trumps.

3.2 Exploring Stage

The exploration stage gives citizens access to the data, via a map-based interface (presented on iPads and also a large interactive wall, used for collaboration activities). The data can be freely explored by selecting:

1. which specific part of data or datasets to look at
2. a region in Lappeenranta (with panning and zooming)
3. the time period (selected by a slider)

Instances of the selected data, based on the made choices, will appear on the map. This is supported by prompts that encourage visitors to focus in to just a small part of a data set, to make meaning from that, and then to do wider explorations. One of the ideas is, to help people find patterns in the data. This ideology is based on principles derived from and tested within the Urban Data School initiative and also expectations of interfaces by users in a study on participatory sensemaking by Filonik et al. [28], who studied this via a dashboard from which users could collaboratively visualize and share meaning from data, finding that visualizations should be 1) dynamic to support playful interactions 2) flexible to allow exploration of relevant data 3) educational and guide the initial inquiries 4) collaborative, allowing visitors to exchange ideas with one another. Therefore, visitors are encouraged to write down questions and predictions and display them, so visitors who will join later on, in different time and/or session, can build upon earlier findings. Visitors can work alone or discuss with others, whichever they prefer. However, collaboration is encouraged, with large interactive map interface.

3.3 Stage to Create

The creation stage provides visitors with artwork creation space to reflect a story they want to tell. Craft materials are provided, inspired by the data sculptures approach of [19]. After representation, they write a story card explaining what they have made and why it is interesting (like in museum exhibition), which visitors can add to museum by leaving their sculptures, or by taking a polaroid picture instead, if visitors prefer to keep the sculpture.

4   Discussion on Action Taking

The question is, does exhibition bring people together around certain topics. Such activities were encouraged and supported in monitoring stage, but not all of the participants were compelled to take action. It was not exactly clear, would additional gamification elements [29] had made people more active, but the general expectation among organizers and active supporters from the city was in this direction. Still in sensei initiative, over 240 participants, aged 7 to 85 years, were involved over a period of 10 months. Ten events and workshops generated over 100 ideas about issues of shared interest, 28 civic tech prototypes and dozens of sense-making artifacts, including data interactions, analysis of datasets and data sculptures [8].

To facilitate volunteering and participation, existing groups (whether pre-existing initiatives or created through earlier Sensei activities) were invited to attend in person and talk about their activities, or at least to leave flyers. Visitors will be able to sign up to participate in the groups or join through social media. New groups forming were able to leave something in the space to attract other people to join, through stigmergic action. E.g. a jar to drop participants contact details into (in anonymous way). This visualizes the traction gaining campaigns.

5   Conclusion

The study described an event to engage citizens of a town with their environmental data (collected during participatory sensing initiative). In any social governance matter, where collective responsibility is considered as a key for success, sensei like methodology to get citizens to participate into technology and data collection activities, makes them more invested to the process and how matters are handled in general in the governance case. In this particular example, the event was staged as an interactive data exhibition, designed to informally build data literacy, to encourage collective sensemaking and, in some cases, to lead to civic action. We suggest future research to look up into opportunities on developing new sustainability innovations on top of civic engagement-based data collection activities as the data is quite unique in nature and could offer seeds for developing e.g. new and novel environmental monitoring services [30–32]. Our research outlines a number of solution for typical challenges for engaging visitors, when playing with the data and in capturing feedback to assess the validity of the design decisions to support the intended outcomes. We recommended on learning from experiences between engineers and representatives of other society groups like artists [33], young students experiences from citizen participation activity [34] and realities of time pressure in innovation processes [27]. Additionally, especially because of the challenges the global covid-19 pandemic has given, e.g. requiring us to endure long term social distancing matters, we would like to suggest researching and experimenting hybrid / almost fully online co-design activities for environmental monitoring innovations, as these will definitely be different from physical events and brainstorming sessions [35].

Acknowledgments. We would like to thank all the volunteers, partners, and authors who wrote and provided helpful comments for this publication writing process. We gratefully acknowledge the support from the Finnish Cultural Foundation for South Karelia Region and the PERCCOM programme. We also give our gratitude for South-East Finland – Russia CBC programme for supporting AWARE project, funded by the European Union, the Russian Federation and the Republic of Finland as the funding has made it possible for publishing this work and disseminate the knowledge.

Competing Interests. Authors have declared that no competing interests exist.

References

1. Foscarini, F.: Citizen engagement. In: Duranti, L., Rogers, C. (eds.) Trusting Records and Data in the Cloud: The Creation, Management, and Preservation of Trustworthy Digital Content, pp. 65–96 (2018). https://doi.org/10.29085/9781783304042.004
2. Palacin-Silva, M., Porras, J.: Shut up and take my environmental data! A study on ICT enabled citizen science practices, participation approaches and challenges. In: Penzenstadler, B., Easterbrook, S., Venters, C., Ahmed, S.I. (eds.) ICT4S2018. 5th International Conference on Information and Communication Technology for Sustainability, vol. 52, pp. 270–288 (2018). https://doi.org/10.29007/mk4k
3. Ghoreishi, M., Happonen, A., Pynnönen, M.: Exploring industry 4.0 technologies to enhance circularity in textile industry: role of internet of things. In: Twenty-first International Working Seminar on Production Economics, 24–28 February 2020, Innsbruck, Austria, pp. 1–16 (2020). https://doi.org/10.5281/zenodo.3471421
4. Happonen, A., Ghoreishi, M.: A mapping study of the current literature on digitalization and industry 4.0 technologies utilization for sustainability and circular economy in textile industries. In: Yang, X.-S., Sherratt, S., Dey, N., Joshi, A. (eds.) Proceedings of Sixth International Congress on Information and Communication Technology. Lecture Notes in Networks and Systems, vol. 217, pp. 697–711. Springer, Singapore (2022). https://doi.org/10.1007/978-981-16-2102-4_63
5. Kilpeläinen, M., Happonen, A.: Awareness adds to knowledge. Stage of the art waste processing facilities and industrial waste treatment development. Curr. Appr. Sci. Technol. Res. **4**, 125–148 (2021). https://doi.org/10.9734/bpi/castr/v4/9636D
6. Happonen, A., Minashkina, D., Nolte, A., MedinaAngarita, M.A.: Hackathons as a company – university collaboration tool to boost circularity innovations and digitalization enhanced sustainability. AIP Conf. Proc **2233**(1), 1–11 (2020). https://doi.org/10.1063/5.0001883
7. Jahkola, O, Happonen, A., Knutas, A., Ikonen, J.: What should application developers understand about mobile phone position data. In: CompSysTech 2017, pp. 171–178. ACM (2017). https://doi.org/10.1145/3134302.3134346
8. Palacin, V., Ginnane, S., Ferrario, M.A., Happonen, A., Wolff, A., Piutunen, S., Kupiainen, N.: SENSEI: harnessing community wisdom for local environmental monitoring in Finland. CHI Conference on Human Factors in Computing Systems, Glagsgow, Scotland UK, pp. 1–8 (2019). https://doi.org/10.1145/3290607.3299047
9. Hagen, L., Kropczynski, J., Dumas, C., Lee, J., Vasquez, F.E., Rorissa, A.: Emerging trends in the use and adoption of E-participation around the world. Proc. Assoc. Inf. Sci. Technol. **52**(1), 1–4 (2016). https://doi.org/10.1002/pra2.2015.14505201008
10. Huffman, T.: Participatory/Action Research/CBPR, The International Encyclopedia of Communication Research Methods, pp. 1–10 (2017). https://doi.org/10.1002/9781118901731.iecrm0180

11. Chudý, F., Slámová, M., Tomaštík, J., Tunák, D., Kardoš, M., Saloň, Š.: The application of civic technologies in a field survey of landslides. Land Degradat. Dev. **29**(6), 1858–1870 (2018). https://doi.org/10.1002/ldr.2957

12. Palacin, V., Gilbert, S., Orchard, S., Eaton, A., Ferrario, M.A., Happonen, A.: Drivers of participation in digital citizen science: case studies on Järviwiki and Safecast. Citizen Science: Theory Pract. 5(1), 1–20 (2020). Article: 22, https://doi.org/10.5334/cstp.290

13. Parra, C., et al.: Synergies between technology, participation, and citizen science in a community-based dengue prevention program **64**(13), 1850–1870 (2020). https://doi.org/10.1177/0002764220952113

14. Balestrini, M., Rogers, Y., Hassan, C., Creus, J., King, M., Marshall, P.: A City in common: a framework to orchestrate large-scale citizen engagement around urban issues. In: CHI 2017: Proceedings of the 2017 CHI Conference on Human Factors in Computing Systems, pp. 2282–2294 (2017). https://doi.org/10.1145/3025453.3025915

15. Rossitto, C.: Political ecologies of participation: reflecting on the long-term impact of civic projects. In: Proceedings of the ACM on Human-Computer Interaction, **5**(CSCW1), 1–27 (2021), Article: 187, https://doi.org/10.1145/3449286

16. Vaddepalli, K., Palacin, V., Porras, J., Happonen, A.:. Connecting digital citizen science data quality issue to solution mechanism table (2020). https://doi.org/10.5281/zenodo.3829498

17. Krumhansl, R., Busey, A., Krumhansl, K., Foster, J. Peach, C.: Visualizing oceans of data: educational interface design. Oceans, San Diego, pp. 1–8 (2013). https://doi.org/10.23919/OCEANS.2013.6741364

18. Capponi, A., Fiandrino, C., Kantarci, B., Foschini, L., Kliazovich, D., Bouvry, P.: A survey on mobile crowdsensing systems: challenges, solutions, and opportunities. IEEE Commun. Surv. Tutor. **21**(3), 2419–2465 (2019). https://doi.org/10.1109/COMST.2019.2914030

19. D'Ignazio, C., Bhargava, R.: DataBasic: design principles tools and activities for data literacy learners. J. Commun. Inform. **12**(3), 83–107 (2016). https://doi.org/10.15353/joci.v12i3.3280

20. Wolff, A., Gooch, D., Cavero Montaner, J.J., Rashid, U., Kortuem, G.: Creating an understanding of data literacy for a data-driven society. J. Commun. Inf. **12**(3), 9–26 (2017). https://doi.org/10.15353/joci.v12i3.3275

21. Landsberger, H.: Hawthorne Revisited. Cornell University, New York (1959)

22. Stevens, J.R.: Digital curation's dilemma: contrasting different uses, purposes, goals, strategies, and values. Int. J. Technol. Knowl. Soc. **9**(4), 1–11 (2014). https://doi.org/10.18848/1832-3669/CGP/v09i04/56399

23. Schuler, D., De Liddo, A., Smith, J., De Cindio, F.: Collective intelligence for the common good: cultivating the seeds for an intentional collaborative enterprise. AI Soc. **33**(1), 1–13 (2017). https://doi.org/10.1007/s00146-017-0776-6

24. Wolff, A., et al.: Engaging with the smart city through urban data games. In: Nijholt, A. (ed.) Playable Cities. Gaming Media and Social Effects, pp. 47–66. Springer, Singapore (2017). https://doi.org/10.1007/978-981-10-1962-3_3

25. Wolff, A., Barker, M., Petre, M.: Creating a Datascape: a game to support communities in using open data. In: C&T 2017 Proceedings of the 8th International Conference on Communities and Technologies, New York, NY, USA pp. 135–138. ACM (2017)

26. Wolff, A., Petre, M., van der Linden, J.: Pixels or plasticine: evoking curiosity to engage children with data. In: Designing for Curiosity workshop at CHI 2017, 7 May 2017, Denver, Colorado (2017)

27. Salmela, E., Happonen, A., Hirvimäki, M., Vimm, I.: Is time pressure an advantage or a disadvantage for front end innovation – case digital Jewelry. J. Innov. Manag. 3(4), 42–69 (2015). https://doi.org/10.24840/2183-0606_003.004_0005

28. Filonik, D., Tomasz, B., Rittenbruch, M., Marcus, F.: Collaborative data exploration interfaces - from participatory sensing to participatory sensemaking. In: Engelke, U., Bednarz, T.P., Heinrich, J., Klein, K., Nguyen, Q.V. (eds.) 2015 Big Data Visual Analytics, Institute of Electrical and Electronics Engineers Inc., Hobart, Australia, pp. 123–125 (2015). https://doi.org/10.1109/BDVA.2015.7314289
29. Santti, U., Happonen, A., Auvinen, H.: Digitalization boosted recycling: gamification as an inspiration for young adults to do enhanced waste sorting. AIP Conf. Proc. **2233**(1), 1–12 (2020). https://doi.org/10.1063/5.0001547
30. Eskelinen, T., Räsänen, T., Santti, U., Happonen, A., Kajanus, M.: Designing a business model for environmental monitoring services using fast MCDS innovation support tools. Technol. Innov. Manage. Rev. **7**(11), 36–46 (2017). https://doi.org/10.22215/timreview/1119
31. Happonen, A., Santti, U., Auvinen, H., Räsänen, T., Eskelinen, T.: Digital age business model innovation for sustainability in University Industry Collaboration Model, E3S Web of Conferences **211**, 1–11 (2020). Article 04005, https://doi.org/10.1051/e3sconf/202021104005
32. Santti, U., Happonen, A., Auvinen, H., Räsänen, T., Eskelinen, T.: Sustainable Business Model Innovation for Digital Remote Monitoring: A Follow up Study on a Water Iot Service, BIOS Forum 2020, St. Petersburg, Russia, 10/2020, pp. 1–7 (2020). https://doi.org/10.5281/zenodo.4290135
33. Happonen, A., et al.: Art-technology collaboration and motivation sources in technologically supported artwork buildup project. Phys. Procedia **78**, 407–414 (2015). https://doi.org/10.1016/j.phpro.2015.11.055
34. Happonen, A., Minashkina, D.: Ideas and experiences from university industry collaboration: Hackathons, Code Camps and citizen participation, LUT Scientific and Expertise Publications report 86, pp. 1–21 (2018). ISBN: 978-952-335-253-7, ISSN: 2243-3384. https://doi.org/10.13140/rg.2.2.29690.44480
35. Salmela, E., Happonen, A.: Applying social media in collaborative brainstorming and creation of common understanding between independent organizations. In: Knowledge Management/Book 2: New Research on Knowledge Management Applications and Lesson Learned, pp. 195–212 (2012). https://doi.org/10.5772/2529

Information Security Resource Allocation Using Evolutionary Game

Jun Li[1], Dongsheng Cheng[2], Lining Xing[2], and Xu Tan[2(✉)]

[1] Academy of Hi-Tech Research, Hunan Institute of Traffic Engineering, Hengyang 421099, People's Republic of China

[2] School of Software Engineering, Shenzhen Institute of Information Technology, Shenzhen 518172, People's Republic of China
tanxu_nudt@yahoo.com

Abstract. Based on the discussion of related concepts and technical theories, the information security resource allocation influencing factors index system is constructed from four aspects: resources, threat sources, vulnerabilities and security measures. With the further analysis of information security factors and their affecting mechanisms, the basic theoretical framework of information security resource allocation is established based on the evolutionary game. Under this framework, the subject relationship in various situations is analyzed. This research work can conduct a reasonable allocation of resources related to information security.

Keywords: Smart city · Information security · Resource allocation · Evolutionary game

1 Introduction

The concept of smart cities, originating from the field of media, refers to using a variety of new technologies or innovative concepts to effectively connect and integrate various systems and services through reasonable resource allocation in cities, so as to optimize urban management and improve life quality of residents [1–3]. Smart cities fully apply all kinds of new technologies (such as Internet of things (IoT), cloud computing, virtual reality, etc.) into all walks of life in cities [4–6]. By establishing the interconnection in broadband ubiquitous networks, integrating application of intelligent technologies and sharing resources widely, smart cities obtain comprehensive and thorough perception abilities to realize fine and dynamic management of cities and effective improvement of life of residents [7–10].

Smart cities have been valued by countries all over the world since they came into being, which provide more convenience for people's life while improving the intelligent level of cities [11–13]. However, smart cities are highly dependent on new technologies including cloud computing and IoT [14–16], which brings a hidden danger of spreading the information risk while applying technologies and poses multi-facetted impacts on information security in cities [17–20]. How to reasonably allocate the current resources

© The Author(s) 2022
Z. Qian et al. (Eds.): WCNA 2021, LNEE 942, pp. 419–425, 2022.
https://doi.org/10.1007/978-981-19-2456-9_43

in cities to avoid the information security risk as far as possible and obtain the maximum benefits has become a practical problem that smart cities have to be faced in their healthy development [21–25].

2 Influencing Factors Index System

Comprehensive analysis on factors influencing resource allocation to information security and establishment of the corresponding index system are the bases for reducing the information security risk in smart cities in the context of big data. From the perspective of information security, the first-level indexes in the index system can be summarized into four aspects, namely resources, threat sources, vulnerability and safety measures by combining with the current situations of smart cities..

2.1 Information Resources

There are many kinds of information resources, but it is evident that the higher the value of resources, the greater the risk may be faced in the actual situations. In accordance with relevant definitions of smart cities and information resources, the influencing factors of resources are sub-classified into three second-level indexes: management personnel, infrastructure and economic investment, that is, manpower, material resources and financial resources. By further analysing the information security risk based on these indexes, the third-level indexes are obtained and the results are shown in Fig. 1.

2.2 Threat Sources

Threat is an objective factor that probably causes the potential risk for information security in smart cities. The influencing factors of a threat source are sub-classified into

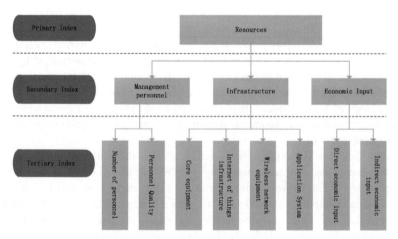

Fig. 1. Index system of factors influencing information security in smart cities based on resource value

two second-level indexes, namely technological and management threats. By further analysing the information security risk based on the indexes, the third-level indexes are obtained and the results are illustrated in Fig. 2.

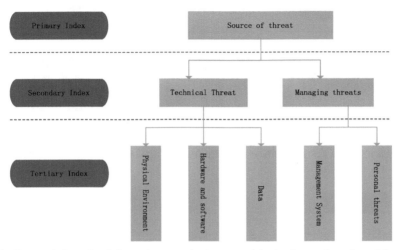

Fig. 2. Factors influencing information security in smart cities in the confirmation of the threat sources

2.3 Vulnerability

Vulnerability is considered mainly because in the context of big data, the defects of the information system in smart cities are threatened and taken advantages of, which renders the system possibly under risk of attack. The influencing factors of vulnerability are sub-classified into two second-level indexes: vulnerability in technology and management. The third-level indexes are obtained by analysing the information security risk based on the above factors, and the results are demonstrated in Fig. 3.

2.4 Safety Measures

Safety measures are a barrier to protect information security in smart cities, which can effectively reduce risks of security accidents and vulnerabilities, and provide technical supports and management mechanisms for some re-sources. The influencing factors of safety measures are sub-classified into two second-level indexes: preventive measures and protective measures, on which basis the information security risk is further analyzed to obtain the three-level indexes. The results are shown in Fig. 4.

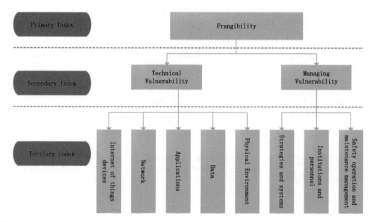

Fig. 3. Factors influencing information security in smart cities in the identification of vulnerability

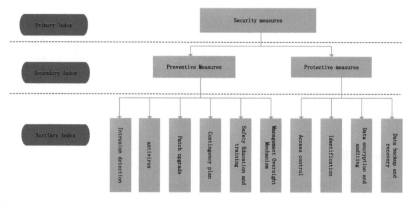

Fig. 4. Factors influencing information security in smart cities based on safety measures

3   Resource Allocation Framework to Information Security

With the constant development and progress in new technologies, such as artificial intelligence, big data, IoT, cloud computing and virtual reality, the development and construction of smart cities has been realized, but there are also great threats and challenges in information security. To effectively respond to these threats and challenges, by fully understanding the factors influencing resource allocation to information security, this study established a reasonable and effective theoretical framework of resource allocation to information security based on the current popular evolutionary game theory. The framework can play its due role in the protection of information security. By analysing the index system of influencing factors in the above section, it can be seen that these common links including software and hardware, data, network, application, external environment and management are involved in all influencing factors in smart cities. In a city, how to plan the limited resources and avoid the restrictions of the above factors, so as to play the maximum efficiency of all resources and well protect the information security is one of the problems that need to be considered. For a city that has communication with the outside world, all internal resources therein are regarded as a whole, in which some external resources can complement, be replaced, and weakly correlated with internal resources. How to allocate the resources reasonably to improve the safeguard effects on information security is also an issue to be considered. In conclusion, the resource allocation to information security in a smart city is to analyse how to allocate internal and external resources of the city. According to the evolutionary game theory, the theoretical framework of resource allocation to information security was obtained, as displayed in Fig. 5.

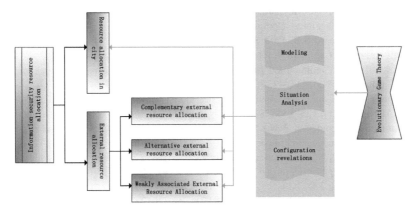

Fig. 5. Theoretical framework of resource allocation to information security

4 Conclusions

On the basis of discussing relevant concepts and technical theories, the research established the index system of factors influencing resource allocation to information security from aspects including resources, threat sources, vulnerability, and safety measures. The factors and mechanisms that influence information security were analysed and the basic theoretical framework of resource allocation to information security was built based on evolutionary game. The resource allocation to information security is divided into internal and external resource allocation in cities, and the latter can be sub-divided into complementary, alternative, and weakly correlated external resource allocation. Moreover, subject relationships under various circumstances were analysed under the framework.

Acknowledgments. This research work is supported by the National Social Science Fund of China (18BTQ055), the Youth Fund of Hunan Natural Science Foundation (2020JJ5149, 2020JJ5150) and the Innovation Team of Guangdong Provincial Department of Education (2018KCXTD031). It is also supported by the Program of Guangdong Innovative Research Team (2020KCXTD040), the Pengcheng Scholar Funded Scheme, and the Basic Research Project of Science and Technology Plan of Shenzhen (SZIITWDZC2021A02, JCYJ20200109141218676).

Conflicts of Interest. The authors declare that they have no conflict of interest.

References

1. Knapp, K.J., Marshall, T.E.: Information security policy: an organizational-level process model. Comput. Secur. **28**(7), 493–508 (2009)
2. Anjaria, K., Mishra, A.: Relating Wiener's cybernetics aspects and a situation awareness model implementation for information security risk management. Kybernetes **47**(1), 69–81 (2017)
3. Webb, J., Ahmad, A., Maynard, S.B., et al.: A situation awareness model for information security risk management. Comput. Secur. **44**, 1–15 (2014)
4. Ahmad, A., Maynard, S.B., Park, S.: Information security strategies: towards an organizational multi-strategy per-spective. J. Intell. Manuf. **25**(2), 357–370 (2014)
5. Bojanc, R.: An economic modeling approach to information security risk management. Int. J. Inf. Manage. **28**(5), 413–422 (2008)
6. Nazareth, D.L., Choi, J.: A system dynamics model for information security management. Inf. Manage. **52**(1), 123–134 (2015)
7. Houmb, S.H., Franqueira, V.N.L., Engum, E.A.: Quantifying security risk level from CVSS estimates of frequency and impact. J. Syst. Softw. **83**(9), 1622–1634 (2010)
8. Feng, N., Li, M.: An information systems security risk assessment model under uncertain environment. Appl. Soft Comput. J. **11**(7), 4332–4340 (2011)
9. Kong, H.K., Kim, T.S., Kim, J.: An analysis on effects of information security investments: a BSC perspective. J. Intell. Manuf. **23**(4), 941–953 (2012)
10. Li, S., Bi, F., Chen, W., et al.: An improved information security risk assessments method for cyber-physical-social computing and networking. IEEE Access **6**(99), 10311–10319 (2018)
11. Basallo, Y.A., Senti, V.E., Sanchez, N.M.: Artificial intelligence techniques for information security risk assessment. IEEE Lat. Am. Trans. **16**(3), 897–901 (2018)
12. Grunske, L., Joyce, D.: Quantitative risk-based security prediction for component-based systems with explicitly modeled attack profiles. J. Syst. Softw. **81**(8), 1327–1345 (2008)
13. Gusm, O.A., Silval, C.E., Silva, M.M., et al.: Information security risk analysis model using fuzzy decision theory. Int. J. Inf. Manage. **36**(1), 25–34 (2016)
14. Baskerville, R.: Integration of information systems and cybersecurity countermeasures: an exposure to risk perspective. Data Base Adv. Inf. Syst. **49**(1), 69–87 (2017)
15. Huang, C.D., Hu, Q., Behara, R.S.: An economic analysis of the optimal information security investment in the case of a risk-averse firm. Int. J. Prod. Econ. **114**(2), 793–804 (2008)
16. Yong, J.L., Kauffman, R.J., Sougstad, R.: Profit-maximizing firm investments in customer information security. Dec. Supp. Syst. **51**(4), 904–920 (2011)
17. Li, J., Li, M., Wu, D., et al.: An integrated risk measurement and optimization model for trustworthy software pro-cess management. Inf. Sci. **191**(9), 47–60 (2012)
18. Benaroch, M.: Real options models for proactive uncertainty-reducing mitigations and appli-cations in cyber-security investment decision-making. Soc. Sci. Electron. Pub. **4**, 11–30 (2017)
19. Gao, X., Zhong, W., Mei, S.: Security investment and information sharing under an alternative security breach probability function. Inf. Syst. Front. **17**(2), 423–438 (2015)
20. Liu, D., Ji, Y., Mookerjee, V.: Knowledge sharing and investment decisions in information security. Dec. Supp. Syst. **52**(1), 95–107 (2012)
21. Gao, X., Zhong, W., Mei, S.: A game-theoretic analysis of information sharing and security investment for complementary firms. J. Oper. Res. Soc. **65**(11), 1682–1691 (2014)

22. Gao, X., Zhong, W.: A differential game approach to security investment and information sharing in a competitive environment. IIE Trans. **48**(6), 511–526 (2016)
23. Wu, Y., Feng, G.Z., Wang, N.M., et al.: Game of information security investment: Impact of attack types and net-work vulnerability. Expert Syst. Appl. **42**(15–16), 6132–6146 (2015)
24. Wang, Q., Zhu, J.: Optimal information security investment analyses with the consideration of the benefits of investment and using evolutionary game theory. In: Proceedings of the International Conference on Information Management, pp. 957–961 (2016)
25. Qian, X., Liu, X., Pei, J., et al.: A game-theoretic analysis of information security investment for multiple firms in a network. J. Oper. Res. Soc. **68**(10), 1–16 (2017)

An Improved Raft Consensus Algorithm Based on Asynchronous Batch Processing

Hao Li, Zihua Liu[✉], and Yaqin Li

School of Mathematics and Computer Science, Wuhan Polytechnic University, 36 Huanhu Middle Road, Dongxihu District, Wuhan, China
liu.zihua@outlook.com

Abstract. The consensus algorithm has been popular in current distributed systems as it is more effective in solving server unreliability. It ensures a group of servers can form a coordinated system, and the entire system continues to work when a part of the service point fails. Raft is a well-known and widely used distributed consensus algorithm, but as it has a built-in purpose of comprehensibility, it is always compromised in terms of performance as a trade-off. In this paper, we mainly aim to improve the traditional Raft consensus algorithm's performance problem, especially in high concurrency scenarios. We introduce a pre-proposal stage on top of the algorithm to achieve efficiency optimization through batch asynchronous log replicated and disk flushing. The experiment proved that the improved Raft could increase the system throughput by 2–3.6 times, and the processing efficiency for parallel requests can be increased by 20% or more.

Keywords: Distributed system · Consensus algorithm · Consistency algorithm · Raft

1 Introduction

The theory of CAP [1] (Consistency, Availability, Partition tolerance) tells us that in any distributed system, the three essential characteristics of CAP cannot be satisfied simultaneously; at least one of them must be given up. Generally, in a distributed system, the partition tolerance is automatically satisfied. Giving up consistency means that the data between nodes cannot be trusted, which is usually unacceptable. Therefore, a possible choice is to give up availability, meaning that the nodes need to be entirely independent to obtain data consistency. When building a distributed system, the main construction goals are to ensure its consistency and partition tolerance, while the former has drawn more interest in recent research.

The consistency problem mainly focuses on how to reach agreement among multiple service nodes. The services of distributed systems are usually vulnerable to various network issues such as server reset and network jitter, making the services unreliable. To solve this problem, a consensus algorithm was created. The consensus algorithm usually uses a replicated state machine to ensure that all nodes have the same log sequence. After all the logs are applied in order, the state machine will eventually reach an agreement.

© The Author(s) 2022
Z. Qian et al. (Eds.): WCNA 2021, LNEE 942, pp. 426–436, 2022.
https://doi.org/10.1007/978-981-19-2456-9_44

The consistency algorithms are widely used in distributed databases [2–4], blockchain applications [5, 6], high-performance middleware [7], and other fields, and they are also the basis for realizing these systems.

Two well-known consensus algorithms are the Paxos [8] and the Raft [9]. The Paxos algorithm has been the benchmark for consensus algorithms in the past decades, but it is somehow obscure, and the implementation detail is missing in the original research, leading to various versions of systems and hard to verify its correctness. The Raft protocol supplements the details of multi-decision stages in the Paxos. It enhances the comprehensibility, decomposes the consistency problem into several consecutive sub-problems, and finally guarantees the system's correctness through the security mechanism.

The distributed consensus problem requires participants to reach a consensus on the command sequence, and a state machine executes the submitted command sequence and ensures the ultimate consistency. In the Raft algorithm, a leader will be selected first, and the leader will execute all requests. Raft's security mechanism ensures that the state machine logs are in a specific sequence according to the logical numbers to reach a consensus, i.e., sequential submission and sequential execution. However, the systems implemented with this procedure have a low throughput rate, a large portion of the requests must be remained blocked, and this reduction in performance will deteriorate, especially in scenarios with high concurrency.

To deal with this problem, an improved Raft consensus algorithm is proposed in this paper. Instead of strict sequential execution of requests, we introduce a pre-proposal stage, in which the asynchronous batch processing is performed to improve the efficiency while retaining the distributed consensus characteristic. The improved Raft algorithm will be deployed on simulated cluster machines for experiments. Finally, the availability and the performance of the proposed method under a large number of concurrent requests will be verified.

2   Related Works

2.1   Replicated State Machine

The consensus algorithm usually uses the replicated state machine structure as its means to achieve fault tolerance. Local state machines on some servers will generate execution copies of the same state and send them to other servers through network, so that the state machine can continue to execute even when some machines are down. A typical implementation is to use the state machine managed by the leader node to execute and send the copy, which can ensure that the cluster can survive externally even when one node is down. Mature open source systems such as Zookeeper [10], TiKV [11] and Chubby [12] are all based on this implementation.

The basis theory of the state machine is: if each node in the cluster is running the same prototype of the deterministic state machine S, and the state machine is in the initial state $S0$ at the beginning, with the same input sequence $I = \{i1,i2,i3,i4,i5,...,in\}$, these state machines will execute the request sequence with the transition path: $s0- > s1- > s2- > s3- > s4- > s5- >...- > sn$, so finally the consistent final state Sn will be achieved, producing the same state output set $O = \{o1(s1),o2(s2),o3(s3),o4(s4),o5(s5),...,on(sn)\}$.

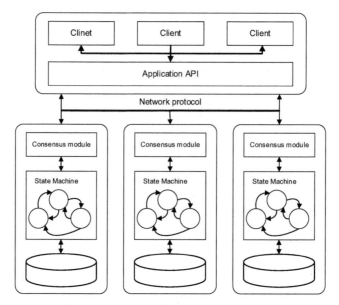

Fig. 1. The replicated state machine structure.

As shown in Fig. 1, the replicated state machine is implemented based on log replication, and the structure usually consists of three parts: a consensus module, a state machine prototype, and a storage engine. The consensus module of each server is responsible for receiving the log sequence initiated by the client, executing, and storing it in the order in which it is received, and then distributing the logs through the network to make the state machines of all server nodes to be consistent. Since the state of each state machine is deterministic, and each operation can produce the same state and output sequence, the entire server cluster acts as one exceptionally reliable state machine.

2.2 Raft Log Compression

The Raft protocol is implemented based on the state machine of log replication. However, in actual systems, the log could not allow unlimited growth. As time increases, the continuous growth of logs will take up more log transmission overhead, as well as more recovery time for node downtime. If there is no certain mechanism to solve this problem, the response time of the Raft cluster will be significantly slower, so log compression is usually implemented in Raft algorithms.

The Raft uses snapshots to implement the log compression. In the snapshot system, if the state Sn in the state machine at a certain time is safely applied to most of the nodes, then Sn is considered safe, and all the states previous to Sn can be discarded, therefore the initial operating state $S0$ is steadily changed to Sn, and other nodes only need to obtain the log sequence starting from Sn when obtaining logs.

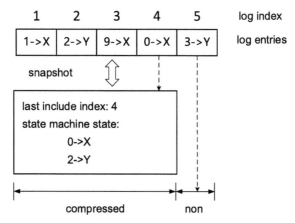

Fig. 2. The Raft log compression implemented by snapshots.

Figure 2 shows the basic idea of the Raft snapshots. A snapshot is created independently by each server node and can only include log entries that have been safely submitted. The snapshot structure contains the index value of the last log entry that was last replaced by the snapshot. Once a node completes a snapshot, it can delete all logs and snapshots before the last index position.

Although each node manages the snapshots independently, Raft's logs and snapshots are still based on the leader node. For followers who are too backward (including nodes that recover from downtime and have large network delays), the leader will send the latest updates through the network and overwrite it.

3   Improved Raft Algorithm

3.1   Premises and Goals of the Improved Algorithm

The premises of the original Raft algorithm is as follows, meaning that its security mechanism should basically guarantees:

- The cluster maintains a monotonically increasing term number (Term).
- The network communication between clusters is not reliable and are susceptible to packet loss, delay, network jitter, etc.
- No Byzantine error will occur.
- There will always be one leader selected in the cluster and there will only be one leader under the same term number.
- Leader is responsible for interacting with client requests. Client requests received by other nodes need to be redirected to the Leader.
- The request to the client meets the linear consistency, and the client can accurately return the interactive information after each operation.

In the improved algorithm, most of the above premises is not changed except for the second one. In actual engineering projects, the communication between computers tends to be stable most of the time (that is, the delay between nodes is much less than the time of a Heartbeat). In addition, general reliable communication protocols such as TCP have a retransmission mechanism, with which lost packets will be retransmitted immediately, so it is possible to recover in a short time even if there is a failure. Therefore, we can change the second premise to: the computer network is not always in a dangerous state. It can be assumed that the communication established between the Leader and the other followers is safe, although node downtime and network partitions still occur, they can be viewed as under control.

3.2 Proposal Process

Each operation of the client that can be performed by the state machine on the server is called a **Proposal**. A complete Proposal process usually consists of an event request (Invocation, hereinafter referred to as Inv) and an event response (Response, hereinafter referred to as Res). A request contains an operation with the type Write or Read, and the non-read-only type Write is finally submitted by the state machine.

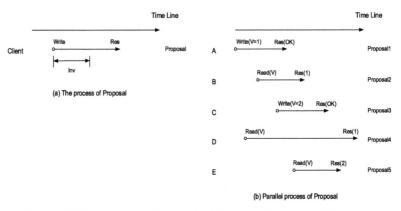

Fig. 3. (a) The process of a Proposal. (b) The parallel process of Proposals.

Figure 3(a) shows the process of a Proposal from client A from initiation to response. From the perspective of Raft, a system that meets linear consistency needs to achieve the following points:

- The submission of Proposal may be concurrent, but the processing is sequential, and the next Proposal can be processed only after a Proposal returns a response.
- The Inv operation is atomic.
- Other proposals occur between the two events of Inv and Res.
- After any Read operation returns a new value, all subsequent Read operations should return this new value.

Figure 3(b) is an example of parallel client requests with linear consistency in Raft. For the same piece of data V, the client A to E initiates a parallel Read/Write request at a certain moment, and Raft receives the Proposal in Real-Time order. As shown in the figure, the request satisfies the following total order relationship:

$$P = \{A, B, C, D, E\} \tag{1}$$

$$R = \{< A, B >, < B, C >, < C, E >, < A, D >, < D, C >\} \tag{2}$$

The $V = 1$ that A initiates the write is successfully written in the Inv period. At this time, B initiates the read between Inv and Res, then $V = 1$ will be read if it can, so as to C and E. The read operation of D is after A and before C, then the value read by D at this time is the data of Inv initiated by A, and $V = 1$ will be returned.

3.3   The Proposed Improved Raft Algorithm

Raft's linear semantics causes client requests to eventually turn into an execution sequence that is received, executed, and submitted sequentially, regardless of the concurrency levels of requests. Under a large number of concurrent requests, two problems will arise. 1. The Leader must process the proposal under the Raft mechanism, so the Leader is a performance bottleneck. 2. The processing rate is much slower than the request rate. A large number of requests will cause a large number of logs to accumulate and occupy bandwidth for a long time and memory.

Problem 1 can be solved with the Mutil-Raft-Group [4]. Mutil-Raft regards a Raft cluster as a consensus group. Each consensus group will generate a leader. Different leaders manage different log shards. In this way, the Leader's load pressure will be evenly divided among all consensus groups, thus preventing the Raft cluster's single Leader from becoming an obstacle. In this paper, we focus on how to solve problem 2.

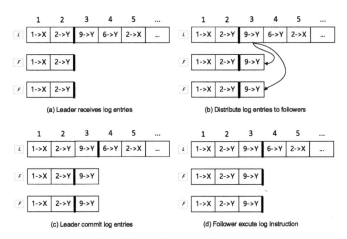

Fig. 4. Log entry commit process

Each proposal will be converted into a log that can be executed by the state machine, as shown in Fig. 4. When the leader node's consistency module receives the log, the

Leader first appends the log to the log collection and then distributes the log items through the RPC method AppendEntries to the remaining follower nodes. Regardless of conditions such as network partition and downtime, the follower node will also copy the log items to its log collection after receiving the request and reply to the leader node ACK to indicate a successful Append. When the Leader receives more than half of the Followers' ACK message, the state machine will submit the log, and the ACK will be sent to other Follower nodes to submit, thereby completing a cluster log submission.

In a highly concurrent scenario, the log items to be processed can be understood as an infinitely growing task queue. The Leader continuously sends Append Entries RPC messages to Follower and waits for half of the nodes to respond. The growth rate of this queue is much greater than that of the submittal time of a log. In this log synchronization mode, consider that the network jitter and packet loss occurs, more logs will be affected, which dramatically impacts system throughput.

Based on the TCP protocol's sliding window mechanism, when multiple consecutive Append Entries RPCs are initiated, the Leader essentially establishes a TCP relationship with the Follower and initiates multiple TCP packets. The sliding window mechanism allows the sender to send multiple packets consecutively before stop-and-wait confirmation instead of stopping to confirm each time a group is sent. The window size determines the number of data packets that can be sent, and when the window is full, the wait will be delayed. The delayed waiting of many TCP data packets will lead to the appearance of LFN (long fat network), which will make the data packets timeout and retransmit. Useless retransmissions generate a lot of network overhead. If the window is large enough, the response can be correctly received by sending multiple data packets continuously and not being retransmitted. If other network overheads are not counted, the network throughput is equivalent to the amount of data transmission per second.

Based on this theory, the synchronous wait of continuous Append Entries is changed to asynchronous in our proposed method so that subsequent ACKs will not be blocked and the network throughput can be improved. However, due to the impact of operating system scheduling during asynchronous callbacks, the message sequence of asynchronous processing may be inconsistent, and direct asynchronous submission may lead to log holes. The solution to this problem is: when the Leader's continuous Heartbeat confirmation can be responded to in time, the network is considered smooth. When an out-of-order sequence occurs, it is within the controllable range, as the logs before the out-of-order log will eventually appear at a certain point in the future. For out-of-order sequences due to scheduling problems, we only need to wait and submit them in order again. If the network fails and is partitioned, the TCP mechanism also ensures that the messages will not be out-of-order.

On this asynchronous basis, the batch is used for log processing. For this reason, we introduce a pre-Proposal stage is to pre-process concurrent Proposals. The Pre-proposal stage is between the client-initiated Proposal and the Leader's processing the Proposal. During this period, a highly concurrent synchronization queue is used to load the Proposal in the order of FIFO (First In First Out). After the Leader starts to process the Proposal, it will sequentially take out the Proposal from the synchronization queue until it encounters the first read-only request in the queue. Then a replica state machine is constructed that is the same as the local state machine. In the replicated state machine, non-read-only logs are submitted in batches, and snapshots are extracted, asynchronous RPCs are sent to make other Follower nodes install snapshots. When more than half of the nodes' ACK responses are received, the replicated state machine is used to replace the original state machine. In order to ensure the consistent reading of the Raft, it is necessary to ensure that the write request has been executed before a read request is executed. For this reason, the synchronization queue needs to be blocked, and the read-related Proposal is processed separately until the next read request. In scenarios that there are more writes than reads, the throughput could be improved more significantly.

4   Experiments and Analysis

The experimental environment is as follows: The server host has 32 GiB of memory, the CPU is Intel Xeon (Cascade Lake) Platinum 8269CY 2.5 GHz with 8 cores. The proposed algorithm is run in the virtual container of this server, 3 nodes are simulated, with each node specifies 4 GiB memory and 2 CPU cores, the operating system is CentOS, and the program code is programmed in Java.

In order to evaluate the efficiency of the improved Raft algorithm, a comparison experiment with traditional Raft [9] was conducted, and the following two aspects were evaluated: 1. The time it takes to process the same level of Proposal before and after the improvement; 2. The impact on the system throughput before and after the improvement.

Multithreading was used to send concurrent requests. In total 17 sets of experiments were carried out for comparison, with different request concurrency levels: from 1000 log entries to up to 13000 log entries. The final results are shown in Fig. 5, Fig. 6 and Table 1.

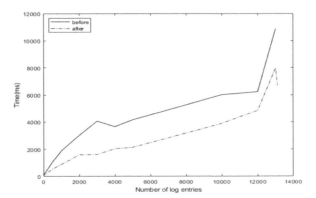

Fig. 5. Performance comparison on the process time of with different number of log entries.

With the increase of concurrency level, the program will inevitably meet the processing bottleneck, that is to say, the point when the program processing speed is far less than the task increments. Figure 5 shows that the bottleneck is around the log concurrency of 12000. If the request number is more than this, the processing capacity of both algorithms will decrease exponentially. Before the bottleneck, it can be clearly seen

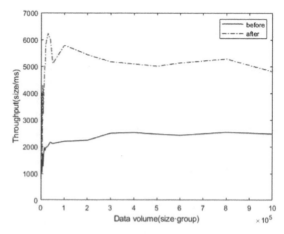

Fig. 6. Performance comparison on throughput of with different size of data volume.

that the proposed algorithm can guarantee more than 20% improvement compared with the traditional algorithm. Even after the bottleneck, the proposed algorithm's process time can adjust to stable because the introduction of the batch process helps alleviate the concurrent task queue. On the contrary, due to the log backlog and task accumulation, the traditional algorithm's processing time will always stay at an exponentially growing trend.

Figure 6 shows that with the increase in the amount of processing data, the throughput of the proposed algorithm system can always be higher than that of the traditional algorithm thanks to the batch processing. Due to many limitations of hardware and software systems, such as the number of disk manipulators, the number of CPU cores, file systems, etc., this improvement is foreseeable to have some limits. Nonetheless, the throughput can be stably guaranteed to be more than two times that of the original algorithm.

Table 1. Performance improvement rate of the optimized algorithm

Improvement rate	Number of log entries(size/ms)						
	1000	2000	4000	5000	10000	12000	13000
Proposal process	0.537	0.472	0.442	0.483	0.353	0.22	0.269
Throughput	1.62	1	1.238	0.592	1.58	1.354	1.353

Table 1 records the improvement rate of the improved algorithm in system throughput and log processing time. It can be seen that the proposed algorithm can at least double the system throughput, and the processing time of the client requests can also be increased by more than 20%.

5   Conclusion

In this paper, the distributed consensus problem is optimized with an improved Raft algorithm. The traditional Raft algorithm executes a client's request to meet linear consistency with sequential execution and sequential submission, which has great impact on performance. In this paper, we intoduces asynchronous and batch processing methods in the pre-Proposal stage to accelerate the processing time and system throughput. After the log submission, snapshot compression of the logs is sent in the sequential queue. Since the network response time is much shorter than the memory calculation, the throughput can be greatly promoted. Experimental results show that this method can increase the system throughput by more than 2 to 3.6 times, and the parallel request processing efficiency can also be increased by more than 1.2 times, which can improve the efficiency of the algorithm while ensuring the correct operation of the algorithm.

Acknowledgments. This research was funded by the National Natural Science Foundation of China (NSFC, Grant No. 61906140, 61705170), the NSFC-CAAC Joint Fund (Grant No. U1833119), and Natural Science Foundation of Hubei Province (Grant No. 2020CFA063).

References

1. Kleppmann, M.: A Critique of the CAP Theorem. arXiv:1509.05393 (2015)
2. Brewer, E.: Spanner, TrueTime and the CAP Theorem (2017)
3. Huang, D., Liu, Q., Cui, Q., et al.: TiDB: a Raft-based HTAP database. Proc. VLDB Endowment **13**(12), 3072–3084 (2020)
4. Taft, R., Sharif, I., Matei, A., et al.: Cockroachdb: the resilient geo-distributed SQL database. In: Proceedings of the 2020 ACM SIGMOD International Conference on Management of Data, pp. 1493–1509 (2020)
5. Huang, D., Ma, X., Zhang, S.: Performance analysis of the raft consensus algorithm for private blockchains. IEEE Trans. Syst. Man Cybernet. Syst. **50**(1), 172–181 (2020)
6. Mingxiao, D., Xiaofeng, M., Zhe, Z., Xiangwei, W., Qijun, C.: A review on consensus algorithm of blockchain. In: IEEE International Conference on Systems, Man, and Cybernetics (SMC), (Banff, AB, Canada), pp. 2567–2572 (2017)
7. Wang, G., et al.: Building a replicated logging system with Apache Kafka. Proc. VLDB Endow. **8**(12), 1654–1655 (2015)
8. Van Renesse, R., Altinbuken, D.: Paxos made moderately complex. ACM Comput. Surv. **47**(3), 36 (2015)
9. Ongaro, D., Ousterhout, J.: In search of an understandable consensus algorithm. In: 2014 USENIX Annual Technical Conference, pp. 305–319 (2014)
10. Frömmgen, A., Haas, S., Pfannemüller, M., et al.: Switching ZooKeeper's consensus protocol at runtime. In: 2017 IEEE International Conference on Autonomic Computing (ICAC), pp. 81–82 (2017)

11. https://github.com/tikv/tikv
12. Ailijiang, A., Charapko, A., Demirbas, M.: Consensus in the cloud: Paxos systems demysti-fied. In: 25th International Conference on Computer Communication and Networks (ICCCN), pp. 1–10 (2016)
13. Lamport, L.: Time, clocks, and the ordering of events in a distributed system. In: Concurrency: The Works of Leslie Lamport, New York, USA, pp. 179–196 (2019)

Distributed Heterogeneous Parallel Computing Framework Based on Component Flow

Jianqing Li[1,2(✉)], Hongli Li[2], Jing Li[3], Jianmin Chen[3], Kai Liu[3], Zheng Chen[3], and Li Liu[3]

[1] Science and Technology on Electronic Information Control Laboratory, Chengdu, China
`lijq@uestc.edu.cn`
[2] School of Electronic Science and Engineering, University of Electronic Science and Technology of China, Chengdu, China
[3] Chengdu Haiqing Technology Co., Ltd., Chengdu, China
`{lijing,chenjm,liukai,chenzheng,liuli}@cdhaiqing.com`

Abstract. Single processor has limited computing performance, slow running speed and low efficiency, which is far from being able to complete complex computing tasks, while distributed computing can solve such huge computational problems well. Therefore, this paper carried out a series of research on the heterogeneous computing cluster based on CPU+GPU, including component flow model, multi-core multi processor efficient task scheduling strategy and real-time heterogeneous computing framework, and realized a distributed heterogeneous parallel computing framework based on component flow. The results show that the CPU+GPU heterogeneous parallel computing framework based on component flow can make full use of the computing resources, realize task parallel and load balance automatically through multiple instances of components, and has the characteristics of good portability and reusability.

Keywords: CPU-GPU heterogeneous processors · Component flow · Multicore multiprocessor · Radar signal processing

1 Introduction

High performance computing (HPC) is the basic technology of information technology, and the key technology to promote information networking. With the diversified development of chip technology, there are so many kinds of high-performance processors, including CPU, GPU, MIC, FPGA, etc.. Each of these processors is suitable for different application scenarios or algorithms [1, 2]. The current simple computing mode of single processor can not meet the complex work requirements [3]. In order to improve the hardware processing capacity, we usually take CPU as the main control and connect GPU, MIC, FPGA and CPU through PCIE bus to accelerate the computing tasks, that is, the heterogeneous computing mode of CPU+X. Among them, the heterogeneous computing mode of CPU+GPU is the most mature and has the best performance [4]. The peak performance of NVIDIA Tesla V100 GPU reaches 15TFlops. Compared with

© The Author(s) 2022
Z. Qian et al. (Eds.): WCNA 2021, LNEE 942, pp. 437–445, 2022.
https://doi.org/10.1007/978-981-19-2456-9_45

the traditional CPU, the GPU-accelerated server can improve the calculation speed by dozens of times under the same computational accuracy [5, 6]. Therefore, this paper studies the heterogeneous computing cluster of CPU+GPU. However, the heterogeneous computing of CPU+GPU brings two new problems [7, 8], including distributed computing resource scheduling strategy and task scheduling strategy between CPU and GPU. For these two problems, we can use multi-core multi processor to solve [9]. The full application of multi-core and multi-processor involves multi-core resource scheduling, multi-task scheduling, inter-processor communication, load balancing, etc.. Optimal scheduling of parallel tasks on multiple processors has been proven to be NP-hard [10]. TDS (Task Duplication Scheduling) [11] divides all tasks into multiple paths according to the dependency topology, and the tasks on each path are executed as a group on one processor. Although this method reduces the delay and shortens the running time, it will increase the energy consumption. In addition, the hardware structure, application and development mode of CPU and GPU processor are different, resulting in poor portability [12]. Sourouri [13] used a simple 3D 7-point stencil computation and statically partition the suitable workload between CPU and GPU to show 1.1–1.2 times of acceleration. Pereira [14] demonstrated a simple static load balancing between CPU and GPU on a single template application, showing up to 1.28 acceleration. Then, Pereira [15] used time tiling on the same pskel framework to reduce the communication requirements between CPU and GPU, but increased redundant computing. Most of them use static load balancing, only consider a single (often repeated) mold, it is difficult to extend to larger applications, with poor reusability.

In view of the above contents, this paper researches on component flow, multi-core multi processor and real-time computing process. Firstly, based on the model of component flow, the model and function of components and component flow suitable for CPU and GPU heterogeneous parallel computing are determined. Then, based on multi-core and multi processor, the task scheduling strategy, data distribution strategy and multi-core parallel strategy are explored. Finally, on the basis of radar signal level simulation, the CPU+GPU heterogeneous computing framework system based on the simulation model is proposed and verified. The results show that the CPU+GPU heterogeneous framework based on component stream can make full use of the computing resources of heterogeneous multiprocessors, improve the computing speed and efficiency of radar signal simulation, realize the automatic distribution and load balancing on multiple computers through components, and has the characteristics of good portability, strong reusability and fast computing speed.

2 Component Flow Model

2.1 Component Flow Model

Developing algorithms directly on CPU and GPU processors will lead to poor reusability and portability of algorithms. Therefore, this paper studies the model based on component flow to realize the algorithm reuse. A component is an abstract model of a computing function, as shown in Fig. 1. The numbers on the left and right represent the serial numbers of the input and output ports respectively. The component model also includes initialization function and processing function, which are automatically called when initialization and data arrive, respectively. Component container is a process running on CPU, which is responsible for data communication between processors, dynamic loading and initialization of local components, and providing versions of operating system. The component flow diagram defines the data flow and temporal relationship between components, and realizes the specific algorithm logic. As shown in Fig. 2, the component flow diagram of an application is used to configure the data input and output relationships and data distribution rules among multiple components, and to configure the resources of each component. Each output port can choose data distribution rules as broadcast, equalization or assignment. Each component can be set to run one or more instances. If there are multiple instances, the number of instances will be adjusted adaptively and dynamically according to the running conditions of components, so as to realize data parallel and load balancing among multiple instances of the same component.

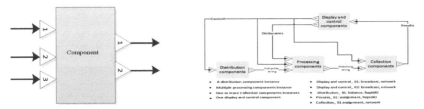

Fig. 1. Component diagram.   Fig. 2. Component flow diagram of an application.

2.2   Task Scheduling Strategy for Multi-core and Multi Processor

The composition of multi-core multi processor task scheduling framework is shown in Fig. 3.

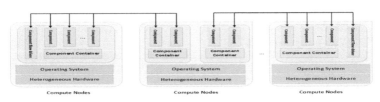

Fig. 3. Multi-core multi processor task scheduling framework

The framework consists of three parts: component flow management software, component container software and component. In Fig. 3, the same filling color belongs to the same component flow task, and the system supports multiple tasks running at the same time. The operation of a component flow needs a component flow driver software for overall control and management, to achieve component flow analysis, resource application and component control. The components in the same component flow are controlled by a component container on a computing node to realize the functions of component loading, task splitting, data distribution, component calling, etc., which will not increase the traffic and delay. In the framework of component-based parallel computing, there are three cases to use multi-core: different cores run different serial component instances, different cores run multiple instances of the same serial component, and multi-core parallelism within a component. In view of the above two cases, CPU establishes thread pool through multitasking for multi-core parallel processing. GPU realizes the data transmission between CPU and GPU through multi thread and multi stream, and improves the processing efficiency of GPU through parallelism.

According to the number of two adjacent components and the data distribution strategy of the output port of the previous component, there are the following few scenarios: 1-to-1, 1-to-N broadcast, 1-to-N balance, N-to-1, M-to-N balance, and N-to-N balance, etc. Some data distribution scenarios are shown in Fig. 4. There are three kinds of location relationships between the two components: running on different processors, loaded by the same process, and running on different cores. Therefore, there are three communication modes: network communication, in-process communication and inter core communication. The priority order is in-process, inter core and network.

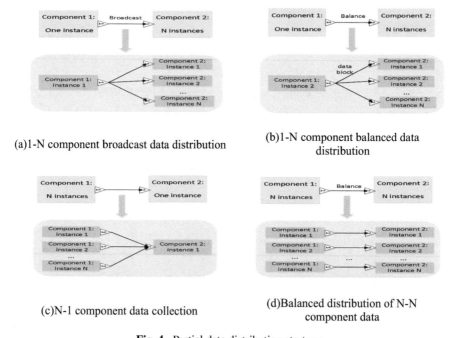

(a)1-N component broadcast data distribution

(b)1-N component balanced data distribution

(c)N-1 component data collection

(d)Balanced distribution of N-N component data

Fig. 4. Partial data distribution strategy

3 Component Flow Framework

The component flow framework and its deployment are shown in Fig. 5, including hardware platform, distributed computing platform and application layer.

Hardware platform includes heterogeneous hardware layer and the operating system layer above it. The former is composed of CPU and GPU processors. The latter runs on

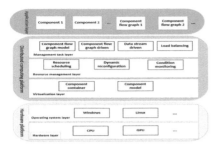

Fig. 5. The component flow framework

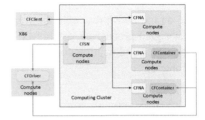

Fig. 6. System composition.

CPU processor and can be windows and Linux operating system. Distributed computing platform includes three parts. The virtualization layer shields the influence of the hardware platform on the components through the component model, which makes the processor hardware universal and simple, and automatically realizes the dynamic component reconfiguration and multi-core parallel. The resource management layer is responsible for the monitoring, scheduling and management of CPU and GPU resources. It abstracts CPU and GPU processors into unified resource pools to achieve automatic deployment, automatic startup, dynamic monitoring and dynamic optimization of resources. Task management layer is responsible for task scheduling and management. It analyses the configuration of component flow graph, applies for computing resources from resource management layer, calls processing functions for real-time parallel computing, and achieves load balancing among multiple instances of the same component. The application layer is the user component developed for users or the component flow diagram used in the actual scene.

The system composition is shown in Fig. 6. The computing cluster is composed of multiple computing nodes to realize the visual monitoring of resource status. CFSM is the system management module. The function is to summarize the resource information of all computing nodes, realize component management, provide component upload, download, delete functions, and provide component flow operation record storage function. CFNA is the node agent module. The function is to manage the component container on the node, collect the resource information of the node and report to CFSM. Cfdriver is component flow driver. It has four functions: (1) parsing component flow and applying for computing resources from CFSM, (2) Deploy the components in the component flow to the applied computing nodes -- start cfcontainer, (3) Build the data transfer network between each cfcontainer and start the component flow calculation, (4) Monitor the running status of component flow. Cfcontainer is the component container. The functions are: loading and initializing components, receiving data and calling component processing functions, uploading the status of each component to cfdriver regularly. Cfclient is the system client. The functions are: (1) provides cluster status monitoring interface, (2) Provide component management function, users can upload, download or delete components in the interface, (3) The component flow operation monitoring function can view the real-time operation record or history record of component flow information.

4   Results Analysis

Based on the above research, the framework based on component flow is applied into a radar signal processing, as showed in Fig. 7, which included the display and control component, amplification component, IQ component and sampling component, and so on (Table 1).

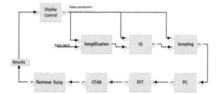

Fig. 7. Flow diagram of radar signal processing.

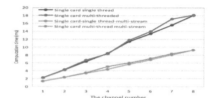

Fig. 8. Performance of multi-channel data processing mode.

Table 1. Performance results of each sub algorithm. (4096 points for segmented FFT transform)

Pulse numbers	IF	IQ	A/D	PC	FFT	CFAR
4	0.127	0.21	0.047	0.116	0.071	0.045
8	0.159	0.233	0.046	0.12	0.071	0.045
16	0.154	0.416	0.047	0.119	0.07	0.047
32	0.308	0.733	0.047	0.115	0.071	0.048
64	0.565	1.353	0.064	0.157	0.078	0.049
12	1.066	2.635	0.105	0.283	0.142	0.048

The performance test results of each sub algorithm in Fig. 7 are shown in Fig. 1. The performance index is the time from the beginning to the end of each sub algorithm process, and the total number of cycles is 10000 (unit: ms). This paper tests the performance of four modes: single card single thread, single card multi thread, single card single thread multi stream, and single card multi thread multi stream, as shown in Fig. 8. For convenience, each data channel takes an input signal of the same length (16 pulses). The performance index is the time from the beginning to the end of all channel data processing, including interface function initialization, input signal data transmission to the video memory, signal process processing, and processing results transmission back to the host memory. Loop "input+process+output" code for 10000 times, and count the average performance. As a comparison, the performance of single channel data cycle test is 2.093 ms.

It can be seen from Fig. 8 that the final performance of using stream mode is better than that of not using stream mode, which indicates that the underlying hardware working mechanism of GPU plays a decisive role in the performance of data processing. The performance of single card single thread mode and single card multi thread mode is almost the same, because when there is no stream mode, API calls use the default null stream, and all CUDA operations in the stream are executed in sequence. When using stream mode, it is faster than using the default null stream, which should be related to the performance improvement of the non pageable memory of the host matching the asynchronous data transmission of the stream. The performance of single thread multi stream is almost the same as that of multi thread multi stream. This is because each step of "input+processing+output" in multi stream test is called asynchronously, so it will not significantly affect the delivery efficiency of related CUDA operations. However, the performance of the latter is slightly better than that of the former, because it is always more efficient for multi CPU threads to compute and deliver CUDA operation commands to the stream.

5   Conclusion

In order to improve the speed and efficiency of the computer, the research is carried out on the CPU+GPU heterogeneous computing cluster. This paper studies the component flow model, uses multi-core multi processor to achieve the dynamic scheduling of tasks, and builds a heterogeneous computing framework system of radar multi signal real-time simulation. This paper abstractly separates the algorithm from the specific hardware environment and operating system through components and component flow, which adapts to the different processor types of CPU and GPU, and realizes the scalability and reconfiguration of the system. The results show that the CPU+GPU heterogeneous framework based on component flow can make full use of heterogeneous multiprocessor computing resources, improve simulation efficiency, and has the characteristics of good portability and reusability.

Acknowledgements. This work was supported by Science and Technology on Electronic Information Control Laboratory Program (Grand No. 6142105190310) and Sichuan Science and Technology Program (Grand No. 2020YFG0390).

References

1. Asano, S., Maruyama, T., Yamaguchi, Y.: Performance comparison of FPGA, GPU and CPU in image processing. In: International Conference on Field Programmable Logic and Applications, pp. 126–131 (2009)
2. Segal, O., Nasiri, N., Margala, M., Vanderbauwhede, W.: High level programming of FPGAs for HPC and data centric applications. In: IEEE High Performance Extreme Computing Conference (HPEC), pp. 1–3 (2014)
3. Dittmann, F., Gotz, M.: Applying single processor algorithms to schedule tasks on reconfigurable devices respecting reconfiguration times. In: Proceedings 20th IEEE International Parallel & Distributed Processing Symposium, p. 4 (2006)
4. Paik, Y., Han, M., Choi, K.H., Kim, M., Kim, S.W.: Cycle-accurate full system simulation for CPU+GPU+HBM computing platform, International Conference on Electronics, Information, and Communication (ICEIC), pp. 1–2 (2018)
5. Rai, S., Chaudhuri, M.: Improving CPU performance through dynamic GPU access throttling in CPU-GPU heterogeneous processors. In: 2017 IEEE International Parallel and Distributed Processing Symposium Workshops (IPDPSW), pp. 18–29 (2017)
6. Di, Y., Weiyi, S., Ke, S., Zibo, L.: A high-speed digital signal hierarchical parallel processing architecture based on CPU-GPU platform. In: IEEE 17th International Conference on Communication Technology (ICCT), pp. 355–358 (2017)
7. Wei, C.: Research on Key Technologies of large scale CFD efficient CPU/GPU heterogeneous parallel computing. University of Defense Science and Technology (2014)
8. Dev, K., Reda, S.: Scheduling challenges and opportunities in integrated CPU+GPU processors. In: 2016 14th ACM/IEEE Symposium on Embedded Systems For Real-time Multimedia (ESTIMedia), pp. 1–6 (2016)
9. Kirk, D.B.: Multiple cores, multiple pipes, multiple threads - do we have more parallelism than we can handle? In: IEEE Hot Chips XVII Symposium (HCS), pp. 1–38 (2005)
10. Jingui, H., Jianer, C., Songqiao, C.: parallel task scheduling in network cluster computing system. Acta Comput. Sin. 27(6), 765–771 (2004)
11. Zaharia, M., Borthakur, D., Sen Sarma, J., Elmeleegy, K., Shenker, S., Stoica, I.: Delay scheduling: a simple technique for achieving locality and fairness in cluster scheduling. In: Proceedings of the 5th European Conference on Computer System, pp. 265–278 (2010)
12. Siklosi, B., Reguly, I.Z., Mudalige, G.R.: Heterogeneous CPU-GPU execution of stencil applications. In: IEEE/ACM International Workshop on Performance, Portability and Productivity in HPC (P3HPC), pp. 71–80 (2018)
13. Sourouri, M., Langguth, J., Spiga, F., Baden, S.B., Cai, X.: Cpu+gpu programming of stencil computations for resource-efficient use of gpu clusters. In: 2015 IEEE 18th International Conference on Computational Science and Engineering, pp. 17–26, October 2015
14. Pereira, A.D., Ramos, L., Ges, L.F.W.: Pskel: a stencil programming framework for cpu-gpu systems. Concurrency and Computation: Practice and Experience, 27(17) (2015)
15. Pereira, A.D., Rocha, R.C.O., Ramos, L., Castro, M., Ges, L.F.W.: Automatic partitioning of stencil computations on heterogeneous systems. In: 2017 International Symposium on Computer Architecture andHigh Performance Computing Workshops (SBAC-PADW), pp. 43–48, October 2017

Design of Multi-channel Pressure Data Acquisition System Based on Resonant Pressure Sensor for FADS

Xianguang Fan, Hailing Mao, Chengxiang Zhu, Juntao Wu, Yingjie Xu, and Xin Wang[(✉)]

School of Aerospace Engineering, Xiamen University, Xiamen 361005, China
xinwang@xmu.edu.cn

Abstract. Resonant pressure sensors have high accuracy and are widely used in meteorological data acquisition, aerospace and other fields. The design and experiment of multi-channel pressure data acquisition system based on resonant pressure sensor, which used for the flush air data sensing(FADS) system, are described. The hardware architecture of DSP and FPGA is applied to the data acquisition system. The digital cymometer and 16-bit analog-to-digital converter are used to measure the output signal of the sensor. It is shown the data acquisition system has favourable performance within the operating temperature range. The maximum experimental error is less than 0.02%FS over the range 2–350 kPa. The period of sampling and fitting is less than 8 ms. The frequency and voltage measurements meet accuracy requirements. The calculated pressure and standard pressure result appears excellent linearity, which reach up to 0.9999.

Keywords: Data acquisition · Resonant pressure sensor · DSP+FPGA · High accuracy

1 Introduction

Atmospheric data parameters include dynamic pressure, static pressure, Mach number, angle of attack, and sideslip angle and other parameters related to the airflow environment of the aircraft during flight [1]. The measurement of atmospheric data is of great significance to the attitude control and structural design of hypersonic vehicles. For example, the design of the air intake and tail nozzle of the aircraft is closely related to the Mach number and the angle of attack. In the overall design of the compression ignition ramjet, the dynamic pressure and the angle of attack are also two important parameters. At present, the measurement of atmospheric data mainly adopts the Flush Air Data Sensing system (FADS) [2], which depends on the design of the pressure sensor array to measure the pressure distribution on the surface of the aircraft head or other local positions, and converts the pressure data through a specific solution algorithm mode 1 [3]. Measure and obtain atmospheric parameters during flight (Fig. 1).

The FADS system mainly uses IPT (Integrated Pressure Transducer) to obtain incoming flow pressure data. IPT is a MEMS pressure sensor, and its working principle has

© The Author(s) 2022
Z. Qian et al. (Eds.): WCNA 2021, LNEE 942, pp. 446–455, 2022.
https://doi.org/10.1007/978-981-19-2456-9_46

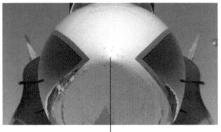

Pressure measuring hole

Fig. 1. Pressure measuring hole for FADS on aircraft nose

undergone the evolution process of piezoresistive, capacitive and resonant [4–6]. The IPT of Honeywell of the United States integrated a piezoresistive pressure sensor with both pressure and temperature sensitive components. It was smart and had an accuracy of 0.03% FS. It was also equipped with EEPROM for the storage of the correction factor of the sensor, without additional pressure and temperature calibration [7]. The accuracy of the pressure sensor integrated in the ADP5 five-hole PTV tube of Simtec Buergel AG in Switzerland was up to 0.05% FS, but it was not calibrated at high Mach numbers. The temperature compensation range was $-35\,°C$–$+55\,°C$. At $-40\,°C$–$+70\,°C$, the performance would decrease. A resonant pressure sensor was integrated in an air data test instrument of GE DRUCK, which had an accuracy of 0.02% FS and an operating temperature of $0\,°C$–$50\,°C$.

With the continuous development of modern aircraft in the direction of high maneuverability and hypersonic speed [8], it is necessary to obtain more accurate atmospheric data parameters during a wider temperature range. So we chosen the resonant pressure sensor. The resonant pressure sensor measures pressure indirectly by detecting the natural frequency of the object [9]. It has the characteristics of high sensitivity and high accuracy, and is suitable for calculation of atmospheric data in flight tests [10].

In order to further study the FADS system, the pressure measurement is required to achieve a stable accuracy of 0.02%FS over the full operating temperature range ($-40\,°C$–$+80\,°C$) and the calculation time of pressure should less than 10 ms. This paper has designed a multi-channel pressure data acquisition system based on a self-developed silicon resonant pressure sensor and a hardware architecture scheme of DSP and FPGA. The data acquisition system shows excellent performance on the ground experimental platform.

2   System Structure

The principle of the multi-channel pressure data acquisition system based on resonant pressure sensor is shown in Fig. 2. It mainly consists of power supply module, ADC data acquisition module, main control module and RS422 communication module. The entire acquisition system realizes the preprocessing and acquisition of the output signal of the resonant sensor, the filtering and fitting of data, and the communication function of the host computer.

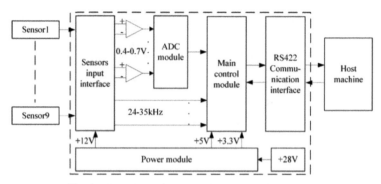

Fig. 2. Overall architecture of the acquisition system

2.1 Sensor

The selected sensor is shown in Fig. 3. Its pressure measurement range is absolute pressure 2 kPa to 350 kPa, working temperature −40 °C to 80 °C. The accuracy and annual stability are better than 0.02%FS. The output signal of the sensor is TTL square wave signal and the voltage signal. TTL square wave signal is related to pressure, and its frequency output range is 25–35 kHz. The voltage signal is related to temperature, and its output range is 400–700 mV. The TTL square wave signal and the voltage signal are fitted into the pressure value through the temperature compensation polynomial (1)

$$P_c = \sum_i^n \sum_j^m C_{ij} f^i V^j \, (n \ge 3, m \ge 2, i = 0 \; to \; n, i = 0 \; to \; m) \tag{1}$$

where Pc is the calculated pressure value, Cij is the fitting coefficient, f is the sensor output frequency, and V is the sensor output voltage [11], m and n are fitting orders, generally, n = 5 and m = 4.

Fig. 3. Resonant pressure sensor and its sensitive core.

2.2 Main Control Module

According to the functional requirements of the data acquisition system, in order to improve the real-time performance of data acquisition and calculation, DSP+FPGA was used as the main control architecture [12]. The structure of the main control module is shown in Fig. 4. The FPGA completes the timing control of the ADC and the frequency measurement of the square wave signal output by the sensor, and the DSP completes the software filtering of the collected data, temperature compensation fitting and RS422 communication with the host computer. This module used TI C674x series 32-bit floating-point DSP. System clock was 456 MHz. The EMIFA bus of the DSP was connected to the FPGA device and FPGA called a dual-port RAM IP core to realize data interaction between FPGA and DSP.

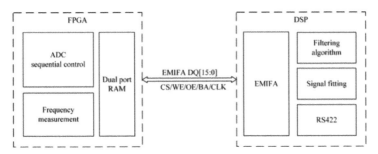

Fig. 4. Main control module.

2.3 Analog-to-Digital Conversion Module

The analog-to-digital conversion uses two 8-channel 16-bit analog-to-digital conversion chips AD7689, which use an external 2.048 V reference voltage. Its input mode is unipolar input. The output voltage signal of the pressure sensor is preprocessed by the two-stage op amplifier and then connected to the analog-to-digital conversion. AD7689 uses a serial port interface and is driven FPGA after passing through a digital isolation chip (Fig. 5).

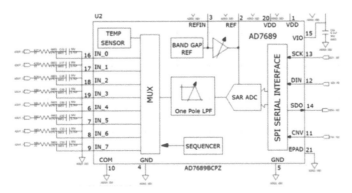

Fig. 5. Analog-to-digital conversion circuit diagram.

3 Software Design

3.1 Principle of Signal Acquisition.

Sensitivity of the sensor is 28.4 Hz/Kpa. In order to ensure the consistency of the measurement accuracy within the output range of the measurement sensor's frequency signal, and eliminate the ± 1 error caused by directly counting the measurement signal, the period method is used to measure the sensor's frequency signal [13, 14]. The principle is shown in Fig. 6. The gating time T is an integer multiple of the measured single fx. The gating time T is Ns clock cycles of fx. The reference clock fs numbered during the gating time T is Nx. Then,

$$T = \frac{N_x}{f_x} = \frac{N_s}{f_s} \tag{2}$$

Ignoring the error of the reference clock itself, the measurement error comes from the ± 1 error generated by counting the reference signal. The relative error σ shows below.

$$\sigma = \frac{1}{T \cdot f_s} \tag{3}$$

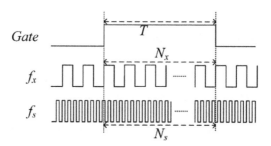

Fig. 6. Principle of frequency acquisition.

When the sampling frequency is 50 Hz, the frequency sampling time should be less than 10 ms. The gating time is 200 clocks of fx, and the reference clock is 50 MHz temperature-compensated crystal oscillator. In the case of sensor output frequency fx = 30000 Hz, we can get:

$$T = \frac{N_x}{f_x} = \frac{200}{30000} = 0.006667 \, s \tag{4}$$

The count value of the reference clock is 333333 or 333334, which converted for 30000.03 Hz or 29999.94 Hz. The error is less than 0.0002%, which meets the measurement requirements.

3.2   Collection Process

The main program flow chart is shown in Fig. 7 below. After the system is powered on, the initialization operation is performed, the DSP enables IO, peripherals, UART and timer modules, and after the host computer collects the command, the FPGA triggers the ADC drive timing, and at the same time starts to measure the frequency, voltage and frequency of the TTL square wave After the measurement is completed, the FPGA writes the data into the dual-port RAM [15], the data writing is completed and the DSP external interrupt is triggered, and the DSP starts to read the data; after the acquisition is completed, the DSP first preprocesses the read data, including data outlier removal and removal After the filtering is completed, the collected signal is converted in the DSP first, and the converted result is brought into the temperature compensation polynomial fitting to synthesize the measured pressure. After the fitting is successful, the DSP sends the data to the RS422 interface. Host computer control system. The DSP completes the calculation in less than 1 ms at the system clock of 456 MHz. Digital cymometer and ADC needs no more than 7 ms. Therefore, a collection calculation period is less than 8 ms, which meets the requirement.

4   Experiments

The multi-channel data acquisition board and host machine is shown in Fig. 8. All channels were connected in parallel to the same sensor for easy connection and testing. In order to verify the acquisition system, a measurement platform was built based on the ground standard pressure source. The test frame is shown in Fig. 9. Pressure controller is a commercial instrument (GE DRUCK PRS8000),which has the accuracy of 0.01%FS. The thermostatic controller (GF ITH-150) is used to stabilize operation environment. After working for 2.5 h, the temperature fluctuation during the measurement is about 0.1 °C. The board's DC power supply is +28 V. The Agilent logic analyzer is used to obtain sensor output parameters. Static measurement is carried out to plot frequency to pressure at different temperatures. The pressure sensor and the board are put inside the thermostatic controller.

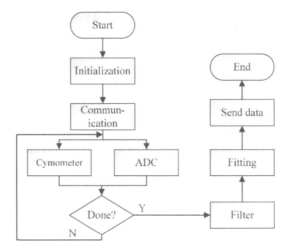

Fig. 7. System acquisition flowchart.

Fig. 8. Multi-channel data acquisition board and host machine.

Fig. 9. Experiment platform. a. DC power; b. Logic analyzer; c. Thermostatic controller; d. Pressure controller; e. Acquisition board; f. Resonant pressure sensor.

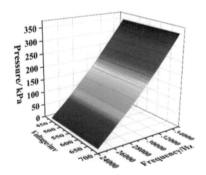

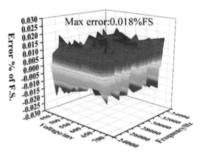

Fig. 10. Fitted pressure surfaces for sensor output frequencies and voltages.

Fig. 11. Full range error under different pressure and temperature points (2 to 350 kPa and −40 to 80 °C).

The setting temperature range of the thermostatic control box is −40 to 80 °C. Pressure sampling is taken every 10 °C for a measuring time of more than 2 h. The data for each point is an average of 100 repeated measurements. The fit of the frequency and voltage is shown in Fig. 10. The uniform surface transition shows that there is a good regularity between the output frequency and the pressure and temperature load. Figure 11 shows the fitting residual. The max error is 0.018%FS, better than 0.02%FS.

The relation between frequency response and applied pressure, which measured at 20 °C, is shown in Fig. 12. The measurement result of the acquisition board is highly in agreement with the performance of the logic analyzer. The frequency error for each measuring point is listed in Fig. 13. The upper and lower margins of error are 0.1718 Hz and −0.0777Hz, which meets the measurement demands of the system.

The system's hysteresis characteristic test curve is shown in the Fig. 14. The forward and reverse fitting results were consistent, which were agreement with the standard pressure. The forward coefficient of determination is 0.999994 and the reverse coefficient of determination is 0.999975.

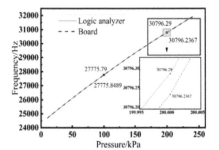

Fig. 12. Frequency under different pressure at room 20 °C.

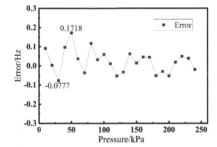

Fig. 13. Error of frequency for each measuring point.

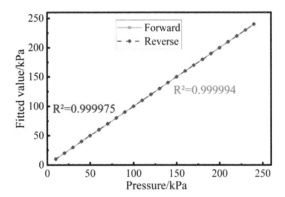

Fig. 14. Forward and reverse fitting results

The coefficient of determination of the 10 repeated experiments is listed in the table below. The coefficient of determination were all better than 0.9999. The exceptional goodness of fit means high measurement accuracy, which indicates our data acquisition system is reliable and stable (Table 1).

Table 1. Coefficient of determination of 10 repeated experiments at room temperature.

Test times	Coefficient of determination(R^2)
1	0.99992
2	0.99994
3	0.99992
4	0.99999
5	0.99992
6	0.99997
7	0.99997
8	0.99997
9	0.99998
10	0.99998

5   Conclusion

This article has demonstrated a multi-channel data acquisition system for measuring the pressure of resonant pressure sensors, whose hardware architecture is based on DSP and FPGA. Digital cymometer and high resolution analog-to-digital converter make the system performed with high measurement accuracy. Experiments showed that the maximum measurement relative error of the sensor output frequency signal is only 0.1718 Hz. The full range error is less than 0.02%FS within the operating temperature range. The measurement is repetitive and there is no hysteresis phenomenon. As such, our multi-channel system is reliable, which can provide accurate data for FADS calculating.

References

1. Angelo, L., Manuela, B.: Safety analysis of a certifiable air data system based on synthetic sensors for flow angle estimation †. Appl. Sci. **11**(7), 3127 (2021)
2. Jiang, X., Li, S., Huang, X.: Radio/FADS/IMU integrated navigation for Mars entry. Adv. Space Res. **61**(5), 1342–1358 (2018)
3. Karlgaard, C.D., Kutty, P., Schoenenberger, M.: Coupled inertial navigation and flush air data sensing algorithm for atmosphere estimation. J. Spacecraft Rockets. **54**, 128–140 (2015)
4. Song, P., et al.: Recent progress of miniature MEMS pressure sensors. Micromachines **11**(1), 56 (2020)
5. Nag, M., Singh, J., Kumar, A., Alvi, P.A., Singh, K.: Sensitivity enhancement and temperature compatibility of graphene piezoresistive MEMS pressure sensor. Microsyst. Technol. **25**(10), 3977–3982 (2019). https://doi.org/10.1007/s00542-019-04392-5

6. Samridhi, M.K., et al.: Stress and frequency analysis of silicon diaphragm of MEMS based piezoresistive pressure sensor. Int. J. Modern Phys. B **33**(07), 1950040 (2019)
7. Hu, B., Liu, X.J.: Design and research of multi-channel temperature calibration system based on the LabVIEW. Adv. Mater. Res. **1362**, 241–246 (2011)
8. Xiaodong, Y., Shi, L., Shuo, T.: Analysis of optimal initial glide conditions for hypersonic glide vehicles. Chin. J. Aeronaut. **27**(02), 217–225 (2014)
9. Radosavljevic, G.J., et al.: A wireless embedded resonant pressure sensor fabricated in the standard LTCC technology. IEEE Sens. J. **9**(12), 1956–1962 (2009)
10. Alcheikh, N., Hajjaj, A.Z., Younis, M.I.: Highly sensitive and wide-range resonant pressure sensor based on the veering phenomenon. Sens. Actuators, A **300**, 111652 (2019)
11. Du Xiaohui, L.W.A.L.: High accuracy resonant pressure sensor with balanced-mass DETF resonator and twinborn diaphragms. J. Microelectromech. Syst. **99**, 1–11 (2017)
12. Haowen, T., et al.: Design and implementation of a real-time multi-beam sonar system based on FPGA and DSP. Sensors **21**(4), 1425 (2021)
13. Pardhu, T., Harshitha, S.: Design and simulation of digital frequency meter using VHDL. In: International Conference on Communications & Signal Processing, pp. 704–710 (2014)
14. Lenchuk, D.V.: Simulation of error analysis in a digital frequency meter for meteorological signals. Telecommun. Radio Eng. **57**(2–3), 18 (2002)
15. Hidaka, H., Arimoto, K.: A high-density dual-port memory cell operation and array architecture for ULSI DRAM's. IEEE J. Solid-State Circuits **27**(4), 610–617 (1992)

Research on Intrusion Detection Technology Based on CNN-SaLSTM

Jiacheng Li[1(✉)], Qiang Du[2], and Feifei Huang[2]

[1] School of Cybersecurity, Chengdu University of Information Technology, Chengdu 610225, Sichuan, China
695791811@qq.com
[2] PetroChina Southwest Oil and Gas Field Communication and Information Technology Center, Chengdu 610051, Sichuan, China

Abstract. As Internet-connected application devices become more and more popular, more and more services need to be done through the network, which also leads to users paying more attention to network security performance. Due to the continuous iterative development of cyber attack means and attack scale, it is difficult to conduct passive security detection systems such as traditional intrusion detection mechanisms to conduct endless attacks. Later, intrusion detection was studied as an active defense technique to compensate for the shortcomings of traditional safety detection techniques. Active defense and response technology has also attracted the attention of researchers at home and abroad. The complex, engineering and large-scale scenarios presented by network attacks prevent the original passive intrusion detection system to meet the users' needs for network security performance. With the continuous expansion of network scale, the continuous increase of network traffic scenarios and the rapid iteration of attack means, the performance of network intrusion detection system has put higher requirements. Therefore, we introduced the CNN, LSTM and self attention mechanisms in deep learning into invasion detection and performed experiments in the tensorflow framework, increasing the accuracy to 97.4%.

Keywords: CNN · LSTM · Self-attention · Intrusion detection

1 Background Introduction

With the continuous development of Internet technology, people also face various security threats while relying on the great convenience of the network. Therefore, network security testing is of great significance to ensuring national security and people's life. How to quickly identify various attacks in real time, especially unpredictable attacks, is an inevitable problem today. Intrusion Detection and Defense Systems (IDS) is an important achievement in information security field. Compared to traditional static security technology [1], such as firewalls and vulnerability scanners, it can identify intrusions that are already occurring or are occurring. The network intrusion detection system [2] is an active cybersecurity defense tool to monitor and analyze key nodes in a network environment in real time and detect for signs of attacks or security violations. Policies in network systems. Behavior and deals with the behavior accordingly. To effectively improve the detection performance of intrusion detection systems in a network environment, many researchers have applied machine learning technology to the research and development of intelligent detection systems. For example, literature [3] applies support vector machines to invasion detection, introduces statistical learning theory into invasion detection studies, literature [4] introduces a naive Bayesian nuclear density estimation algorithm into invasion detection, literature [5] introduces random forest to deal with attack detection disequilibrium and short attack response time. However, most traditional machine learning algorithms are shallow learning algorithms. They aim to emphasize feature engineering and feature selection and do not solve the classification of massive invasive data in actual networks. As the network data grows rapidly, its accuracy will constantly decline. Deep learning [6] is one of the most widely used technologies in the AI field. Many scholars have applied it to intrusion detection and achieved better accuracy. Deep learning is a kind of machine learning. Its concept comes from the study of artificial neural networks. Its structure is actually a multi-layer perceptron with multiple hidden layers. Convolutional neural networks (CNN) require fewer parameters and are well suited to processing data with statistical stability and local correlations. In Ref [7], applying convolutional neural networks to sparse attack type r2l invasion detection improves the u2r detection rate, but requires further improvement on the detection of sparse attack type r2l. Long short-term memory (LSTM) is specifically used for learning time-series data with long dependencies. It has great advantages in learning long-term dependencies and timing in higher advanced feature sequences. Long short-term memory neural network (LSTM) is a special recurrent neural network and is one of the classical deep learning methods. Literature [8] applied LSTM to intrusion detection, effectively solving the problem of gradient disappearance and gradient explosion in data training, and effectively solving the problem of input sequence features. However, the model is still not accurate enough for feature extraction in small and medium-sized datasets. It takes advantage of the advantages of convolutional neural networks in processing locally relevant data and feature extraction, as well as long-and short-term memory neural networks in capturing data sequences and long-term dependencies. Combined with the attention [9] self attention mechanism, it has the advantages of processing the serialized data and classification. In this paper a CNNsalstm based intrusion detection model to further improve accuracy and reduce misuse rate.

2   Related Theories

2.1   Long and Short-term Neural Memory Network

Commonly known as LSTM, is a special RNN [10], that can learn about long dependence. They were introduced by Hochreiter & schmidhuber [11] and improved and popularized by many. They work well on a variety of issues and are now widely used. RNN is good at processing sequence data, but exhibits gradient extinction or gradient explosion as well as long-term dependence in the course of RNN training. The LSTM has been carefully designed to avoid long-term dependence. Keep in mind that long-term historical information is actually their default behavior, not what they are trying to learn. All recurrent neural networks have the form of recurrent module chains of neural networks.

In the standard RNN, repeat modules will have very simple structures, such as a single tanh layer (Fig. 1).

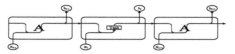

Fig. 1. Single layer neural network with repeated modules in standard RNN

LSTM also has this chain structure, but the structure of the repeat modules is different. Compared to the simple layers of neural networks, LSTM have four layers, which interact in special ways (Fig. 2).

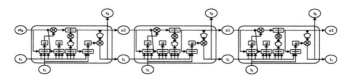

Fig. 2. Four interactive neural network layers included in the repeating module in LSTM

The long, short-term neural memory model actually adds three gates to the hidden layer of the RNN model, namely the input gate, the output gate, the forgetting gate, and a cell state update, as shown in the figure below (Fig. 3).

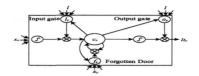

Fig. 3. Long short-term memory module

By forgetting the gate, we screen the cell states in the upper layer, leaving the desired information and discarding useless information. The formula is as follows:

$$f_t = \sigma(w_f * [h_t, x_t] + b_f) \tag{1}$$

They are the weight matrices and bias terms of the forgetting gate, are the activation functions of the sigmoid, and [,] is connecting the two vectors into one vector. The input gate determines the importance of the information and sends the important information to the place where the cell state is updated to complete the cell state update. This process consists of two parts, the first part uses the sigmoid function to determine new information needed to be added to the cell state, and the second part uses the tanh function to general new candidate vectors. The calculation formula is as follows:

$$\begin{cases} f_t = \sigma(w_i * [h_{t-1}, x_t] + b_i) \\ \tilde{c}_t = tanh(w_c * [h_{t-1}, x_t] + b_c) \end{cases} \tag{2}$$

Among them, it is the weight and bias of the input gate, which is the weight and bias of the cell state. After the above treatment, the cell state is updated to the cell state c, formula as follows:

$$c_t = f_t * c_{t-1} + i_t * \tilde{c}_t \tag{3}$$

Among them, * represents multiplied elements, represents deleted information, and * represents new information.

The output gate controls the output of the cell state of the present layer and determines which cell state enters the next layer. The calculation formula is as follows:

$$\begin{cases} o_t = \sigma(w_o * [h_{t-1}, x_t] + b_o) \\ h_t = o_t * tanh(c_t) \end{cases} \tag{4}$$

According to the LSTM network invasion method, the initial detection dataset was first digitized, standardized, normalized, then the preprocessed dataset was input into the trained LSTM model, and finally the results into the softmax classifier to get good classification results. Although the proposed method can extract more comprehensive features and improve the accuracy of network intrusion detection when processing sequence data, the proposed method has a high false alarm rate.

2.2 Convolutional Neural Network

Convolutional neural networks is a hierarchical computational model. As the number of network layers increases, increasingly complex abstract patterns can be extracted. The emergence of convolutional neural networks was inspired by bioprocessing, as the connectivity between neurons is similar to the tissue structure of the animal visual cortex. The typical architecture of CNN is: input the → conv → pool → fullcon, which combines the idea of local receptive fields, shared weights, and spatial or temporal subsampling. This architecture makes CNN well-suited for processing data with statistical stability and local correlations, and makes it highly deformable upon translation, scaling, and tilt. It is a deep feedforward neural network. Each network has a multiple neuron population. Each neuron receives only the upper-layer of the output. After the layer is calculated, the results are output to the next layer. Elements of homric neurons are not connected. The proposed algorithm can obtain the output from a multi-layer network trained with the input data. Convolutional neural network includes input layer, convolutional layer, pooling layer, fully connected layer, and the structure in Fig (Fig. 4).

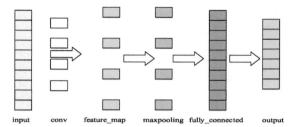

Fig. 4. Convolutional neural network structure

Input Layer. It can be represented as the beginning of the entire neural network. In the field of data processing, the input to convolutional neural networks can be viewed as a data matrix.

Convolutional Layer. As the most important part of the convolutional neural network, each convolutional layer comprises several convolutional units, each of whose parameters are optimized by a backpropagation algorithm. The purpose of the convolution operations is to extract the different features of the input. The first convolutional layer can only extract low-level features such as edges, lines, and angles. More multiple layers of the network can iteratively extract more complex features from low-level features. Convolutional layers perform more thorough analysis of each small block to obtain more abstract features. Convolutional neural networks first extract local features and then fuse local features at a higher level, which can not only obtain global features, but also reduce the number of neuronal nodes. However, the number of neurons is still very large at this time, so by setting the same weight of each neuron, the number of network parameters is greatly reduced. For the m th convolutional layer, its output is y_m, then the output of the Kth convolution kernel is y_m:

$$y_k^m = \delta\left(\sum_{y_i^{n-1}} \in m_k y_i^{m-1} * W_{ik}^m + b_k^m\right) \tag{5}$$

Pooling Layer. You can reduce the size of the data matrix very efficiently. The two most commonly used methods are maximal pooling and average pooling, which further reduce the number of nodes in the fully connected layer. The task of reducing the entire neural network parameters is finally implemented.

Fully Connected Layer and Output Layer. Features of the data were extracted and classified by the full connectivity layer. The output layer completes the detailed prime classification of the risk factors according to the professional type to obtain the probability distribution problem.

2.3 Attention

The attention mechanism was first proposed in the field of image recognition. The idea is that when humans deal with certain things or images, they allocate more energy to specific parts of the key information. Once concentrated, the information can be accessed more efficiently. When processing a large amount of input information, the neural network can also learn from the attention mechanism of the human brain, and select only some key input information for processing, thus improving the efficiency of the neural network. When using neural networks, we can usually encode using convolutional or recurrent networks to obtain an output vector sequence of the same length (Fig. 5).

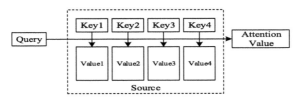

Fig. 5. The essence of the Attention mechanism: addressing

The essence of the attention mechanism is an addressing process [12], as shown above: given a task-related query vector Q, calculates the attention value by calculating the attention distribution of the key and attaching it to the value. This process is actually the embodiment of the attention mechanism in reducing the complexity of the neural network model: there is no need to input all the N input information into the neural network for calculation. Simply select some task-related x information and input it into the neural network. The attention mechanism can be divided into three steps: one is the information input; the other is to calculate the attention distribution α; three is the attention distribution α, used to calculate the weighted average of the input information. When using neural networks, we can usually encode using convolutional or recurrent networks to obtain an output vector sequence of the same length, as shown in Fig (Fig. 6):

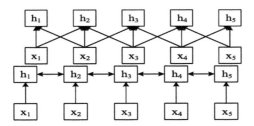

Fig. 6. Variable length sequence coding based on convolutional network and recurrent network

As can be seen from the figure above, both convolutional and recurrent neural networks are actually "local coding" for the variable length sequence: the convolutional neural network is obviously based on n-gram local coding; for recurrent neural networks, short-range dependence can be established only due to the disappearance of the gradient (Fig. 7).

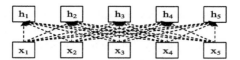

Fig. 7. Self-attention model

In this case, we can use attention mechanisms to generate weights for different connectivity "dynamics". This is the self-attention model. Since the weights of the self attention model are dynamically generated, the longer information sequence can be processed. Overall, why are self-attention models so powerful: attention mechanisms are used to "dynamically" generate weights of different links to process longer sequence of information. The self-attention model was calculated as follows: Let $X = [x1, \cdots, xN]$ represent N input information; obtain the query vector sequence, key vector sequence and value vector sequence through linear transformation:

$$Q = w_Q X \quad K = w_K X \quad V = w_V X \tag{6}$$

From the above formula, Q in self-Attention is a transformation of self-input, and attention calculates the formula as:

$$h_i = att((K, V), q_i)$$

$$= \sum_{j=1}^{N} a_{ij} v_j$$

$$= \sum_{j=1}^{N} softmax(s(k_j, q_j)) v_j \tag{7}$$

In self-attention models, the scaled dot product is usually used as a function of attention scoring, and the output vector sequence can be written as:

$$H = V \; softmax(x = \frac{K^T Q}{\sqrt{d_3}}) \tag{8}$$

2.4   Data Pre-processing

In this paper, the KDD99 [13] dataset is used as our training and test dataset. The dataset is nine-week network connectivity data collected from a simulated USAF LAN, divided into training data with identification information and test data without identification information. The test and training data have different probability distributions. The test data contained some types of attack that did not appear in the training data, which makes intrusion detection more realistic. Each connection in the dataset included 41 functions and 1 attack type. The training dataset contains a normal identification type and 36 training attack types, with training data contains 22 attack patterns, and only 14 attacks in the test dataset (Fig. 8).

Intrusion category	Description	Details
Normal	Normal record	Normal
DOS	Denial of service attack	Back, land, neptune, pod, Smurf, teardrop
Probing	Scanning and detection	Ipswee, ap, portsweep, satan
R2L	Unauthorised remote access	ftp_write, guess_passwd, imap, multihop phf, warezclient, warezmaster
U2R	Illegal access to local super users	Buffer_overflow, loadmodule, perl, rootkit

Fig. 8. Details of five labels

TCP basic connection characteristics (nine kinds) basic connection characteristics include basic connection attributes, such as continuous time, protocol type, number of transmitted bytes, etc. TCP connection content features (13 kinds in total) are extracted from the content features that may reflect intrusion data, such as the number of login failures. Network statistics have time-based traffic (9 kinds, from 23 to 31). Due to the strong temporal correlation of network attack events, there is a certain connection between the current connection records and the previous connection records. Statistical calculation can better reflect the relationship between connections. Host based network

Description	Feature	Data attributes
Basic feature of individual TCP connections	Duration	continuous
	protocol_type	symbolic
	service	symbolic
	flag	symbolic
	src_bytes	continuous
	dst_bytes	continuous
	land	symbolic
	wrong_fragment	continuous
	urgent	continuous
Content feature within a connection suggested by domain knowledge	hot	continuous
	num_failed_logins	continuous
	logged_in	symbolic
	num_compromised	continuous
	root_shell	continuous
	su_attempted	continuous
	num_root	continuous
	num_file_creations	continuous
	num_shells	continuous
	num_access_files	continuous
	num_outbound_cmds	continuous
	is_host_login	symbolic
	is_guest_login	symbolic
	count	continuous
	srv_count	continuous
	serror_rate	continuous
	srv_serror_rate	continuous
	rerror_rate	continuous
	srv_rerror_rate	continuous
	same_srv_rate	continuous
	diff_srv_rate	continuous
	srv_diff_host_rate	continuous
Traffic features computed in and out a host	dst_host_count	continuous
	dst_host_srv_count	continuous
	dst_host_same_srv_rate	continuous
	dst_host_diff_srv_rate	continuous
	dst_host_same_src_port_rate	continuous
	dst_host_srv_diff_host_rate	continuous
	dst_host_srv_serror_rate	continuous
	dst_host_srv_serror_rate	continuous
	dst_host_rerror_rate	continuous
	dst_host_srv_rerror_rate	continuous

Fig. 9. Details of forty one features

traffic statistics (32–41 in total) time based traffic statistics only show the relationship between the current connection and the last two seconds, as shown in the following figure (Fig. 9). Original intrusion data record: x = {0, icmp, ecr_i, SF, 1032, 0, 0, 0, 0, 0, 0, 0, 0, 0, 0, 0, 0, 0, 0, 0, 0, 0, 0, 0, 511, 511, 0.00, 0.00, 0.00, 0.00, 1.00, 0.00, 0.00, 255, 255, 1.00, 0.00, 1.00, 0.00, 0.00, 0.00, 0.00, 0.00, smurf} There are 41 functional parts and a label.

2.5 Character Numeric

First, we should remove the duplicates. In the actual data collected, many intrusion records are the same, so the deduplication technology [14] can be used to reduce the amount of input ID data and eliminate information redundancy. The KDD99 dataset has been counter processed, and filtering is not required in this paper. However, some functions in the KDD99 dataset are number functions and some are characters. All data captured from different ID sources were then converted into digital format using normalization to simplify data processing. Value rules for symbol features are as follows: Use attribute mapping. For example, property 2 is the protocol type protocol_type. It has three values: TCP, UDP and ICMP, are represented by its location. TCP is 1, UDP 22 and ICMP 3. Similarly, the mapping relationship can establish the relationship between the symbol values and the corresponding values through the 70 symbol values and 11 symbol values used by the attribute element service. Labbel processed as follows (Fig. 10).

Intrusion type	Description	Label
Normal	Normal record	0
Dos	Denial of service attack	1
Probe	Scanning and detection	2
R2L	Unauthorised remote access	3
U2R	Illegal access to local super users	4

Fig. 10. Description of five labels

2.6 Normalization

Because some elements have values of 0 or 1, some values to avoid the influence of large range values, too large; and small effects of the values disappear, need to normalize the value of each feature to convert between [0,1].

$$y = (x - xmin/xmax - xmin) \tag{9}$$

After normalization

x = {0.0, 3.38921626955e−07, 0.00128543131293, 0.0, 0.0, 0.0, 0.0, 0.0, 0.0, 0.0, 0.0, 0.0, 0.0, 0.00195694716243, 0.00195694716243, 0.0, 0.0, 0.0, 0.0, 1.0, 0.0, 0.0, 0.125490196078, 1.0, 1.0, 0.0, 0.03, 0.05, 0.0, 0.0, 0.0, 0.0, 0} (Fig. 11).

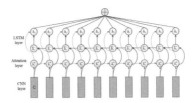

Fig. 11. CNN-SALSTM network structure

3   Model Establishment

3.1   Based on CNN-SALSTM Network Structure

Step 1. Data preprocessing. One-click encoding of network protocols, network service type, and network connection state text type data. Meanwhile, continuous numerical data such as the connection time in the grouping characteristics are normalized according to Eq. 10

$$x_n = \frac{x - x_{min}}{x_{max} - x_{min}} \tag{10}$$

Step 2. Advanced feature extraction. The basic features of the pre-processed packets are sent to lenet for advanced feature extraction, output advanced features via one-dimensional convolution operations. Each volume layer is followed by a BN layer and leakyrelu activation function to speed up the network and avoid collapse as much as possible.

Step 3. The self-attention mechanism highlights the high-weight features. Based to its upper subvector, each vector multiplied its three matrices WQ, wk and WV generated by its upper subvector to obtain a vector. A vector yields a probability then multiplied by the result of the CNN convolution and passed to the next layer.

Step 4. Classified the network connections. Entering-level features into LSTM, yields the classification results of the network data through the softmax function.

3.2   Evaluation Method

Precision, recall and F-measure were used in this experiment to judge the classification effect of the model. TP represents the number of samples correctly identified as an attack, and FP represents the number of samples incorrectly identified as an attack. TN represents the number of samples correctly identified as normal, while FN indicates the number of samples incorrectly identified as normal. Accuracy represents the proportion of network data classified as common attack types. The calculation formula is as follows:

$$Precision = \frac{TP}{TP + FP} \tag{11}$$

Recall represents the proportion of network data classified as an attack to all attack data. The calculation formula is:

$$Recall = \frac{TP}{TP + FN} \tag{12}$$

Measure is the weighted average of both Precision and Recall. It is used to synthesize the scores of Precision and Recall. The calculation formula is:

$$F - Measure = \frac{(1 + \beta^2) \times Precision \times Recall}{\beta^2 \times (Precision + Recall)} \tag{13}$$

β is used to adjust the proportion of accuracy and recall. When $\beta = 1$, $F - Measure$ is the F1 score.

3.3 Experimental Parameter Setting and Result Analysis

The software environment used in this paper is the Python 3.7, tensorflow 2.1 and keras2.24. experimental hardware conditions of Intel Core i7–8700 CPU and 16g ram.The model was trained using the Adam optimizer and the category_ cross-entropy loss function.Adam's learning rate is 0.0001, epoch is 2000, batch_ size is 128, momentum in batch normalization is 0.85, and alpha in leakyrelu is 0.2. Dropout is set to 0.4, and LSTM recurrent_ Dropout is set to 0.01. The experiment is selected from the KDD99 training set 300,000 pieces of data are used to train the model, and the remaining 194021 pieces are used to test the model. The Sklearn toolkit is used to encode the 22 types of attacks in the training set. The results are shown in Fig. 12. The invasion detection accuracy of CNN+LSTM and CNN+SA+LSTM is as follows.

Model	Precision	Recall	F1
CNN+LSTM	0.9536	0.9518	0.9575
CNN+SA+LSTM	0.9742	0.9813	0.9736

Fig. 12 .

For experiments, CNN used a 3×3 convolutional kernel with a step length of 2, after each BN layer and a dropout layer. In Table 2, label0 represents normal network traffic and label1–label22 represents 22 different attack types. From the experimental results, the CNN+SA+LSTM hybrid model has a higher accuracy than the LSTM and CNN+LSTM models, and the convergence rate is significantly better than the CNN+LSTM model. The iterative procedure of model training is shown in Figs. 13 and 14.

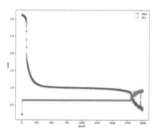   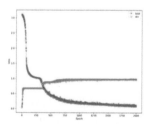

Fig. 13. CNN+LSTM Model accuracy graph   **Fig. 14.** CNN+SaLSTM Model accuracy graph

4 In Conclusion

For the current research status of intrusion detection, a neural network model based on intrusion detection with CNN and self-attention LSTM is proposed to solve the problems of unbalanced invasion data and inaccurate feature representation. Convolutional neural networks were used to extract the features of the raw data. Features that have great effects on classification results are given higher weight by attention autommachines. Then, the processed high-level features were predicted as input parameters for the LSTM network. In this paper, KDD99 training set was used for model training and testing for comparative analysis of CNN+LSTM and CNN+salstm models. Experiments show that the CNN+salstm model-based invasion detection and F1 metrics are better and accurate than the pure CNN+LSTM model.

References

1. Anonymous: Static and dynamic security technology. Comput. Commun. 000(005), 48–49 (1999)
2. Zhang, Y., Layuan, L.: Design and implementation of network intrusion detection system. J. Wuhan University of Technol. (Transp. Sci. Eng. Ed.) **28**(005), 657–660 (2004)
3. Anonymous. Network intrusion detection based on support vector machine. Comput. Res. Dev. (06), 799–807 (2003)
4. Zhong, W., Zhou, T.: Application of Naive Bayes classification in intrusion detection. Comput. Inf. Technol. (12), 24–27 (2007)
5. Guo, S., Gao, C., Yao, J., et al.: Intrusion detection model based on improved random forest algorithm. J. Software **16**(008), 1490–1498 (2005)
6. Guo, L., Ding, S.: Research progress of deep learning. Comput. Sci. **042**(005), 28–33 (2015)
7. Li, Y., Zhang, B.: An intrusion detection algorithm based on deep CNN. Comput. Appl. Software **037**(004), 324–328 (2020)
8. Wang, Y., Feng, X., Qian, T., et al.: Disguised user intrusion detection based on CNN and LSTM deep network. J. Comput. Sci. Expl. **012**(004), 575–585 (2018)
9. Mou, C., Xue, Z., Shi, Y.: Command sequence detection method based on BiLSTM and attention. Commun. Technol. **052**(012), 3016–3020 (2019)
10. Liu, L., Yu, X.: Recurrent Neural Network (RNN) and its application research. Sci. Technol. Vis. **290**(32), 60–61 (2019)
11. Hochreiter, S., Schmidhuber, J.: Long short-term memory. Neural Comput. **9**(8), 1735–1780 (1997)
12. Hinden, R.M., Deering, S.E.: IP Version 6 Addressing Architecture (1998)
13. Stolfo, S.J.: KDD cup 1999 dataset (1999)
14. Rizal, Y.: Data deduplication technology. Star (2011)

The Impacts of Cyber Security on Social Life of Various Social Media Networks on Community Users. A Case Study of Lagos Mainland L.G.A of Lagos State

Oumar Bella Diallo[✉] and Paul Xie

Computer Science and Technology, Zhejiang Gongshang University,
18 Xuezheng St, Jianggan District, Hangzhou 310018, Zhejiang, China
dialosal@gmail.com

Abstract. This study investigates the impacts of cyber Security on social life on community users Lagos L.G.A. The survey research was designed and adopted to describe this study. The sample for this research consists of one hundred and twenty undergraduate students randomly selected from respondents in their different homes using a simple technique. A structured questionnaire titled was developed and validated. It has a reliability coefficient of 0.72 using the test and re-test method. Descriptive statistics of frequency count and simple percentages were used to analysis the research question. This study is based on finding solution, it was affirmed that result from the demographic variable of respondents by age, sex, most visited social networks, duration of visitation, hours spent on social networks daily and people with the gadget that can access the internet. Results show that communities aware of the effects of cyber security on social media platforms are people between the ages of 21–25 years, male and female community dwellers have access to social network platforms, WhatsApp represents the most visited social platforms by people. The research questions show that social media platforms significantly influence the social life of community users in Lagos mainland L.G.A.

Keyword: The impacts of Cyber security on social life

1 Introduction

The internet is the fastest growing infrastructure in everyday single day the life in today's world. The internet is basically the network of networks used across for communication and data sharing. The term "Cyber" describes a person, thing, or idea that is associated with the computer and information age. It is relevant to computer systems or computer networks. A computer network is basically the collection of communicating nodes that helps in transferring data across. The nodes at any given time could be computers, laptops, smartphones, etc. The term crime is denoted as an unlawful act punishable under the law. Cybercrime was defined as a type of crime committed by criminals who use a computer

© The Author(s) 2022
Z. Qian et al. (Eds.): WCNA 2021, LNEE 942, pp. 469–477, 2022.
https://doi.org/10.1007/978-981-19-2456-9_48

as a tool and the internet as a connection to achieve different objectives such as illegal downloading of music and films, piracy, spam mailing, and so on. Cybercrime evolves from the erroneous use or abuse of internet services. According to (Mariam Webster), cybercrime includes any criminal act involving computers or networks (Chatterjee 2014).

Focusing on the case of Lagos mainland L.G.A in Lagos State (Nigeria), this study aims to investigate the impacts of cybercrime on the social life of various social media networks on community users. The study's objective is to find out the variety social media and networking sites community users have access. In addition, to determine how community users got involved in various cybercrime activities and how people prevent themselves from cyber-attack.

2   Materials and Methods

2.1   Design of Research

The design of research that was implemented for this study is a survey research design descriptive. A descriptive survey study is the best method for describing a population that is too large to observe directly.

2.2   Population of the Study

The study was conducted in Lagos mainland L.G.A of Lagos Stateandfocus mainly on community dwellers.

2.3   Sample Techniques

In this survey research, one hundred and twenty (120) people were selected randomly in Lagos mainland L.G.A of Lagos State using a simple random sampling technique 60 male and 60 female. The samples were selected randomly from their different homes. From the above explanation, all the samples were randomly selected according to their population. And the total number that was randomly selected from the L.G.A will make up the total samples that were required for this study.

2.4   Research Instrument

The questionnaire was used as a research instrument for the survey research. The questionnaire was divided into two (2) sections. Section A sought information about Age, Sex/gender; most visited social networks, duration of visitation, hours spent on social networks daily, etc. It was designed to tick the box that corresponds with their opinions on the question asked to express their mind about the subject matter (the question being asked). Section B was explicitly designed to determine the awareness level of students using social media platforms on cyber security.

2.5 Validity of the Instrument

The instrument was given to the expert (project supervisor) for vetting, after which the instruments were collected back with corrections and the proper check was affected before the final copy was produced.

2.6 Instrument Reliability

The instrument reliability was done through the test-retest method. The questionnaires were administered twice on twenty (20) respondents drawn from Alimosho L.G.A, which was out of the sample within two weeks interval. The data collected were correlated using Cronbach's alpha to obtain a standard data range (0.72) that was considered high enough for a study.

2.7 Administration of Instrument and Data Collection

The instruments were administered to the respondents in their different homes, personally by the researcher and were collected back immediately.

2.8 Analysis Method of Data

The data were analyzed using the statistics descriptive of frequency counts and simple percentages.

3 Results

This section is concerned with the presentation and analysis of data on Age, Sex, most visited social networks, duration of visitation, hours spent on social networks daily and students with a gadget that can access the internet.

3.1 Frequency Distribution of Demographic Variables

Table 1. The distribution frequency the respondents by age

	Frequency	Percentage (%)	Valid (%)	Cumulative (%)
15–20 yrs	15.8	15.8	15.8	
21–25 yrs	54	45.0	45.0	60.8
26–30 yrs	38	31.7	31.7	92.5
31–Above	9	7.5	7.5	100.0
Total	120	100.0	100.0	

The result from Table 1 shows that the number of respondents between the ages of 21–25 years is more than other respondents between the ages of 15–20 years, 26–30 years, 31 years and above. Out of the 120 respondents, there were 54 respondents representing 45.0% between the ages of 21–25 years. Since the respondents who make up the highest percentage are between the age ranges of 21–25 years, this means that the number of respondents aware of the effects of cyber security on social media platforms are people between the ages of 21–25 years.

Table 2. Frequency distribution of respondents by sex

Sex				
	Frequency	Percentage	Valid (%)	Cumulative (%)
Male	60	50.0	50.0	50.0
Female	60	50.0	50.0	100.0
Total	120	100.0	100.0	

The result from Table 2 showed that there is an equal result in the gender of the respondents as arranged in the sampling techniques. Out of the 120 questionnaire distributed, there were 60 respondents representing 50.0% males, while there were also 60 respondents representing 50.0% females.

Table 3. Descriptive statistics of frequency count on most visited social networks

M.V.S.N				
	Frequency	Percentage	Valid percent	Cumulative percent
Facebook	20	16.7	16.7	16.7
Twitter	15	12.5	12.5	29.2
WhatsApp	38	31.7	31.7	60.9
Instagram	26	21.7	21.7	82.6
B.B.M	5	4.2	4.1	86.7
2go	3	2.5	2.5	89.3
Google	13	10.8	10.8	100.0
Total	120	100.0	100.0	

Table 4. Descriptive statistics of frequency count on the duration of the visit of social networks site

D.O.V				
	Frequency	Percentage	Valid percent	Cumulative percent
Everyday	68	56.7	56.7	56.7
once a week	27	22.5	22.5	79.2
twice a week	23	19.2	19.2	98.4
Never	2	1.6	1.6	100.0
Total	120	100.0	100.0	

The result from Table 4 above showed that students visit WhatsApp more than other social networks. Out of the 120-questionnaire distributed, there were 38 respondents representing 31.7% WhatsApp users, 26 respondents representing 21.7% Instagram users, 20 respondents representing 16.7% Facebook users, five respondents representing 4.2.7% B.B.M., three respondents representing 2.5% 2go users. In comparison, there were 13 respondents representing 10.8% Google users. Since the respondents who make up the highest percentage of most visited social networks platforms choose WhatsApp, it means that WhatsApp represents the most visited social network platforms by community dwellers.

Table 5. Descriptive statistics of frequency count of students with a mobile phone or any media gadget that can access the internet

M.P.G				
	Frequency	Percentage	Valid (%)	Cumulative (%)
Yes	102	85.0	85.0	85.0
No	18	15.0	15.0	100.0
Total	120	100.0	100.0	

The result from Table 5 above showed that social networks are being visited every day by the respondents as it has the highest percentage of choice. Out of the 120 questionnaires distributed, there were 68 respondents representing 56.7%, daily users, 27 respondents representing 22.5% are once a week visitors, 23 respondents representing 19.2% visit twice a week, and there were only two respondents representing 1.2% that never visit therefore since the respondents who make up the highest percentage are those who visit every day, almost all the people with Lagos mainland L.G.A of Lagos state visit one or two social network sites every day.

The result from Table 6 below showed that the majorities of students have a mobile phone or social media gadget that can access the internet. Out of the 120 questionnaires distributed, there were 102 respondents representing 85.0% students with social media gadgets that can access the internet, while 18 respondents representing 15.0% students, don't have access to the internet. Since the respondents who make up the highest percentage are those with social media gadgets that can access the internet, it means that most community dwellers are aware of the effects of cyber security on social media platforms.

3.2   Analysis of Data Related to the Issues Raised by the Study

HOW DO COMMUNITY PEOPLE GET INVOLVED IN VARIOUS CYBER-CRIME ACTIVITIES?

Table 6. Table showing how community people get involved in various cybercrime activities

S/N	ITEMS	SA	A	D	SD
8	I do click on any available link I come across whenever I am using the internet	21 (17.5%)	45 (37.5%)	33 (27.5%)	21 (17.5%)
9	I visit almost all social media platform everyday	38 (31.7%)	17 (14.1%)	50 (41.7%)	15 (12.5%)
10	I quickly respond to likes and frequently comment on any post on any social media platform	16 (13.3%)	61 (50.8%)	26 (21.7%)	17 (14.2%)
11	With my phone, I do respond to any promotional messages that are sent to me through text messages	30 (25.0%)	21 (17.5%)	55 (45.8%)	14 (11.7%)
12	I always find it easier to shop online with my credit card on any promotional items than visiting a store with a cash	11 (9.2%)	36 (30.0%)	64 (53.3%)	9 (7.5%)
13	I accept every internet free pop up gift and distributes to friends online	45 (37.5%)	17 (14.2%)	47 (39.1%)	11 (9.2%)

The table above shows the percentage summation of those who answered "Strongly agree", "Agree", "Disagree", "strongly disagree", as analysed in the table above.

After the answers on the six items were added, the average percentage was found by dividing the total percentage on the items by six as presented in the table below.

HOW DO PEOPLE PREVENT THEMSELVES FROM CYBER-ATTACKS?

Table 7. Table showing how people prevent themselves from cyber-attack

S/N	Items	SA	A	D	SD
34	I always confirm any financial information from my local banks before I attend to it	29 (24.2%)	60 (50.0%)	21 (17.5%)	10 (8.3%)
35	I always reject or do away with any promotional links I come across during any internet engagements	44 (36.7%)	41 (34.2%)	24 (20.0%)	11 (9.1%)
36	I attend any promotional interview I come across through internet	16 (13.3%)	51 (42.5%)	45 (37.5%)	8 (6.7%)
37	Most people limit the time spent on the internet in other to avert any cyber insecurity or theft	23 (19.2%)	47 (39.1%)	35 (29.2%)	15 (12.5%)

The table above shows the percentage summation of those who answered "Strongly agree", "Agree", "Disagree", "strongly disagree", as analysed in the table above.

After the answers on the four items were added, the average percentage was found by dividing the total percentage on the items by four as presented in the table below.

4 Discussion

The result from demographic variables by age, sex, Most visited social networks, and community dwellers with a mobile phone or any media gadget that can access the internet from Table 1, 2, 3, 4, 5 and 6 show that the numbers of people in the community who are aware of the effects of cyber security on social media platforms are people between the ages of 21–25 years, the gender of the respondents are equal which signify that both male and female community dwellers have access to various social network platforms. From the most visited social network platform, Whatsapp represents the most visited social network platform by the people. The result from how often people visit various social media platforms shows that almost all community dwellers of Lagos mainland L.G.A do visit one or two social network sites every day and above on social network sites on a daily basis, while statistics show that large numbers of community dwellers have social media gadget that can access the internet which signifies that majority of them are aware of the effects of cyber security on social media platforms.

The result obtained from Table 6 indicates that social media platforms have no significant influence on how people get involved in various cybercrime activities. This is against (Global Risks 2013) report, which affirmed that the ability of individuals to share information with an audience of millions is at the heart of the particular challenge that social media presents to businesses. In addition to giving anyone the power to disseminate commercially sensitive information, social media also offers the same ability to spread false information, which can be just as damaging. The rapid spread of false information through social media is an emerging risk. In a world where we're quick to give up our personal information, companies have to ensure they're just as fast in identifying threats, responding in real-time, and avoiding a breach of any kind. Since these social media easily attract people, the hackers use them as bait to get the information and the data they require.

The result obtained from Table 7 indicates that social media platforms have a significant influence on how people prevent themselves from cyber-attack. This supports (Okeshola 2013) report, which affirmed that inspecting your mails before opening is a very useful way of detecting unusual or strange activities. Email spam and cyberstalking can be detected by carefully checking the email header, which includes the sender's real email address, internet protocol address, and the date and time it was sent. It has been discovered that cybercriminals can be extremely careless; therefore, it is recommended that the system be reviewed on a regular basis to detect unusual errors. Individuals should also ensure that proper security controls are in place and that the most recent security updates are installed on their computers. Lakshmi (2015) defines formalised formalised formalised formalised formalised formalised formalised formalised formalised formally.

References

Andreas and Michael: Reading in mass communication and Nigeria satellite. Makurdi: Benue State University (2000)

A.P.R.A.: "Cyber Security Survey Results" in Australian Prudential Regulation Authority (A.P.R.A.) (2016)

Armstrong, R., Mayo, J., Siebenlist F.: Complexity Science Challenges in Cybersecurity (2009). See http://sendsonline.org/wp-content/uploads/2011/02/DOE

Asghari, H., van Eeten, M., Bauer, J.M.: 13. Economics of cybersecurity. Handbook on the Economics of the Internet, p. 262 (2016)

Awake: The benefits of Facebook "friends": exploring the relationship between college students' use of online social networks and social capital. J. Comput.-Med. Commun. 12(3) (2012). article 1

Baron, S.J.: Introduction to Mass Communication: Media Literacy and Culture, 2nd edn. McGraw hill Companies, New York (2012)

Chatterjee, B.B.: Last of the rainmacs? Thinking about pornography in cyber space. Crime and the Internet, by David S. Wall (2014). ISBN 0–203–164504, Page no.-74

Bittner, R.J.: Mass Communication: An Introduction, 3rd edn. Prentice Hall Incorporation, New Jersey (1989)

Wall, D.: "Cyber crimes and Internet", Crime and the Internet, by David S. Wall (2002). ISBN 0-203-164504 ISBN 0-203-164504, Page no.1

Ewepu, G.: Nigeria loses N127bn annually to cyber-crime — N.S.A (2016). http://www.vangua rdngr.com/2016/04/nigeria-loses-n127bn-annually-cyber-crime-nsa

Facuconner: Mass communication research: issues and methodologies. A.P. Express Publishers, Nsukka (1975)

Gartner, H.: Forecast: The Internet of Things, Worldwide (2013). See http://www.gartner.com/ newsroom/id/2636073. Accessed 24 Apr 2016

Hassan, A.B., Lass, F.D., Makinde, J.: Cybercrime in Nigeria: causes, effects and the way out, A.R.P.N. J. Sci. Technol. 2(7), 626–631 (2012)

Analyzing the Structural Complexity of Software Systems Using Complex Network Theory

Juan Du[✉]

Wulanchabu Vocational College, Wulanchabu Inner Mongolia, Ulanqab 012000, China
best_ky123@163.com

Abstract. Software systems have nearly been used in all walks of life, playing an increasingly important role. Thus, how to understand and measure complex software systems has become an ever-important step to ensure a high-quality software system. The traditional analysis of software system structure focuses on a single module. However, the traditional software structural metrics mainly focus on analyzing the local structure of software systems and fail to characterize the properties of software as a whole. Complex network theory provides us with a new way to understand the internal structure of software systems, and many researchers have introduced the theory of complex networks into the examination of software systems by building software networks from the source code of software systems. In this paper, we combine software structure analysis and complex network theory together and propose a SCANT (Software Complexity Analysis using complex Network Theory) approach to probe the internal complexity of software systems.

Keywords: Software · Complex network · Software complexity · Metrics

1   Introduction

Large software systems are usually composed of lots of small constitute elements (e.g., methods, fields, classes, and packages); any small error in one element may lead to catastrophic consequences [1]. Thus, how to ensure a high quality software system has become a problem faced by many people in the field of software engineering. Generally, we cannot control what we cannot measure. Therefore, how to understand and measure complex software systems has become an ever-important step to ensure a high-quality software system [2].

The complexity of a specific software system usually originates from its internal structure. In recent years, some researchers proposed some approaches to explore the complexity of software systems from the perspective of the internal structure of software systems. Up to now, many promising achievements have been reported. Generally, the studies on software structure analysis can be divided into two groups, i.e., i) traditional software structure metrics, and ii) software structure metrics based on complex network theory.

© The Author(s) 2022
Z. Qian et al. (Eds.): WCNA 2021, LNEE 942, pp. 478–486, 2022.
https://doi.org/10.1007/978-981-19-2456-9_49

The traditional software structural metrics mainly focus on analyzing the local structure of software systems and fail to characterize the properties of software as a whole. With the development of complex networks, some researchers have introduced the theory of complex networks into the examination of software systems by building software networks from the source code of software systems. Complex network theory provides us with a new way to understand the internal structure of software systems. At present, the number of studies on software network analysis is still not very large, the construction of software networks is not accurate enough, and the metrics used in software network analysis and the data set used in the experiment are not comprehensive enough.

In this paper, we combine software structure analysis and complex network theory together and propose a SCANT (Software Complexity Analysis using complex Network Theory) approach to probe the internal complexity of software systems. Specifically, we build much more accurate software network models from the source code of a specific software system, and then introduce a set of statistical parameters in complex network theory to characterize the structural properties of the software system, with the aim of revealing some common structural laws enclosed in the software structure. By doing so, we can shed some light on the essence of software complexity.

2 Related Work

The traditional analysis of software system structure focuses on a single module. The McCabe metrics [3] are mainly based on graph theory and program structure control theory, using directed graph to represent the program control flow, so as to represent the complexity of the network according to the ring complexity in the graph. The Halstead metrics [4] are used to measure the complexity of a software system by counting the number of operators and operands in the program. The C&K metric suit [5] is based on the theory of object-oriented metrics and mainly includes six metrics. The MOOD metric suit [6] proposed by Abreu et at. indirectly reflect some basic structural mechanisms of the object-oriented paradigm.

With the development of complex networks, some researchers have introduced the theory of complex networks into the examination of software systems by building software networks from the source code of software systems. In their software networks, software elements such as attributes, methods, classes, and packages are represented by nodes, and the couplings between elements such as inheritance, method call, and implements are represented by undirected (or directed) edges. Based on the software network representation of the software structure, they introduced the complex network theory to characterize the structural properties of a specific software system, and further to improve its quality. Complex network theory provides us with a new way to understand the internal structure of software systems, and many related work has been reported.

3 The Proposed SCANT Approach

Our SCANT approach is mainly composed of four three, i.e., i) software network model construction, ii) calculating the values of statistical parameters, and iii) analyzing the parameter values to reveal the structural characteristics.

3.1 The Software Network Model

The software systems studied in this work are all open source software systems developed by using Java programming language. The topological information in software systems will be analyzed and extracted. In this work, we extract various software elements.

Since most statistical parameters in complex network theory do not consider the weight on the edges (or links), i.e., they only can be applied to un-weighted software networks. Thus, to apply the statistical parameters in complex network theory to characterize the software structure, in this work, we construct an un-weighted software network at the class level, i.e., Un-weighted Class Relationship Network (UCRN for short), to represent classes and the relationships between them. In UCRN, nodes represent the software elements at the class level (i.e., classes and interfaces), edges between nodes represent the relationship between classes, and the direction of edges represents the relationship direction between classes. In UCRN, we consider the following seven types of relationships [7], i.e., Inheritance relationship, Implementation relationship, Parameter relationship, Global Variable Relationship, Method Call Relationship, Local Variable Relationship, and Return Type Relationship.

If there is one of the seven kinds of relationships between two classes, then we establish a directed edge in the UCRN network between the nodes denoting the two classes. This edge is used to describe the coupling relationship. Thus, UCRN is essentially an un-weighted directed network which can be defined as

$$UCRN = (V, L), n \in V, l \in L,$$
$$l = < n_i, n_j >, n_i, n_j \in V \tag{1}$$

where V denotes the class (or interface) set in the software system, and L denotes the coupling relationship set between all pairs of nodes. Generally, if one class uses the service provided by another class, then a directed edge connecting the two classes will be established in the UCRN. We do not consider the weight on the edges. Thus, the weight on the edges will be the same, i.e., 1.

3.2 The Statistical Parameters

Here we introduce some statistical parameters widely used in complex network theory to characterize the structural properties of software systems. These statistical parameters are borrowed from [8].

Definition 1. Betweenness Centrality.
Betweenness is a very important parameter in complex network theory, and it is usually used to reflect the importance of nodes. The betweenness centrality of node i in a network can be described as the ratio of the number of all shortest paths passing through node i to the number of the shortest paths in the whole network. Till now, the betweenness centrality has been widely applied in a wide range of networks such as biological networks, transportation networks, and social networks. Betweenness centrality can be formally described as

$$B(v) = \sum_{s \neq v \neq t} \frac{\phi_{st}(v)}{\phi_{st}}, \tag{2}$$

where ϕ_{st} is the number of shortest paths between nodes s and t, and $\phi_{st}(v)$ denotes the number of shortest paths between nodes s and t which also passes node v.

Definition 2. Closeness Centrality.
Closeness centrality refers to the degree of closeness between a specific node and other nodes in the network. The higher the closeness centrality of a node is, the closer it is to other nodes. The closeness centrality of a node is the reciprocal of the average of the shortest path lengths between the node and all other nodes in the network and thus can be defined as

$$C(i) = \frac{n}{\sum_j d(j, i)}, \tag{3}$$

where $d(j, i)$ is the shortest path length between nodes i and j, and n is the number of nodes in the whole network.

Definition 3. Degree Distribution.
The degree of a node is the number of edges that the node used to be connected to other nodes. Degree distribution is a general description of the degree of nodes in a graph (or network), which is the probability distribution or frequency distribution of the degrees of the nodes in the network.

If a graph (or network) is composed of n nodes with n_k nodes whose degree is k, then the degree distribution $P(k) = \frac{n_k}{n}$. For directed graph (or network), $P(k)$ has two versions, i.e., in-degree distribution and out-degree distribution.

Definition 4. Clustering Coefficient.
Clustering coefficient is used to measure the degree to which nodes in a graph (or network) tend to cluster together, i.e., the aggregate density of nodes in a graph (or network). The clustering coefficient of a node in a network mainly refers to the proportion of the number of connections between the node and adjacent nodes to the maximum number of edges that can be connected between these nodes. The clustering coefficient of node i, C_i, can be computed according to the following formula

$$C_i = \frac{2e_i}{k_i(k_i - 1)} = \frac{\sum_{jm} a_{ij}a_{im}a_{mj}}{k_i(k_i - 1)}, \tag{4}$$

where e_i is equal to the number of nodes whose clustering coefficient is equal to the edges actually connected by its neighbours. $\frac{k_i(k_i-1)}{2}$ is the maximum possible number of edges. Then the clustering coefficient of the network is the average of the clustering coefficients of all the nodes in the network, i.e.,

$$C = \langle C_i \rangle = \frac{1}{N} \sum_{i \in V} C_i, \tag{5}$$

where N is the number of nodes in the graph (or network), and V is the nodes set.

Definition 5 Average Shortest Path Length.

For an un-weighted network, the shortest path length is the minimum number of edges from one node to another node in the network; for the weighted network, the shortest path length is the minimum value of the sum of the edge weights from one node to another node. The average shortest path length of a network is defined as the average of the shortest path lengths between any two nodes in the network. The average shortest path length of a network can be defined as

$$L = \frac{2}{N(N-1)} \sum_{i \neq j} d_{ij}, \tag{6}$$

where d_{ij} is the number of edges on the shortest path between nodes i and j, and N denotes the number of nodes in the network.

4 Software Structure Analysis

In this section, we use a set of four open source software systems as case studies to probe their topological properties.

4.1 Subject Systems

We selected a set of four open-source Java systems as our research subjects. These systems are selected from different domains with different scales. Specifically, the subject systems contain ant, jedit, jhotdraw, and wor4j. Table 1 shows some simple statistics of the four subject software systems. Specifically, *System* is the name of the subject system, *Version* shows the version of the corresponding software system, *Directory* is our analysed directory, *LOC* is the lines of code, and *#C* is the number of classes and interfaces.

Table 1. Statistics of the subject systems.

System	Version	Directory	*LOC*	*#C*
ant	1.6.1	src/main	81515	900
jedit	5.1.0	src	112492	1082
jhotdraw	6.0b.1	src	28330	544
wro4j	1.6.3	src	33736	567

4.2 Results and Analysis

In this section, we constructed the software networks for all subject systems, and then used the statistical parameters to characterize the topological properties of these subject systems.

Node Centrality Analysis. Network centrality metrics are mainly used to find the nodes which play an important role in the complex network. In this section, two centrality metrics are used, i.e., betweenness centrality and closeness centrality.

Betweenness centrality is one of the most important centrality metrics in complex network theory. It is widely used to characterize the importance of nodes. As shown in Fig. 1, we can find that, nearly in all the subject systems, about 90% of the nodes have a betweenness value less than 0.05, which means only 5% of classes contain important information and play important role in the implementation of the key functionalities of the software system; a large part of the classes do not perform important role. Betweenness centrality reflects the degree of interdependence between each class node and other class nodes. The higher betweenness centrality of class nodes is, the more important it is to the software network.

In the actual development process, the class call is usually a call chain, and the important class will generally be more called and called other classes, such as the core function class is usually called by various types of software to perform the corresponding action. Therefore, the key class in the software system, the performance of the betweenness centrality is that the betweenness centrality value is larger.

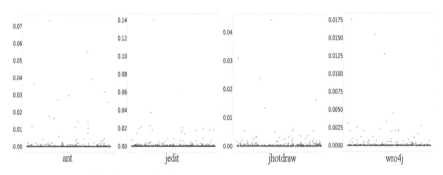

Fig. 1. The distribution of betweenness centrality values

As shown in Fig. 2, there is no class nodes whose closeness value is larger than 0.5, and in the four subject software systems, the closeness centrality values of most nodes are close to 0. The fact that the closeness value of some class is equal to 0 indicates that there are some isolated nodes in the network without any connections to other nodes. The larger the closeness centrality value of the class node is, the closer the class is related to all other class nodes, which means these class nodes have a best position in the network and can perceive the dynamics of the whole software network including the flow direction of information. Generally, key classes usually use the services provided by many more classes to complete core functionality. Thus, in the software network, we may find that some key class are more closely related to other class nodes.

Clustering coefficient analysis. Figure 3 shows the distribution of clustering coefficient values. Obviously, the clustering coefficient values of most class nodes in ant, jedit, jhotdraw, and wro4j are close to 0, which means that most of the nodes whose neighbors

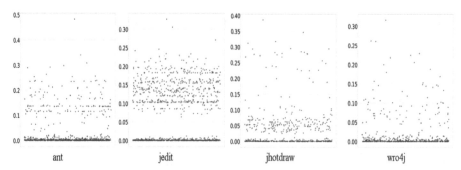

Fig. 2. The distribution of closeness centrality values.

are not closely coupled with each other; only a few class nodes have high clustering coefficient values.

For all the subject software systems, only a few class nodes have a relatively high clustering coefficient, i.e., only a few classes will use many other classes or be used by many other classes. This is in line with the characteristics of key classes of software systems. In the practical development process, classes that provide core functionalities (i.e., key classes) are usually called by many other classes to execute core functionalities. Generally, developers will write some small classes to provide some single-functionality classes, and then key classes will use the services provided by these classes to provide complex functionalities. Thus, the neighbours of key classes are usually coupled closely, which is reflected by a larger value of clustering coefficient.

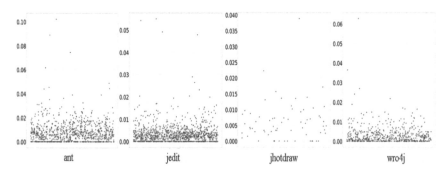

Fig. 3. The distribution of clustering coefficient values.

Degree Distribution. Figure 4 shows the degree distribution of nodes in the software network. As shown in Fig. 4, we can observe that the number of nodes decreases as the degree increases, and the more nodes in the software network, the more obvious this trend is.

It can be observed from Fig. 4 that when the degree is less than 10, the number of nodes accounts for almost 90% of the nodes in the software network; when the degree is greater than 50, the number of nodes is almost close to 0. Therefore, most of the nodes in the software network are only connected to a few nodes, and a few nodes are connected to most of the nodes, which is in line with the typical characteristics of scale-free networks. It indicates that in the software system, most of the classes only call a very small number of classes or are called by a very small number of classes, and only a few classes are called a large number of other classes or are called by a large number of classes.

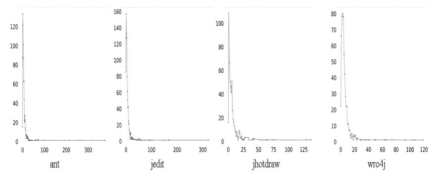

Fig. 4. The degree distribution.

Average Path Length Analysis. As shown in Table 2, although the software scales are different across systems, the average shortest path length is roughly equal to 3. The maximum average shortest path length is 3.379, and the minimum average shortest path length is 2.806. Therefore, software networks have small-world property.

Table 2. The average path length of software networks.

Subject systems	ant	jedit	jhotdraw	wro4j
Average shortest path length	3.178	3.290	3.235	3.379

5   Conclusions

In this work, we used un-weighted software networks to represent software structure and introduced some statistical parameters in complex network theory to characterize the structural properties of software systems. We used a set of four open-source software systems as subject systems to reveal some topological properties of software systems. Specifically, we analyzed the distribution of many statistical parameters, such as centrality metrics (i.e., betweenness and closeness), clustering coefficient, and average shortest path length.

The results show that the software networks proposed in this work also belong to small-world and scale-free networks. The analysis of these important structural properties in software networks is of great significance to the field of software metrics.

References

1. Fenton, N.E., Ohlsson, N.: IEEE Tran. Softw. Eng. **26**, 797 (2000)
2. Fenton, N.E., Neil, M.: IEEE Trans. Softw. Eng. **25**, 675 (1999)
3. McCabe, T.J.: IEEE Trans. Softw. Eng. **SE-2**, 308 (1976)
4. Felician, L., Zalateu, G.: IEEE Trans. Softw. Eng. **15**, 1630 (1989)
5. Shatnawi, R.: IEEE Trans. Softw. Eng. **36**, 216 (2010)
6. Harrison, R., Counsell, S.J., Nithi, R.V.: IEEE Trans. Softw. Eng. **24**, 491 (1998)
7. Li, H., et al.: IEEE Access **9**, 28076 (2021)
8. Battiston, F., Nicosia, V., Latora, V.: Eur. Phys. J. Spec. Top. **226**, 401 (2017)

Cluster-Based Three-Dimensional Particle Tracking Velocimetry Algorithm: Test Procedures, Heuristics and Applications

Qimin Ma, Yuanwei Lin, and Yang Zhang[✉]

Department of Fluid Machinery and Engineering, Xi'an Jiaotong University, 28 Xianning West Rd., Xi'an 710049, China
zhangyang1899@mail.xjtu.edu.cn

Abstract. Particle tracking velocimetry (PTV) algorithm based on the concept of particle cluster is investigated and improved. Firstly, an artificial test flow is constructed, and a dimensionless parameter C_{PTV} is introduced to characterize the difficulty for the PTV reconstruction. Secondly, the heuristics that particle-cluster based algorithms must follow are summarized, and a three-dimensional cluster-based PTV incorporating the Delaunay Tessellation is proposed and tested by using the artificial flow. The criteria property of C_{PTV} is then analysed and verified. Combining the proposed algorithm with a three-dimensional particle detection system, two particle flows are successfully reconstructed, therefore verifying the practicality of the algorithm.

Keywords: Flow visualization · Particle tracking algorithm · Particle cluster · Artificial test flow

1 Introduction

Due to the thriving demands for the non-intrusive flow measurements and the progresses of volumetric photography techniques, three-dimensional particle image velocimetry (PIV) and particle tracking velocimetry (PTV) are considered effective ways to achieve complex flow reconstruction at satisfying spatiotemporal resolutions [1, 2]. In comparison with PIV based on the Eulerian viewpoint [3–5], PTV is based on the Lagrangian viewpoint [6, 7] and has three distinctive features: firstly, PTV restores the local large velocities without smoothing them by spatial averaging; secondly, PTV is able to restore particle trajectories from the sequence of inter-particle matching relations, which is important to certain special occasions; thirdly, resolution of PTV depends on the particle intensity instead of the minimum size of the interrogation window. However, if particle intensity is so high that the particle images are overlapping or adhering with each other, PIV is considered a better choice than PTV [8–10].

The idea of PTV is to correlate particle coordinates from consecutive frames to obtain inter-frame particle displacements. Such displacements combined with the frame interval lead to the velocity of the corresponding flow field [11]. [12] and [13] came up with the

© The Author(s) 2022
Z. Qian et al. (Eds.): WCNA 2021, LNEE 942, pp. 487–496, 2022.
https://doi.org/10.1007/978-981-19-2456-9_50

earliest PTV based on the concept of particle clusters, where the clusters are composed of particles from the same frame and within the fixed interrogation window. In comparison with the optimization or hybrid algorithms [14–17], the cluster-based algorithm has simpler structure and fewer preset parameters, which can be easily adapted to the three-dimensional practice. The fundamental idea is to match the clusters according to self-defined geometrical characteristics, so that the corresponding particles as the cluster centers can be matched. [18] proposed a PTV using Delaunay tessellation (DT-PTV), in which the cluster refers to the DT triangle that is formed flexibly without using any fixed interrogation window. [19] extended DT-PTV to three-dimensional domain, in which the cluster refers to the DT tetrahedron. However, the degree of freedom of either triangle or tetrahedron is so low that when particle intensity is high, clusters become geometrically similar to each other, which is detrimental to PTV judgement. Then the Voronoi Diagram (VD, the dual of DT) was adopted to propose a VD-PTV [20] and its quasi-three-dimensional version [21]. Then the geometrical change of cluster responds sensitively to the inter-frame flow variation, thus leading to a satisfactory matching accuracy.

This paper introduces an improved cluster-based PTV with higher parametric independence than the aforementioned ones, so as to better meet the practice of flow reconstruction involving the three-dimensional particle detection systems [22]. The paper is organized as follows: in Sect. 1, the artificial flow with a wide range of testing challenges is constructed, following which a dimensionless number incorporating the challenges for PTV is proposed; in Sect. 2, the heuristics for the cluster-based PTV are suggested, followed by an improved double-frame PTV and its simple verification; in Sects. 3, the criteria feature of the dimensionless number is tested and analysed by the artificial flow; Finally in Sect. 4, the improved algorithm is applied to two actual particle flows.

2 Artificial Test Flow

The double-frame artificial particle flow is generated as follows. Firstly, a certain number of particles are randomly distributed in the "imaging field" to form the first frame. Secondly, particles move along the flow that is determined by linear superposition of basic flows, namely, shear, dipole expansion and rotation, which correspond to the different components of the rate of strain, thereby giving birth to the second frame. The artificial flow is easy to generate while providing challenges tough enough to test PTV. This is important because it is the flow intensity, rather than the complexity of flow pattern (or structure), that brings substantial challenges to PTV. The governing equations of basic flows are shown in (1), and the examples are shown in Fig. 1.

$$\frac{dx}{dt} = f_{shr,x}(x, y, z) + f_{vor,x}(x, y, z) + f_{dip,x}(x, y, z) \tag{1-1}$$

$$\frac{dy}{dt} = f_{shr,y}(x, y, z) + f_{vor,y}(x, y, z) + f_{dip,y}(x, y, z) \tag{1-2}$$

$$\frac{dz}{dt} = f_{shr,z}(x, y, z) + f_{vor,x}(x, y, z) + f_{dip,x}(x, y, z) \tag{1-3}$$

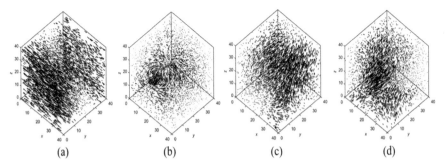

Fig. 1. Artificial test flows. (a) Shear. (b) Dipole expansion. (c) Rotation. (d) Superposition. The units of the coordinates are in pixel for simplicity.

$$f_{shr,x}(x, y, z) = C_{shr}(y - y_0) \tag{1-4}$$

$$f_{shr,y}(x, y, z) = 0 \tag{1-5}$$

$$f_{shr,z}(x, y, z) = 0 \tag{1-6}$$

$$f_{vor,x}(x, y, z) = \sum (-C_{vor,i,y} \frac{y_j - y_{vor,i}}{r_{i,j}^p} + C_{vor,i,z} \frac{z_j - z_{vor,i}}{r_{i,j}^p}) \tag{1-7}$$

$$f_{vor,y}(x, y, z) = \sum (-C_{vor,i,z} \frac{z_j - z_{vor,i}}{r_{i,j}^p} + C_{vor,i,x} \frac{x_j - x_{vor,i}}{r_{i,j}^p}) \tag{1-8}$$

$$f_{vor,z}(x, y, z) = \sum (-C_{vor,i,x} \frac{x_j - x_{vor,i}}{r_{i,j}^p} + C_{vor,i,y} \frac{y_j - y_{vor,i}}{r_{i,j}^p}) \tag{1-9}$$

$$f_{dip,x}(x, y, z) = -\sum C_{abs,i} \frac{x_j - x_{dip,i}}{r_{i,j}^p} + \sum C_{exp,i} \frac{x_j - x_{dip,i}}{r_{i,j}^p} \tag{1-10}$$

$$f_{dip,y}(x, y, z) = -\sum C_{abs,i} \frac{y_j - y_{dip,i}}{r_{i,j}^p} + \sum C_{exp,i} \frac{y_j - y_{dip,i}}{r_{i,j}^p} \tag{1-11}$$

$$f_{dip,z}(x, y, z) = -\sum C_{abs,i} \frac{z_j - z_{dip,i}}{r_{i,j}^p} + \sum C_{exp,i} \frac{z_j - z_{dip,i}}{r_{i,j}^p} \tag{1-12}$$

where f_{shr} is the spatial distribution of shear, C_{shr} is the intensity of shear; f_{vor} is the spatial distribution of rotation, $C_{vor,i,x}$, $C_{vor,i,y}$, $C_{vor,i,z}$ are the intensities of rotation in three dimensions; f_{dip} is the spatial distribution of dipoled expansion, C_{abs} and C_{exp} are the absorbing and expanding intensites for a pair of dipoles; p is the influencing index of $r_{i,j}$, which defines the decay of the flow intensity with distance. In generating the flow, all these intensity parameters are randomly selected in [0, 1].

Generating an artificial flow also requires the pre-input of the following controlling parameters: particle number in the first frame N_{ptc}, side length of the rectangular "imaging field" L, the maximum displacement parameter C_{dsp}, numbers of vortices and/or dipoles, proportion of the randomly occuring particles in the second frame compared to the first frame μ_1, and that of the missing particles μ_2 C_{dsp} determines the maximum displacement of the entire flow field. Specifically, after generating the flow field, the displacements of all particles should be normalized not to exceed the maximum value L/C_{dsp}. μ_1 and μ_2 simulate the failure of particle number conservation across two frames: overlapping of the particles in the second frame, particles escaping out of the illuminating sheet, and image noises mistakenly recognized. The particle intensity is represented by the average distance between the neighboring particles d_m:

$$d_m = \frac{L}{\sqrt[3]{N_{ptc}}} \tag{2}$$

The inter-frame particle displacements are indicated by their average value f_m.

The particle coordinates of two frames will be the input for PTV to match. By comparing the matched result by PTV with the genuinely generated result, one can obtain the accuracy of PTV, as well as the way those parameters influnce the performance of PTV. The accuracy of PTV is defined as:

$$Acc = \frac{N_c}{N_{ptc}} = \frac{N_{c,m}}{N_{ptc}} + \frac{N_{c,d}}{N_d} \frac{N_d}{N_{ptc}} \tag{3}$$

where N_c is the number of particles in the first frame which are correctly matched or correctly determined as no-match; $N_{c,m}$ is the number of particles in the first frame which are correctly matched, $N_{c,d}$ is the number of particles in the first frame which are correctly determined as no-match; N_d is the number of genuinely missing particles in the second frame. Generally speaking, if f_m gets smaller or d_m gets larger, it would be easier for PTV to reconstruct the flow. Therefore, influences of f_m and d_m are collected as $C_{PTV} = \frac{f_m}{d_m}$, indicating that C_{PTV} may be a criteria to describe the difficulty for the PTV reconstruction. Since it is unable to define "the difficulty for the PTV reconstruction" by equations, the verification of the criteria property of C_{PTV} would be conducted with the help of the following principle:

$$\forall C_{PTV} \in P, \forall (f_m, d_m) \in \left\{ (f_m, d_m) \middle| \frac{f_m}{d_m} = C_{PTV} \right\} \tag{4-1}$$

$$\exists\, g \text{ and } f, \quad A_{cc} = g(C_{PTV}) = f(f_m, d_m) \tag{4-2}$$

In Sect. 3, the criteria property of C_{PTV} is to be tested.

3   Heuristics and Improvement

In order to match clusters across the frames, the assumption of small deformation is applied. Specifically, it indicates that across the frames, the cluster's feature changes so mildly that the differences among clusters in the same frame are greater than that between the same cluster in different frames. Based on this assumption, the characteristic index of the cluster (as a vector) should meet the followimg heuristics: (1) the index is sensitive to the selection of particles in the same frame. This heuristic is usually easy to satisfy by choosing a sufficient amount of irrelevant characteristic values to form the index. (2) the index is insensitive to the translation and rotation of the cluster, i.e., the selection of the reference system. (3) the index is insensitive to the deformation of clusters over time, which can be achieved by selecting the high-order terms of the basic geometrical parameters of the cluster. (4) the way the elements of the index are arranged should be unique, to avoid traversing all possible arrangements of the elements while comparing two clusters. (5) the index should be insensitive to the missing particles. Particle missing and occuring is inevitable in practical situations, so the influence of no-match particles should be treated seriously rather than be neglected.

DT based three-dimensional PTV [21] meets the abovementioned heuristics, and the present work is to focus on its last preset parameter: the searching radius R_s. To find candidate particles in the second frame which are in a certain range around the target particle from the first frame, a searching radius R_s was always used to traverse all particles, to check if their distance to the given coordinate are smaller than R_s. However, a fixed R_s may include redundant candidates to threat the PTV accuracy and eat up a good amount of time. Moreover, R_s must be estimated according to the everage feature of flow field, and is very likely to fail on the inhomogenuous velocity field. In the improved algorithm, therefore, the particle coordinates of the target particle and those in the second frames are superposed in the same space and then processed with the Delauney Tesselation. Then these particles become the knots in a DT grid. The searching area is defined by the connection of the DT grid and specified by an integral number, the contact level C_l: a particle is considered a matching candidate for the target particle if they are connected by grid lines through a number of $(n-1)$ knots under $C_l = n$. DT grid is not influenced by the size of image area or particle intensity, and it appears that contact level C_l higher than 2 would have no practical use, while $C_l = 2$ would be of use only if the situation is extreme. Therefore, C_l is usually set to be 1 to suit the assumption of small deformation, which in fact reduces the number of preset parameters and makes the algorithm more concise. As shown in Fig. 2, influence of the improvement on the accuracy of PTV is small when the particle number is over 2000; meanwhile, the computing time decreases significantly. Therefore, In the cases where the computing speed is stressed on, the inproved version has an obvious advantage. Tests using other flow types shown in Fig. 1 have obtained similar results, which therefore are not shown here.

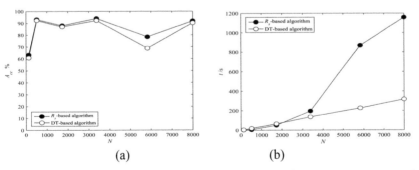

(a)   (b)

Fig. 2. Comparison between the original and the improved algorithms on (a) accuracy and (b) computing time. The artificial rotation flow is used, and N denotes the particle number in the first frame.

4 Analysis and Test of C_{PTV}

Figure 3 shows the variation of accuracy with the dimensionless parameter C_{PTV}. C_{PTV} varies in a wide range of value by randomly changing f_m or/and d_m in artificial flows. There is an explicitly monotonic relationship between C_{PTV} and A_{cc}, with the scattered data collapsing stably on a regression curve for three basic flows. Therefore, C_{PTV} is showing a good property of criteria. This is an interesting phenomenon, because the increase of f_m and the decrease of d_m, although they bring about the same degree of challenge for PTV, actually indicate quite different changes of flow states (in contrast with the former one, the latter one changes nothing to the flow structure).

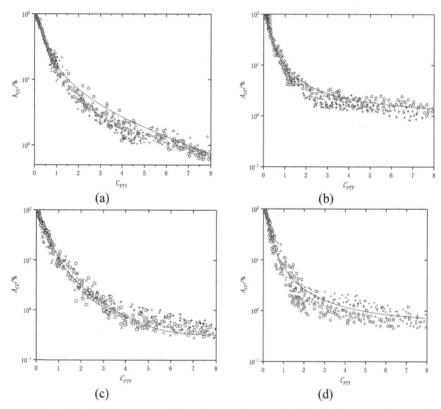

Fig. 3. Ests of the criteria property of C_{PTV} by using (a) rotation, (b) dipoled expansion, (c) shear, and (d) complex flow by (a)–(c), blue circle: f_m and d_m simultaneously change, black cross: only d_m changes, red triangle: only f_m changes.

By combining the conclusion with the basic idea of the cluster-based PTV, one question raises: what is exactly a "small deformation" for PTV? Obviously it is not the "tiny deformation" that can be ignored as in the field of materal mechanics. In fact, the deformation is significant even if C_{PTV} equals 0.5, while the accuracy of PTV is still satisfactory. But why does the algorithm fails as soon as the C_{PTV} gets larger than 0.5? A conjecture is introduced that as the C_{PTV} is increasing, the particles in a cluster becomes more likely to pass through the planes determined by other particles in the cluster, and such passing-through will change the connecting relationship among the particles in that cluster and its neighbours. In other words, the topological property of the grid is changed by the passing-through. Then any method applied to extract the characteristic index of the cluster will fail, since the characteristic index simply no longer represents the same particle when C_{PTV} is over a certain threshold.

Assume that a cluster is made of a center particle on the origin and three vertice particles on three axes at a distance of d_m from the origin, and the three vertices determine a plane. Then let all the particles move in random directions at a certain distance of f_m. The motion of these four particles are independent of each other. Let p_0 be the possibility that the displacement of the center particle does not pass through the plane determined by the three vertices after motion, and the relationship between p_0 and $C_{PTV} = f_m/d_m$ is shown in Fig. 4, from which one can see that the results do collapse on the function. Therefore, (1) the dimensionless parameter C_{PTV} has the criticia property because it determines the possibility that the topological property of grid changes after the inter-frame displacement, and if the property drastically changes, PTV would not be able to conduct any successful match across frames. (2) The mathematical principle that C_{PTV} affects PTV accuracy determines not only the algorithm improved and tested here, but

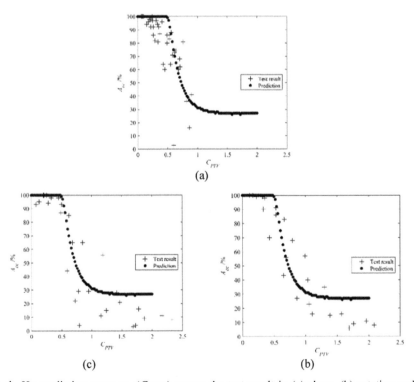

Fig. 4. He predicting curve p_0 (C_{PTV}) versus the test result in (a) shear, (b) rotation and (c) dipoled expansion.

also all the cluster-based PTV, and no matter what methods are used to form clusters, the passing-through will give them strong interruption, and their average accuracy curve will not be higher than p_0 (C_{PTV}). Considering there is no standard for PTV testing, this curve can be regarded as one that makes sense to most of the algorithms.

5   Application of the Algorithm

The improved algorithm is applied to the analysis of the output data of three-dimensional particle detection recognition system to verify the practicability of the algorithm. The test is a shear flow in a water tunnel with transparent walls. The tunnel is illustrated by four surrounding neon lamps. In the illuminating volume, a V3V system captures the instantaneous coordinates of tracer particles, which is used as the input data of PTV. The tracer particles are glass beads with a diameter of 10–20 μm and 1.05 times heavier than water. On one side of the tunnel, a sealed drawer plate is assembled. After the tunnel is filled with water, the plate is drawn out horizontally to generate a shear flow. C_l is set

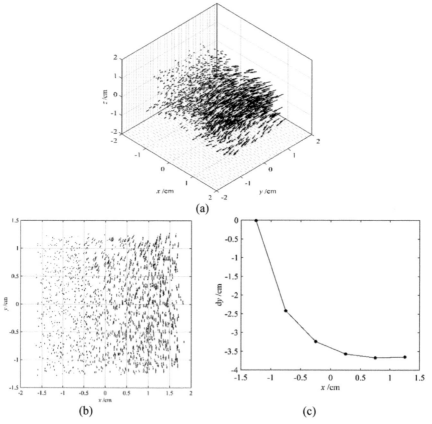

Fig. 5. PTV reconstruction of a shear flow. (a) Three-dimensional result, (b) projection of the result on x-y plane and (c) profile of the y-direction displacement along x.

as 1. The double-frame reconstruction of the flow field is shown in Fig. 5. The first and the second frames contain 1883 and 1878 particles, respectively, and there are a total of 1679 correct matches. As shown in Fig. 5(c), The profile of the y-direction displacement along x is well restored, so the algorithm meets the expected shear in this experiment.

6   Conclusion

An artificial flow was constructed that can pose sufficient challenges to PTV. The artificial flow allows for comprehensive testing of several factors that affect the performance of PTV. By analyzing the sufficient conditions for the cluster-based algorithm to take effect, it has been concluded that the applicability of the PTV algorithm depends on whether the small deformation assumption is satisfied. The five heuristics that the cluster-based algorithm should satisfy were proposed, so that PTV based on VD becomes fully parameter-independent. The improved algorithm was tested using artificial and actual flow fields to verify its effectiveness and practicality. The criteria property of the dimensionless parameter C_{PTV} was also verified, i.e., it can be considered as a standard for PTV design and test.

Acknowledgments. This work is funded by National Natural Science Foundation of China (11402190) and China Postdoctoral Science Foundation (2014M552443).

References

1. Hassan, Y.A., Canaan, R.E.: Full-field bubbly flow velocity measurements using a multiframe particle tracking technique. Exp. Fluids **12**, 49–60 (1991)
2. Boushaki, T., Koched, A., Mansouri, Z., Lespinasse, F.: Volumetric velocity measurements (V3V) on turbulent swirling flows. Flow Meas. Instrum. **54**, 46–55 (2017)
3. Adrian, R.J.: Twenty years of particle image velocimetry. Exp. Fluids **39**, 159–169 (2005)
4. Westerweel, J., Elsinga, G.E., Adrian, R.: Particle image velocimetry for complex and turbulent flows *Annu.* Rev. Fluid Mech. **45**, 409–436 (2013)
5. Ishima, T.: Fundamentals of Particle Image Velocimetry (PIV). J. Combust. Soc. Jpn **61**(197), 224–230 (2019)
6. Schanz, D., Gesemann, S., Schröder, A.: Shake the Box: lagrangian particle tracking at high particle image densities. Exp. Fluids **57**, 70 (2016)
7. Zhalehrajabi, E., Lau, K.K., Kushaari, K.Z., Horng, T.W., Idris, A.: Modelling of urea aggregation efficiency via particle tracking velocimetry in fluidized bed granulation. Chem. Eng. Sci. **223**(21), 115737 (2020)
8. Schröder, A., Geisler, R., Staack, K.: Eulerian and Lagrangian views of a turbulent boundary layer flow using time-resolved tomographic PIV. Exp. Fluids **50**, 1071–1091 (2010)
9. Cerqueira, R.F.L., Paladino, E.E., Ynumaru, B.K., Maliska, C.R.: Image processing techniques for the measurement of two-phase bubbly pipe flows using particle image and tracking velocimetry (PIV/PTV). Chem. Eng. Sci. **189**, 1–23 (2018)
10. Takahashi, A., Takahashi, Z., Aoyama, Y., Umezu, M., Iwasaki, K.: Three-dimensional strain measurements of a tubular elastic model using tomographic particle image velocimetry. Cardiovasc Eng. Technol. **9**, 395–404 (2018)

11. Ruhnau, P., Guetter, C., Putze, T., Schnörr, C.: A variational approach for particle tracking velocimetry. Meas. Sci. Technol. **16**, 1449–1458 (2005)
12. Okamoto, K.: Particle tracking algorithm with spring model. J. Visual. Soc. Jpn. **15**, 193–196 (1995)
13. Ishikawa, M., Murai, Y., Wada, A., Iguchi, M., Okamoto, K., Yamamoto, F.: A novel algorithm for particle tracking velocimetry using the velocity gradient tensor. Exp. Fluids **29**, 519–531 (2000)
14. Ohyama, R.I., Takagi, T., Tsukiji, T., Nakanishi, S., Kaneko, K.: Particle tracking technique and velocity measurement of visualized flow fields by means of genetic algorithm. J. Visual. Soc. Jpn. **13**, 35–38 (1993)
15. Labonte, G.: New neural network for particle-tracking velocimetry. Exp. Fluids **26**, 340–346 (1999)
16. Ohmi, K., Li, H.Y.: Particle-tracking velocimetry with new algorithm. Meas. Sci. Technol. **11**, 603–616 (2000)
17. Brevis, W., Nino, Y., Jirka, G.H.: Integrating cross-correlation and relaxation algorithms for particle tracking velocimetry. Exp. Fluids **50**, 135–147 (2010)
18. Song, X., Yamamoto, F., Iguchi, M., Murai, Y.: A new tracking algorithm of PIV and removal of spurious vectors using Delaunay tessellation. Exp. Fluids **26**, 371–380 (1999)
19. Zhang, Y., Wang, Y., Jia, P.: Improving the Delaunay tessellation particle tracking algorithm in the three-dimensional field. Measurement **49**, 1–14 (2014)
20. Zhang, Y., Wang, Y., Yang, B., He, W.: A particle tracking velocimetry algorithm based on the Voronoi diagram. Meas. Sci. Technol. **26**, 075302 (2015)
21. Cui, Y.T., et al.: Three-dimensional particle tracking velocimetry algorithm based on tetrahedron vote. Exp. Fluids **59**, 31 (2018)
22. Kalmbach, A., Breuer, M.: Experimental PIV/V3V measurements of vortex-induced fluid-structure interaction in turbulent flow-a new benchmark FSI-PfS-2a. J. Fluids Struct. **42**, 369–387 (2013)

Robust Controller Design
for Steer-by-Wire Systems in Vehicles

Nabil El Akchioui[1]([✉]), Nabil El Fezazi[2,3], Youssef El Fezazi[2], Said Idrissi[2,4], and Fatima El Haoussi[2]

[1] Faculty of Sciences and Technology, LRDSI Laboratory,
Abdelmalek Essaâdi University, Al Hoceima, Morocco
n.elakchioui@uae.ac.ma

[2] Faculty of Sciences Dhar El Mehraz, Department of Physics, LISAC Laboratory,
Sidi Mohammed Ben Abdellah University, Fez, Morocco
youssef.elfezazi@usmba.ac.ma

[3] National School of Applied Sciences, ERMIA Team,
Abdelmalek Essaâdi University, Tangier, Morocco

[4] Polydisciplinary Faculty of Safi, LPFAS Laboratory,
Cadi Ayyad University, Safi, Morocco

Abstract. The steer-by-wire (SbW) technology enables to facilitate better steering control as it is based on an electronic control technique. The importance of this technology lies in replacing the traditional mechanical connections with steering auxiliary motors and electronic control and sensing units as these systems are of paramount importance with new electric vehicles. Then, this research paper discusses some difficulties and challenges that exist in this area and overcomes them by presenting some results. These results meet the SbW's robust performance requirements and compensate oscillations from the moving part of the steering rack in the closed-loop system model: modeling, analysis and design. Thus, the issue of robust control for nonlinear systems with disturbances is addressed here. Finally, the results are validated through detailed simulations.

Keywords: SbW technology · Electronic control · Electric vehicles · Robust performance · Nonlinear systems

1 Introduction

The auto industry has implemented many modern and advanced systems in an attempt to raise the quality of driving, especially in off-road, as well as increase the safety and comfort of users of these vehicles [11,13,17]. Parallel to these developments, we see a significant shift from classical to modern systems [9] and SbW is another very promising application in terms of practicality, safety, and functionality [4,14]. For that reason, several automobile manufacturers have

© The Author(s) 2022
Z. Qian et al. (Eds.): WCNA 2021, LNEE 942, pp. 497–508, 2022.
https://doi.org/10.1007/978-981-19-2456-9_51

introduced SbW systems in vehicles to improve operational efficiency and fuel economy [3,8,19,24,36]. Then, SbW is a technology that replaces the traditional systems for steering with electronic controllers [7,10,18,20,31,32]. This technique enables to facilitate better steering control as it is based on what we call electronic control [12,15,27].

The primary objective of these vehicles is to obtain control capabilities that are not mechanically related to the vehicle's engine, but are sensed through advanced devices and transmitted by electrical signals based on effective mechanisms [26]. Then, the accuracy, performance and efficiency of the machinery in these vehicles is directly related to the positioning systems on roads and tracks [16,22] where DC motors are often used in this case. The steering wheel (SW) rotation is transmitted in the classic steering system through an intermediate shaft that is connected via the rack/pinion torque to front wheels (FWs) [38]. In SbW technology, the main component, the intermediate element, is dispensed and in turn many modern sensors and efficient actuators are connected to the SW and FW parts [30]. Then, the dynamic model obtained for this technology represents the close relationship between the current steering mechanism, the electrodynamics of the DC motor, and the torque of the rack/pinion part as shown in Fig. 1 [18,23].

Finally, this paper discusses the robust control problem using a technology called SbW. The primary objective of the considered strategy is to maintain stability, traceability and resistance to interference under complex working and road conditions. A novel scheme is developed here for modern vehicles that is equipped with the active steering system under consideration to cope well with difficult and varied road conditions. Then, in this research paper we discuss difficulties and challenges that exist in this area and give some results to overcome them. These results meet the SbW's robust performance requirements and compensate oscillations from the moving part of the steering rack. Finally, the obtained graphs are presented to see the achieved high performance, the resulting strong stability, and the durability that this type of system requires.

2   Modeling and Problem Statement

Based on the great development of vehicles production, it has become urgent to rely on SbW auto technology in order to replace the traditional parts with new technologies. The FW rotation satisfies the following dynamic equation [2]:

$$\ddot{\delta}_f = -\frac{B_w}{J}\dot{\delta}_f + \frac{1}{J}\tau_m - \frac{1}{J}\tau_a - \frac{F_c}{J}sign(\dot{\delta}_f) \tag{1}$$

where

   J is the DC motor inertia moment;
   B_w is the constant DC motor viscous friction;
   δ_f is the FW steering angle;
   τ_a is the self-aligning torque;

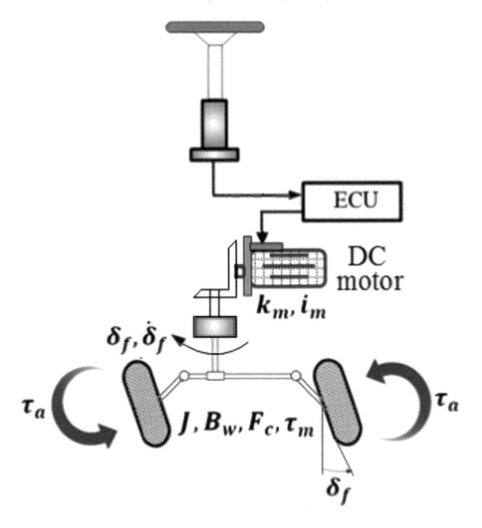

Fig. 1. Schematically model of SbW.

τ_m is the DC motor torque;

F_c is the constant Coulomb friction;

$F_c sign(\dot{\delta}_f)$ is the Coulomb friction in the steering system.

During a handling maneuver, the forces acting on the FW and rear wheel (RW) is illustrated in Fig. 2 (bicycle model [1,2]). Also, the pneumatic trail is the distance between the center of the tire and where the lateral force is applied as shown in the same figure.

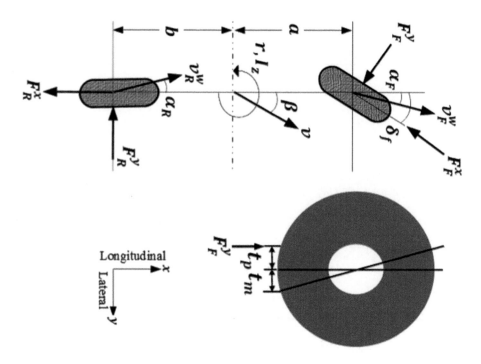

Fig. 2. Bicycle model.

The equations to calculate the both torque are given at small sideslip angles (approximately less than 6°) by (2) [20, 23, 30].

$$\tau_a = F_F^y(t_p + t_m), \ F_F^y = -C_F^\alpha \alpha_F, \ F_R^y = -C_R^\alpha \alpha_R, \ \tau_m = k_m i_m \tag{2}$$

where

F_F^y is the FW lateral force;
F_R^y is the RW lateral force;
F_F^x is the FW longitudinal force;
F_R^x is the RW longitudinal force;
v is the vehicle velocity at the center of gravity (CoG);
v_F^w is the FW velocity;
v_R^w is the RW velocity;
C_F^α is the FW cornering coefficient;
C_R^α is the RW cornering coefficient;
α_F is the FW sideslip angle;
α_R is the RW sideslip angle;
t_p is the pneumatic trail;
t_m is the mechanical trail;
k_m is the constant DC motor;
i_m is the armature current.

Also, the sideslip angles of the FW and RW are given by the Eq. (3) [5, 20, 35].

$$\alpha_F = -\delta_f + \beta + \frac{a}{v}r, \ \alpha_R = \beta - \frac{b}{v}r \tag{3}$$

where

β is the vehicle sideslip angle;
r is the yaw rate at the CoG;
a is the FW distance from the vehicle CoG;
b is the RW distance from the vehicle CoG.

On the other side, the yaw rate dynamics at the CoG and the dynamics of the sideslip angle are:

$$v(\dot{\beta} + r) = \frac{1}{m}(F_F^y + F_R^y), \ I_z\dot{r} = aF_F^y - bF_R^y \tag{4}$$

where

m is the vehicle mass;
I_z is the vehicle inertia moment.

Using (2), (3), (1), and (4), we have:

$$\ddot{\delta}_f = -\frac{B_w}{J}\dot{\delta}_f + \frac{k_m}{J}i_m - \frac{C_F^\alpha(t_p + t_m)}{J}\delta_f + \frac{C_F^\alpha(t_p + t_m)}{J}\beta + \frac{C_F^\alpha(t_p + t_m)a}{Jv}r$$
$$- \frac{F_c}{J}sign(\dot{\delta}_f)$$
$$\dot{\beta} = \frac{C_F^\alpha}{mv}\delta_f - \frac{C_F^\alpha + C_R^\alpha}{mv}\beta + (-1 + \frac{C_R^\alpha b - C_F^\alpha a}{mv^2})r$$
$$\dot{r} = \frac{C_F^\alpha a}{I_z}\delta_f + \frac{C_R^\alpha b - C_F^\alpha a}{I_z}\beta - \frac{C_F^\alpha a^2 + C_R^\alpha b^2}{I_z v}r \tag{5}$$

Remark 1. The new wire-based steering system, that dispenses with the mechanical column between the handwheel and front wheels and replaces it by modern devices, incorporates various types of non-linearity and disturbances, such as Coulomb friction, tyre self-aligning torque and so on [6]. Then, the SbW auto systems show considerable advantages over conventional steering arrangements; however there are also a number of limitations. For this reason, a controller is developed and presented in this paper to ensure the reliability and the robustness of these systems [21, 28, 29, 33, 34].

Remark 2. In the implementation of the vehicles control technique that are equipped with the active steering system SbW, due to the fact that the actual steering angle is generated via the front wheel steering motor, the steering controller drive the actual steering angle to exactly track the reference angle provided by the yaw control [25, 37].

Figure 3 gives an overview of a simplified DC motor circuit and a rotor mechanical model [23].

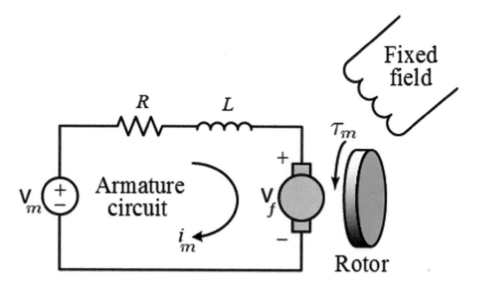

Fig. 3. DC motor sub-system model.

Then, the electrical circuit mathematical model is expressed by the Eq. (6) using $V_f = K_f \dot{\delta}_f$.

$$\dot{i}_m = -\frac{K_f}{L}\dot{\delta}_f - \frac{R}{L}i_m + \frac{1}{L}V_m \qquad (6)$$

where

V_f is the electromotive force;
K_f is the electromotive force constant;
L is the armature inductance;
R is the armature resistance;
V_m is the voltage at the armature terminals.

Combining the Eqs. (5) and (6) in a state-space form, a dynamics system model for steering is obtained and presented in the following equations:

$$\dot{x}(t) = Ax(t) + Bu(t) + D_w w(t)$$
$$y(t) = C_y x(t)$$
$$z(t) = C_z x(t)$$

where

$$x = \begin{bmatrix} \delta_f \\ \dot{\delta}_f \\ i_m \\ \beta \\ r \end{bmatrix}, \quad u = V_m, \quad w = \text{sign}(\dot{\delta}_f), \quad y = \delta_f, \quad z = \begin{bmatrix} \dot{\delta}_f \\ i_m \end{bmatrix},$$

$$A = \begin{bmatrix} 0 & 1 & 0 & 0 & 0 \\ \frac{C_F^\alpha(t_p+t_m)}{J} & -\frac{B_w}{J} & \frac{k_m}{J} & -\frac{C_F^\alpha(t_p+t_m)}{J} & -\frac{C_F^\alpha(t_p+t_m)a}{Jv} \\ 0 & -\frac{K_f}{L} & -\frac{R}{L} & 0 & 0 \\ \frac{C_F^\alpha}{mv} & 0 & 0 & -\frac{C_F^\alpha+C_R^\alpha}{mv} & -1+\frac{C_R^\alpha b-C_F^\alpha a}{mv^2} \\ \frac{C_F^\alpha a}{I_z} & 0 & 0 & \frac{C_R^\alpha b-C_F^\alpha a}{I_z} & -\frac{C_R^\alpha a^2+C_F^\alpha b^2}{I_z v} \end{bmatrix},$$

$$B = \begin{bmatrix} 0 \\ 0 \\ \frac{1}{L} \\ 0 \\ 0 \end{bmatrix}, \quad D_w = \begin{bmatrix} 0 \\ -\frac{F_c}{J} \\ 0 \\ 0 \\ 0 \end{bmatrix}, \quad C_y = \begin{bmatrix} 1 \\ 0 \\ 0 \\ 0 \\ 0 \end{bmatrix}^T, \quad C_z = \begin{bmatrix} 0 & 0 \\ 1 & 0 \\ 0 & 1 \\ 0 & 0 \\ 0 & 0 \end{bmatrix}^T.$$

Remark 3. Considering the necessity for a reliable motor, an effective way to model the friction of the DC motor is determined in this paper. Then, basic and main friction models are derived and a mathematical model that is linear of the DC motor is generated using Newton's mechanics.

3   Main Results

Now, some results are given to illustrate the applicability of the proposed approach. Then, the parameters of the SbW model are listed in Table 1 where $u_0 = V_m = 12\ V$.

Table 1. Parameter values of the SbW model.

Parameter	Value	Parameter	Value
J	$0.0004\ Kg.m^2$	a	$0.85\ m$
B_w	$0.36\ N.m.s/rad$	b	$1.04\ m$
k_m	$0.052\ N.m/A$	C_F^α	$10000\ N/rad$
t_p	$0.0381\ m$	C_R^α	$10000\ N/rad$
t_m	$0.04572\ m$	v	$13.4\ m/s$
F_c	$2.68\ N.m$	L	$0.0019\ H$
m	$800\ Kg$	K_f	$0.0521\ V.s/rad$
I_z	$3136\ Kg.m^2$	R	$0.39\ \Omega$

Graphically, to note the developments resulting from the proposed approach, Figs. 5 and 6 provide a clear view of the evolution of the state and input variables. On the other side, the disturbance used in these simulations is given in Fig. 4.

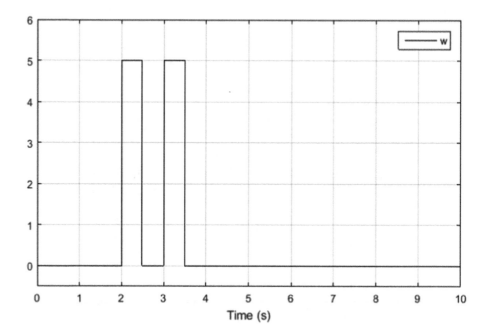

Fig. 4. Disturbance used in the simulations.

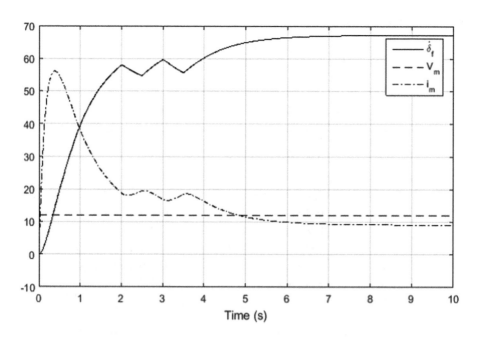

Fig. 5. Evolution of the state and input variables (a).

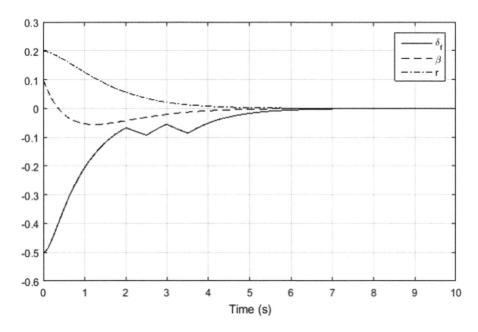

Fig. 6. Evolution of the state variables (b).

Based on the above, the control technique that is presented exhibits good steering performance and excellent stability, and behaves with strong force against parameter changes and external varying road disturbance. Also, the simulations show that the Coulomb friction model gives strong results compared to the viscous friction model. Then, the adopted controller has the ability to track the vehicle's movement path under the successive disturbances of the road, in terms of steering angle tracking.

Finally, the simulation results give a clear view that the FW angle can be convergent to the reference angle in SW ideally and quickly with SbW technology despite significant perturbations.

Remark 4. The effectiveness of the proposed method is verified using these results. Despite the excellent and great work that has been done to develop this technology, there are several important things to consider in this regard that will be touched upon in upcoming works.

4   Conclusion

Vehicles based on SbW technology are able to provide a more comfortable and safer driving by performing the primary function of isolating occupants from off-road conditions. SbW technology is simply a technology that completely eliminates the vehicle's primary mechanical link that controls its steering. This link is between the steering wheel and the front wheels. To better discuss the advantages of this technique, a complete and thorough description is given in this paper and then a linear mathematical model is presented to meet the challenges at hand. Among these challenges is ensuring robust vehicles stability under complex working and road conditions. Simulation results are given at the end of this paper to confirm that stability of the system and its robustness can be obtained despite the disturbance. On the other side, the FW angle can move well and perfectly time towards the SW reference angle.

References

1. Anwar, S.: Generalized predictive control of yaw dynamics of a hybrid brake-by-wire equipped vehicle. Mechatronics **15**(9), 1089–1108 (2005)
2. Anwar, S., Chen, L.: An analytical redundancy-based fault detection and isolation algorithm for a road-wheel control subsystem in a steer-by-wire system. IEEE Trans. Vehicular Technol. **56**(5), 2859–2869 (2007)
3. Balachandran, A., Gerdes, J.C.: Designing steering feel for steer-by-wire vehicles using objective measures. IEEE/ASME Trans. Mechatron. **20**(1), 373–383 (2014)
4. Bertoluzzo, M., Buja, G., Menis, R.: Control schemes for steer-by-wire systems. IEEE Ind. Electron. Mag. **1**(1), 20–27 (2007)
5. Chang, S.-C.: Synchronization in a steer-by-wire vehicle dynamic system. Int. J. Eng. Sci. **45**(2–8), 628–643 (2007)
6. Chen, T., Cai, Y., Chen, L., Xu, X., Sun, X.: Trajectory tracking control of steer-by-wire autonomous ground vehicle considering the complete failure of vehicle steering motor. Simul. Model. Pract. Theory **109**, 102235 (2021)
7. Chen, H., Zhu, L., Liu, X., Yu, S., Zhao, D.: Study on steering by wire controller based on improved H_∞ algorithm. Int. J. Online Eng. **9**(S2), 35–40 (2013)
8. El Fezazi, N., Tissir, E.H., El Haoussi, F., Bender, F.A., Husain, A.R.: Controller synthesis for steer-by-wire system performance in vehicle. Iranian J. Sci. Technol. Trans. Electr. Eng. **43**(4), 813–825 (2019)
9. Hang, P., Chen, X., Fang, S., Luo, F.: Robust control for four-wheel-independent-steering electric vehicle with steer-by-wire system. Int. J. Automot. Technol. **18**(5), 785–797 (2017)
10. Huang, C., Du, H., Naghdy, F., Li, W.: Takagi-sugeno fuzzy H_∞ tracking control for steer-by-wire systems. In: 1st IEEE Conference on Control Applications, Sydney, Australia, pp. 1716–1721 (2015)
11. Huang, C., Li, L.: Architectural design and analysis of a steer-by-wire system in view of functional safety concept. Reliability Eng. Syst. Saf. **198**, 106822 (2020)
12. Huang, C., Naghdy, F., Du, H.: Delta operator-based model predictive control with fault compensation for steer-by-wire systems. IEEE Trans. Syst. Man Cybern. Syst. **50**(6), 2257–2272 (2018)
13. Huang, C., Naghdy, F., Du, H., Huang, H.: Fault tolerant steer-by-wire systems: An overview. Annu. Rev. Control. **47**, 98–111 (2019)
14. Kim, K., Lee, J., Kim, M., Yi, K.: Adaptive sliding mode control of rack position tracking system for steer-by-wire vehicles. IEEE Access **8**, 163483–163500 (2020)
15. Kirli, A., Chen, Y., Okwudire, C.E., Ulsoy, A.G.: Torque-vectoring-based backup steering strategy for steer-by-wire autonomous vehicles with vehicle stability control. IEEE Trans. Veh. Technol. **68**(8), 7319–7328 (2019)

16. Lan, D., Yu, M., Huang, Y., Ping, Z., Zhang, J.: Fault diagnosis and prognosis of steer-by-wire system based on finite state machine and extreme learning machine. Neural Comput. Appl. https://doi.org/10.1007/s00521-021-06028-0 (2021)

17. Li, R., Li, Y., Li, S. E., Zhang, C., Burdet, E., Cheng, B.: Indirect shared control for cooperative driving between driver and automation in steer-by-wire vehicles. IEEE Trans. Intell. Transp. Syst. (2020). https://doi.org/10.1109/TITS.2020.3010620

18. Mohamed, E.S., Albatlan, S.A.: Modeling and experimental design approach for integration of conventional power steering and a steer-by-wire system based on active steering angle control. Am. J. Vehicle Design **2**(1), 32–42 (2014)

19. Mortazavizadeh, S.A., Ghaderi, A., Ebrahimi, M., Hajian, M.: Recent developments in the vehicle steer-by-wire system. IEEE Trans. Transp. Electrification **6**(3), 1226–1235 (2020)

20. Shah, M.B.N., Husain, A.R., Dahalan, A.S.A.: An analysis of CAN-based steer-by-wire system performance in vehicle. In: 3rd IEEE Conference on Control System, Computing and Engineering, Penang, Malaysia, pp. 350–355 (2013)

21. Sun, Z., Zheng, J., Man, Z., Fu, M., Lu, R.: Nested adaptive super-twisting sliding mode control design for a vehicle steer-by-wire system. Mech. Syst. Signal Process. **122**, 658–672 (2019)

22. Tashiro, T.: Fault tolerant control using disturbance observer by mutual compensation of steer-by-wire and in-wheel motors. In: IEEE Conference on Control Technology and Applications, pp. 853–858 (2018)

23. Virgala, I., Frankovský, P., Kenderová, M.: Friction effect analysis of a DC motor. Am. J. Mech. Eng. **1**(1), 1–5 (2013)

24. Wang, H., Kong, H., Man, Z., Cao, Z., Shen, W.: Sliding mode control for steer-by-wire systems with AC motors in road vehicles. IEEE Trans. Industr. Electron. **61**(3), 1596–1611 (2013)

25. Wang, H., Shi, L., Li, Z.: Robust hierarchical sliding mode control for steer-by-wire equipped vehicle yaw stability control. In: 11th Asian Control Conference, pp. 239–243 (2017)

26. Wu, X., Li, W.: Variable steering ratio control of steer-by-wire vehicle to improve handling performance. Proc. Inst. Mech. Eng. Part D J. Automobile Eng. **234**(2–3), 774–782 (2020)

27. Yang, H., Liu, W., Chen, L., Yu, F.: An adaptive hierarchical control approach of vehicle handling stability improvement based on Steer-by-Wire Systems. Mechatronics **77**, 102583 (2021)

28. Yang, Y., Yan, Y., Xu, X.: Fractional order adaptive fast super-twisting sliding mode control for steer-by-wire vehicles with time-delay estimation. Electronics **10**(19), 2424 (2021)

29. Ye, M., Wang, H.: Robust adaptive integral terminal sliding mode control for steer-by-wire systems based on extreme learning machine. Comput. Electr. Eng. **86**, 106756 (2020)

30. Yih, P., Ryu, J., Gerdes, J.C.: Modification of vehicle handling characteristics via steer-by-wire. IEEE Trans. Control Syst. Technol. **13**(6), 965–976 (2005)

31. Zakaria, M.I., Husain, A.R., Mohamed, Z., El Fezazi, N., Shah, M.B.N.: Lyapunov-krasovskii stability condition for system with bounded delay-an application to steer-by-wire system. In: 5th IEEE Conference on Control System, Computing and Engineering, Penang, Malaysia, pp. 543–547 (2015)

32. Zakaria, M.I., Husain, A.R., Mohamed, Z., Shah, M.B.N., Bender, F.A.: Stabilization of nonlinear steer-by-wire system via LMI-based state feedback. 17th Springer In: Asian Simulation Conference, Melaka, Malaysia, pp. 668–684 (2017)

33. Zhang, J., Wang, H., Ma, M., Yu, M., Yazdani, A., Chen, L.: Active front steering-based electronic stability control for steer-by-wire vehicles via terminal sliding mode and extreme learning machine. IEEE Trans. Veh. Technol. **69**(12), 14713–14726 (2020)
34. Zhang, J., et al.: Adaptive sliding mode-based lateral stability control of steer-by-wire vehicles with experimental validations. IEEE Trans. Veh. Technol. **69**(9), 9589–9600 (2020)
35. Zhao, W., Qin, X., Wang, C.: Yaw and lateral stability control for four-wheel steer-by-wire system. IEEE/ASME Trans. Mechatron. **23**(6), 2628–2637 (2018)
36. Zheng, B., Altemare, C., Anwar, S.: Fault tolerant steer-by-wire road wheel control system. IEEE American Control Conference, Portland, USA, pp. 1619–1624 (2005)
37. Zheng, H., Hu, J., Liu, Y.: A bilateral control scheme for vehicle steer-by-wire system with road feel and steering controller design. Trans. Inst. Meas. Control. **41**(3), 593–604 (2019)
38. Zou, S., Zhao, W.: Synchronization and stability control of dual-motor intelligent steer-by-wire vehicle. Mech. Syst. Signal Process. **145**, 106925 (2020)

Printed by Printforce, the Netherlands